LTE Communications and Networks

LTE Communications and Networks

Femtocells and Antenna Design Challenges

Edited by Masood Ur Rehman and Ghazanfar Ali Safdar

School of Computer Science and Technology,
University of Bedfordshire, Luton, UK

WILEY

This edition first published 2018

Registered Offices
John Wiley & Sons, Inc., 111 River Street, Hoboken, NJ 07030, USA
John Wiley & Sons Ltd, The Atrium, Southern Gate, Chichester, West Sussex, PO19 8SQ, UK

Editorial Office
The Atrium, Southern Gate, Chichester, West Sussex, PO19 8SQ, UK

For details of our global editorial offices, customer services, and more information about Wiley products visit us at www.wiley.com.

Wiley also publishes its books in a variety of electronic formats and by print-on-demand. Some content that appears in standard print versions of this book may not be available in other formats.

Library of Congress Cataloging-in-Publication Data

Names: Ur-Rehman, Masood, editor. | Safdar, Ghazanfar Ali, 1973– editor.
Title: LTE communications and networks : femtocells and antenna design challenges / edited by Masood Ur Rehman, Ghazanfar Ali Safdar.
Description: First edition. | Hoboken, NJ : John Wiley & Sons, 2018. | Identifiers: LCCN 2017056167 (print) | LCCN 2018006050 (ebook) | ISBN 9781119385240 (pdf) | ISBN 9781119385257 (epub) | ISBN 9781119385226 (cloth)
Subjects: LCSH: Long-Term Evolution (Telecommunications) | Femtocells. | Antennas (Electronics)–Design and construction.
Classification: LCC TK5103.48325 (ebook) | LCC TK5103.48325 .L7346 2018 (print) | DDC 621.3845/6–dc23
LC record available at https://lccn.loc.gov/2017056167

Cover design by Wiley
Cover image: © Jan Jirous/Shutterstock

Set in 10/12pt Warnock by SPi Global, Pondicherry, India

Printed in Great Britain by TJ International Ltd, Padstow, Cornwall

10 9 8 7 6 5 4 3 2 1

To my parents, Khalil Ur Rehman and Ilfaz Begum
and my siblings, Habib Ur Rehman, Waheed Ur Rehman and
Tahera Kalsoom.
Masood Ur Rehman

To my late parents, Safdar Hussain and Sadiq Sultana
and my two little angels, Taha and Taqi and my wife, Misbah.
Ghazanfar Ali Safdar

Contents

List of Contributors *xv*
Preface *xvi*

1 Introduction *1*
Ghazanfar Ali Safdar and Masood Ur Rehman
1.1 Evolution of Wireless and Cellular Communication *2*
1.1.1 1 G *3*
1.1.2 2 G *3*
1.1.3 2.5 G *3*
1.1.4 2.75 G *4*
1.1.5 3 G *4*
1.1.6 3.5 G *4*
1.1.7 4 G/LTE *5*
1.2 LTE Architecture *5*
1.2.1 Communications Perspective Challenges in LTE Networks *8*
1.2.1.1 Signalling System *8*
1.2.1.2 Backward Compatibility *9*
1.2.1.3 BS Efficiency *9*
1.2.2 LTE Radio Frame *10*
1.3 LTE Antennas *11*
1.4 LTE Applications *11*
1.4.1 Communications *11*
1.4.2 Public Safety *12*
1.4.3 Device-to-Device Communications *12*
1.4.4 Video Streaming *12*
1.4.5 Voice over LTE (VoLTE) *12*
1.4.6 Internet of Things *13*

1.4.7 Wearable Systems *13*
1.4.8 Cloud Computing *13*
1.5 Book Organization *14*
References *16*

Part I LTE Femtocells *19*

2 LTE Femtocells *21*
Ghazanfar Ali Safdar
2.1 Introduction *21*
2.1.1 Cross-Tier Interference *22*
2.1.2 Co-Tier Interference *24*
2.1.3 Downlink Interference Modelling *24*
2.1.4 Uplink Interference Modelling *25*
2.2 Platform for Femtocell Deployment *26*
2.3 LTE Architecture Overview *26*
2.3.1 LTE Downlink Transmission *27*
2.3.2 LTE Uplink Transmission *27*
2.4 LTE Femtocell Interference Analysis *28*
2.4.1 Scenario 1: Cross-Tier Interference Analysis *28*
2.4.2 Scenario 2: Effects of Femtocell Access Mode Deployment *28*
2.4.3 Scenario 3: Co-Tier Interference Analysis *29*
2.4.4 Scenario 4: Effects of Varying FAP Transmit Power Levels on MUEs *29*
2.5 Interference Mitigation: Current State of the Art *31*
2.5.1 Spectrum Access/Frequency Assignment *31*
2.5.2 Power Control *32*
2.5.3 Antenna Schemes *33*
2.6 Cognitive Femtocells: A Smart Solution to a Complex Problem *33*
2.7 Summary *35*
References *36*

3 Interference Mitigation in Cognitive Radio-Based LTE Femtocells *38*
Ghazanfar Ali Safdar
3.1 Introduction *39*
3.2 Femtocells *41*

3.2.1 Femtocells – Interference versus Deployment *43*
3.2.2 Femtocells – Typical Interference Mitigation Techniques *46*
3.2.2.1 Spectrum Access/Frequency Assignment Schemes *46*
3.2.2.2 Power Control (PC) Schemes *46*
3.2.2.3 Antenna Schemes *48*
3.3 Interference Mitigation in Femtocells using Cognitive Radio *49*
3.3.1 Cognitive Interference Mitigation *51*
3.3.1.1 Cognitive Interference Mitigation – PC *52*
3.3.1.2 Cognitive Interference Mitigation – Spectrum Access *54*
3.3.1.3 Cognitive Interference Mitigation – Antenna Schemes *64*
3.3.1.4 Cognitive Interference Mitigation – Joint Schemes *66*
3.3.2 Cognitive Interference Mitigation versus Conventional Interference Mitigation *70*
3.4 Summary *74*
References *75*

4 Coverage Area-Based Power Control for Interference Management in LTE Femtocells *84*
Ghazanfar Ali Safdar
4.1 Introduction *85*
4.2 Coverage Radius Based Power Control Scheme (PS) *88*
4.2.1 Radius Limit Setting *89*
4.2.2 Initial Coverage Radius *89*
4.2.3 Self-Update *89*
4.2.4 Final Radius *89*
4.3 System Model *90*
4.4 Performance Analysis *92*
4.4.1 Results and Discussion *93*
4.4.1.1 SINR Cross-Tier (Single Cell) *93*
4.4.1.2 SINR Co-Tier (Single Cell) *94*
4.4.1.3 Downlink Throughput (Single Cell) *95*
4.4.1.4 Co- and Cross-Tier SINR (Single Cell versus Multicell) *96*
4.4.1.5 Droppage in SINR (Single Cell versus Multicell) *97*
4.4.1.6 Coverage Area Bounds and Impact on SINR (Single Cell versus Multicell) *99*
4.5 Summary *100*
References *101*

5 Energy Management in LTE Femtocells *104*
Kapil Kanwal, Ghazanfar Ali Safdar, Masood Ur Rehman and Xiaodong Yang
5.1 Introduction *105*
5.2 Architecture of LTE Networks *105*
5.2.1 Communications Perspective Challenges in LTE Networks *106*
5.2.1.1 Signalling System *106*
5.2.1.2 Backward Compatibility *107*
5.2.1.3 BS Efficiency *107*
5.2.2 Importance of Energy Management in LTE Networks *108*
5.3 Classification of ES Schemes *108*
5.3.1 Static Power Consumption *109*
5.3.2 Dynamic Power Consumption *109*
5.4 Energy Efficient Resource Allocation *113*
5.4.1 Hybrid FBS and MBS Based Schemes *113*
5.4.2 Link Adaptation Schemes *114*
5.4.3 Cross Layer Resource Allocation Schemes *115*
5.4.4 MBSFN Resource Allocation Scheme *115*
5.5 Bandwidth Expansion Schemes *117*
5.5.1 CoMP Based Coverage Expansion *117*
5.5.2 Time Compression (TCoM) Scheme *118*
5.5.3 Bandwidth Expansion Mode (BEM) Scheme *119*
5.5.4 Component Carrier Based Schemes *121*
5.5.5 Scheduling Based Schemes *122*
5.6 Load Balancing Schemes *123*
5.6.1 Distance Aware Schemes *123*
5.6.2 Coverage Expansion Based Schemes *125*
5.6.3 Distributed Schemes *125*
5.6.4 Shared Relay Based Schemes *127*
5.6.5 CRN Adopted Switching Off of a BS *128*
5.6.6 Reduced Early Handover (REHO) Scheme *129*
5.7 Comparative Analysis *130*
5.8 Open Research Issues *135*
5.9 Summary *139*
References *140*

6 Spectrum Sensing Mechanisms in Cognitive Radio Based LTE Femtocells *150*
Tazeen S. Syed and Ghazanfar Ali Safdar
6.1 Fundamentals of Signal Processing *151*

6.1.1 Channel Model *151*
6.1.1.1 Additive Gaussian Noise Channel *151*
6.1.1.2 Linear Filter Channel *152*
6.1.1.3 Band Limited Channel *153*
6.1.2 Modulation Technique *153*
6.1.3 Error Probability *154*
6.2 Spectrum Sensing Techniques *155*
6.2.1 Primary Transmitter Detection *155*
6.2.1.1 Energy Detector *156*
6.2.1.2 Matched Filter Detection *158*
6.2.1.3 Cyclostationary Feature Detection *159*
6.2.1.4 Waveform Detection *160*
6.2.1.5 Wavelet Detection *161*
6.2.1.6 Hybrid Sensing *162*
6.2.1.7 Multi-Taper Spectrum Sensing *163*
6.2.2 Collaborative/Cooperative Detection *163*
6.2.3 Interference Temperature Detection *166*
6.2.4 Primary Receiver Detection *166*
6.3 History Assisted Spectrum Sensing *166*
6.4 Model- and Statistics-Based Spectrum Sensing Classification *167*
6.5 Challenges and Issues *172*
6.6 Summary *176*
References *177*

Part II Antennas for LTE Femtocells *185*

7 Antenna Consideration for LTE Femtocells *187*
Masood Ur Rehman
7.1 Antenna Fundamentals *187*
7.1.1 Input Impedance and Matching *188*
7.1.2 Bandwidth *189*
7.1.3 Radiation Pattern *190*
7.1.4 Directivity and Gain *191*
7.1.5 Efficiency *193*
7.1.6 Polarization *193*
7.2 Antenna Requirements for LTE Femtocells *196*
7.2.1 Frequency Bands *197*
7.2.2 Form Factor and Size Limitation *201*
7.2.3 Impedance Matching, Directivity, Gain and Efficiency *201*

7.2.4 Directionality *202*
7.2.5 Polarization *203*
7.2.6 Human Body Effects and Specific Absorption Rate (SAR) *204*
7.2.7 Multiple Input Multiple Output (MIMO) *205*
References *206*

8 Multiband Antennas for LTE Femtocells *209*
Masood Ur Rehman and Xiaodong Yang
8.1 Fundamentals of Multiband Antennas *209*
8.1.1 Multiband Techniques *210*
8.1.1.1 Higher Order Resonances *210*
8.1.1.2 Multiple Resonant Structures *211*
8.2 Types of Multiband Antennas *211*
8.3 Multiband Antenna Design: Case Studies *214*
8.3.1 Multi-Slot Antenna *215*
8.3.1.1 Antenna Geometry *215*
8.3.1.2 Antenna Performance Evaluation *215*
8.3.2 Patch-Loop Combination Antenna *220*
8.3.2.1 Antenna Configuration *220*
8.3.2.2 Antenna Performance *220*
8.4 Open Research Issues *227*
References *227*

9 Reconfigurable Antennas for LTE Femtocells *230*
Masood Ur Rehman and Waqas Farooq
9.1 Fundamentals of Reconfigurable Antennas *230*
9.1.1 Types of Reconfigurable Antennas *231*
9.1.1.1 Use of Switches *232*
9.1.1.2 Structural and Mechanical Changes *232*
9.1.1.3 Material Changes *234*
9.2 Realization of Reconfigurable Antennas *234*
9.3 Rectangular Patch Reconfigurable LTE Femtocell Antenna *237*
9.3.1 Design Conception *237*
9.3.2 Frequency Reconfiguration Mode *239*
9.3.3 Antenna Performance Evaluation *240*
9.4 Circular Patch Reconfigurable LTE Femtocell Antenna *246*
9.4.1 Frequency Reconfiguration Mode *248*

9.4.2 Antenna Performance Evaluation *248*
9.5 Open Research Issues *253*
References *254*

10 Multimode Antennas for LTE Femtocells *259*
Oluyemi Peter Falade, Xiaodong Chen and Masood Ur Rehman
10.1 Multimode Antennas: Fundamentals and Types *260*
10.2 Design of a Compact Multimode LTE Femtocell Antenna for Handheld Devices *261*
10.2.1 Numerical Analysis *263*
10.2.2 Experimental Investigation *266*
10.3 Design of a Multifunctional Compact Antenna for LTE Femtocells and GNSS Systems *268*
10.3.1 Numerical Analysis *273*
10.3.2 Experimental Investigation *279*
10.4 Summary *284*
10.5 Open Challenges and Issues *284*
References *284*

11 Human Body Effects on LTE Femtocell Antennas *289*
Masood Ur Rehman and Qammer Hussain Abbasi
11.1 Interaction of the Human Body with Antennas *290*
11.2 Numerical Modelling of the Human Body *291*
11.2.1 Evaluation and Comparison of Numerical Models of Human Body *294*
11.2.1.1 On-Body Transmission *294*
11.2.1.2 Effects on Antenna Radiation Pattern *297*
11.2.1.3 Electric Field Distribution *299*
11.2.1.4 Specific Absorption Rate (SAR) *300*
11.3 Evaluation of Human Body Effects on LTE Femtocell Antennas *305*
11.3.1 On-Body Antenna Placement *308*
11.3.2 Antenna-Body Separation *310*
11.3.3 On-Body LTE Channel Characterization *312*
11.3.4 On-Off Body LTE Channel Characterization *313*
11.3.5 Body-to-Body LTE Channel Characterization *315*
11.4 Open Research Issues *316*
References *317*

12 The Road Ahead for LTE Femtocells *322*
Masood Ur Rehman and Ghazanfar Ali Safdar
12.1 Future Prospects and Challenges *323*
12.1.1 Spectrum Sharing *324*
12.1.2 Intelligent/Efficient Spectrum Sensing Schemes *324*
12.1.3 Primary/Secondary User Issue *325*
12.1.4 Energy Saving *325*
12.1.5 Security *326*
12.1.6 Pilot Power/Coverage Radius Issue *326*
12.1.7 Signalling Overhead *326*
12.1.8 Proximity Services *326*
12.1.9 The Internet-of-Things (IoT) *327*
12.1.10 The Age of Big Data *328*
12.1.11 5G and Femtocells *328*
12.1.12 Antenna Design and Channel Modelling *328*
References *330*

Index *332*

List of Contributors

Qammer Hussain Abbasi
School of Engineering
University of Glasgow,
Glasgow, UK

Xiaodong Chen
School of Electronic Engineering and Computer Science
Queen Mary, University of London, London, UK

Oluyemi Peter Falade
Tempfad Limited, Leyton,
London, UK

Waqas Farooq
School of Computer Science and Technology
University of Bedfordshire,
Luton, UK

Kapil Kanwal
School of Computer Science and Technology
University of Bedfordshire,
Luton, UK

Masood Ur Rehman
School of Computer Science and Technology
University of Bedfordshire,
Luton, UK

Ghazanfar Ali Safdar
School of Computer Science and Technology
University of Bedfordshire,
Luton, UK

Tazeen S. Syed
School of Computer Science and Technology
University of Bedfordshire,
Luton, UK

Xiaodong Yang
School of Electronic Engineering
Xidian University,
Xi'an, China

Preface

Long Term Evolution (LTE) technology has brought about a revolution in the field of wireless communications. It has attracted huge attention due to its essential features of being an easily deployable network, offering high data rates and low latencies over long distances. Almost all new cellular and portable communication devices are now LTE enabled. It is also being used as a basis for the upcoming 5G technology and Internet-of-Things (IoT) concept, which will allow connectivity anywhere and anytime. It is growing fast to fulfil the ever-increasing demand from millions of users worldwide with applications ranging from communications to infotainment, healthcare to surveillance and transportation to manufacturing. Sales of LTE-enabled smart phones alone were expected to grow from 450 million units in 2015 to over 900 million units in 2017.

With huge benefits on offer, the LTE faces challenges of spectrum cognition, interference mitigation and power control. Efficient solutions to these challenges are necessary to enhance the performance of this technology. Femtocells are envisioned as a step forward to smart and low-interference LTE systems. Moreover, the performance of the overall wireless devices is dictated by the working of embedded antennas. Design of the LTE antennas is getting more complex day-by-day due to the advent of new design methodologies, innovative material technologies, miniaturization of devices and performance degradations caused by the user.

The current developments and expected future growth of the LTE demands availability of a comprehensive reference that deals with these systems in the context of femtocells and antennas. This book is an effort to fill this gap by educating the reader on the most important aspects of LTE femtocells and lays the foundations for future advancements. It brings together multidisciplinary contributions in the field of wireless and mobile communications, signal processing and antenna design to identify technical challenges and present recent results related to the development, integration and enhancement of LTE systems in portable devices. Both state-of-the-art and advanced topics including application of cognitive radio with efficient sensing mechanisms, interference mitigation and power management schemes for the LTE systems are discussed. Moreover, a comprehensive account of design challenges and approaches, performance enhancement techniques and effects of a user's presence on the LTE antennas is presented. Particular focus is put on the promising technologies of multiband, multimode and reconfigurable antennas for efficient design of portable LTE devices. Although the book is intended to be practical, theoretical details are revisited where it is required.

This is the first dedicated book that gives such a broad treatment to LTE systems in the context of femtocells and antenna design, covering wide range of issues related to the topic. The organization of the book makes it a valuable reference for the LTE system designers, as well as an introductory text for researchers, lecturers and students.

Masood Ur Rehman
Ghazanfar Ali Safdar

1

Introduction

Ghazanfar Ali Safdar and Masood Ur Rehman

School of Computer Science and Technology, University of Bedfordshire, Luton, UK

Wireless communication has involved relentless years of research and design and comprises cellular telephony, broadcast and satellite television, wireless networking to today's 3rd Generation Partnership Project (3GPP) and Long Term Evolution (LTE) technology. However, cellular telephony networks surpass the others in terms of usage [1]. Although cellular networks were designed to provide mobile voice services and low rate mobile data services, data services have excelled voice and findings show that global data traffic has grown by 280% since 2008 and is expected to double annually in the next 5 years [2]. Importantly, it already exceeded those expectations by 2010 by nearly tripling and it is further predicted that by 2020 nearly 1 billion people will access the Internet using a wireless mobile device [3].

The introduction of new or the upgrade of existing wireless standards such as the Institute of Electrical and Electronics Engineers (IEEE) Worldwide Interoperability for Microwave Access (WiMAX) and 3GPP's LTE have been developed to meet traffic and high data rates. Most of the methods to increase spectrum capacity in practice today are aligned towards; (1) improving the macro layer by upgrading radio access, (2) densifying the macro layer by reducing inter-site distances and (3) the use of low power nodes to complement the macro layer [4]. Macro layer deployment is a typical approach of deploying Base Station (BS) in proximity to each other covering large distances with reduced handover frequency. Although it is the backbone of

LTE Communications and Networks: Femtocells and Antenna Design Challenges,
First Edition. Edited by Masood Ur Rehman and Ghazanfar Ali Safdar.

most wireless networks, it has proven to be inefficient as it does not guarantee a high-quality link in situations where the BS and Mobile Station (MS) are relatively far away. Moreover, a BS serving hundreds of contentious users all vying for resources is old fashioned [5]. Researchers indicate that 50% of all voice calls and most of the data traffic, more than 70%, originate indoors [6]. However, indoor users may suffer from a reduced Received Signal Strength (RSS) due to low signal penetration through the walls or attenuation leading to total loss of signal in situations where the distance between transmitter and receiver is large. There is a need to provide solutions for poor indoor coverage to satisfy consumers. According to [5] the solutions to poor indoor coverage can be classified into two types, Distributed Antenna Systems (DAS) and Distributed Radios.

Distributed Antenna Systems comprise a group of Remote Antenna Units (RAU) spaced apart, providing not only enhanced indoor signal quality by significantly reducing transmission distance but also reducing transmit power (the power of the reference signal) [7]. Some of the challenges involved in deploying DAS are the choice of antennas and selecting a suitable location [8, 9]. Distributed radios involve the introduction of smaller cells to complement the deficiencies of the larger macrocell and the gains include an efficient spatial reuse of spectrum [10]. These small cells, which include picocells and microcells, are overlaid in the macrocell to provide voice and data service. Due to the two-tier nature of its architecture, it is prone to interference that may result in a low Signal to Interference plus Noise Ratio (SINR), throughput and in some cases a total disruption of service. As a result, there is a need to provide interference avoidance and mitigation schemes. Recently, a new distributed form of radio, LTE femtocells, has emerged that promises to be a viable solution to indoor cellular communication.

1.1 Evolution of Wireless and Cellular Communication

Communication has been essential for humanity to interact with one another where distance, quality of communication and high demand have always been important factors. Thus, it has evolved over the recent decades to overcome such factors in which newer and more obstacles have arrived in order to meet these challenges. Mobile communication has gradually evolved in shape of different generations as described next.

1.1.1 1 G

1G stands for the first generation of wireless mobile communication, which was first implemented in North America in the early 1980s. The technology was also known as Analogue Mobile Phone Systems (AMPS) based on an analogue system; that is, where information is transmitted by controlling a continuous transmission signal, such as amplifying signal strength or varying its frequency in relation to actual data. This system mainly provided services such as voice over a set radio frequency. In order for users to communicate, they would have to maintain a large distance from communicating points and use sufficiently large handsets. A mobile user would have to connect to the mobile base station that connects to the MTSO (Mobile Telecommunication Switching Office) that contains an MSC (Mobile Switching Centre) for routing mobile calls. The MTSO is then connected to the PSTN (Public Switch Telephony Network), which is a collection of unified voice-oriented public telephone networks [11].

1.1.2 2 G

2G stands for the second generation of wireless mobile communication and finished its establishment in the late 1990s. It was based on the Global System for Mobile Communication (GSM). GSM is a digital cellular phone system and it uses a variation of TDMA (time-division multiple access). 2G introduced digital traffic and voice encoded into digital signals. From its predecessor, it evolved and brought features such as SMS (short messaging service) and the quality of service for voice communication considerably improved [11].

1.1.3 2.5 G

2.5G GPRS (General Packet Radio Service) is a bridge between the second and third generations of wireless technology. GPRS supports MMS (multimedia messaging service), WAP (Wireless Application Protocol) Access and connects to the Internet. The first major step in the advancement of GSM networks to 3G (3rd Generation) of wireless mobile technology is GPRS. The service has added value to the GSM network by transmitting data by overlaying a packet based air interface on the existing circuit-switch-based GSM network. The voice traffic with this carrier is circuit switched, whereas the data is packet switched [12].

1.1.4 2.75 G

2.75G is based on an Enhanced Data rates for GSM Evolution (EDGE) and was the major breakthrough before the evolution of 3G. EDGE technology allows fast transmission of data and information and one of its major advantages is that the existing GSM networks can also support this technology and be upgraded. EDGE is preferred over GSM due to its flexibility and the provision of capacity, global roaming and data size as compared to GPRS [12].

1.1.5 3 G

3G stands for the third generation of mobile technology, which was introduced in 2005. It is based on set standards that are used for mobile devices meeting the terms of the ITU (International Telecommunication Union). 3G features CDMA (code division multiple access), a channel access method where a single channel can be used by multiple users to transmit data on the same frequency. The most common form of 3G usually identified as UMTS (Universal Mobile Telecommunications System/Standards) is WCDMA (Wideband Code Division Multiple Access). It can use both voice and data services consecutively and offers faster data rates compared to EDGE. Data is sent through packet switching while video traffic is managed through circuit switching. 3G provides services like web browsing, multimedia, navigation and smartphone applications that require higher data rates. It has backward compatibility with 2G mobile technology, which means a user is able to use services such as voice and SMS alongside data [13].

1.1.6 3.5 G

3.5G is an improvement of UMTS and also known as CDMA2000 and High Rate Packet Data (HRPD) or Evolution Data Optimised (EV-DO). With 3.5G technology, there is improved capacity featuring high-speed packet access, almost five times faster than an average 3G mobile technology. HSPA (High-Speed Packet Access) extends and improves the performance and working of existing WCDMA systems. Although there are some technical differences between CDMA2000 and UMTS, which includes the fact that CDMA2000 is backward compatible with IS-95. Interim Standard 95 was the first CDMA-based digital technology; that is, IS-95 devices can communicate with CDMA2000 BS whereas UMTS is not compatible with 2G GSM. Furthermore, UMTS uses the same carrier frequency for all types of traffic such as voice and data whereas CDMA2000 separates the traffic to multiple carriers [13].

1.1.7 4 G/LTE

LTE is a standard introduced by the 3GPP (3rd Generation Partnership project). There are a number of factors that LTE has helped to overcome with its following characteristics.

- High throughput – high data rates, which can be achieved in uplink and downlink
- Low Latency – unnoticeable delays between an input being processed and the corresponding output providing real time characteristics; for example, establishing a connection to a nearby network within a few milliseconds
- Improved Quality of Service
- Smooth handover across heterogeneous networks
- High network capacity to accommodate user demands for high bandwidth.

LTE is based on OFDMA (Orthogonal Frequency Division Multiple Access) in which the system transmits large amount of data; that is, large bandwidths up to 20 Mbps. Multiple access is achieved in OFDMA by assigning subnets of subcarriers to individual users. Table 1.1 briefly compares different generations of mobile technologies.

1.2 LTE Architecture

The LTE systems usually provide low latency, high data rate and packet optimized radio access. Compared to 3G, LTE additionally provides international roaming and compatibility with other legacy

Table 1.1 Comparison of mobile technologies.

Technology	1G	2G/2.5G/2.75G	3G/3.5G	LTE/4G
Features	• Makes use of analogue radio signals • Analogue voice service • No data service	• Uses digital radio signals • Voice encoded to digital signals • *GSM* Supports digital voice service, SMS messaging, improved voice clarity, Comparatively secure • *GPRS* Supports MMS, Internet	• Fast data transfer rate, improved spectral efficiency, greater network capacity • Enhanced audio video streaming, video conferencing support, • Web browsing at higher speeds, IPTV support	• Converged data and voice over IP • Entirely packet switched network, • Higher bandwidth to provide multimedia services at lower cost • Enhanced audio, video streaming, IP telephony, HD mobile TV
Standards	MTS, AMTS, IMTS	2G: GSM 2.5: GPRS 2.75: EDGE	IMT-2000 3.5G-HSDPA 3.75G: HSUPA	Single unified standard LTE, LTE adv. Mobile WiMAX
Web Standards	—	www	www (IPv4)	www (IPv4)
Technology	Analogue cellular Technology 14.4 Kbps	Digital narrow band circuit data, Packet data 171.2 Kbps (peak). 20–40 Kbps	Digital Broadband Packet data 3.1 Mbps (peak) 500–700 Kbps 3.5G:14.4 Mbps (peak) 1–3 Mbps	Digital Broadband Packet, Very high throughput 100–300 Mbps (peak) 3–5 Mbps 100 Mbps (Wi-Fi)

Service	Voice Calls	2G: Digital voice, SMS 2.5: Higher capacity packetized	Integrated high-quality audio, video and data	Dynamic information access, wearable devices with AI capabilities
Switching	Circuit switching	2G: Circuit 2.5G: Circuit for access network and air interface, packet for core network	Packet except for circuit for air interface	Packet switching, message switching
Handoff	Horizontal only	Horizontal only	Horizontal and vertical	Horizontal and vertical
Shortfall	Low capacity, unreliable handoff, poor QoS for voice, less secure	Reliant on location and proximity, required strong digital signals to help mobile phones	Requires higher network capacity to accommodate growing consumers	Being deployed

networks [14–16]. The 4G systems make use of OFDMA and Single Channel Frequency Division Multiple Access (SC-FDMA) schemes to support flexible bandwidth [17–23]. LTE architecture is generally based on Evolved Packet Core (EPC), Universal Terrestrial Radio Access (UTRA) and Universal Terrestrial Radio Access Network (UTRAN), each of which communicates with core network air interfaces and radio access network [24, 25].

Figure 1.1 illustrates the overall architecture of the LTE networks showing both EPC and evolved UTRAN (E-UTRAN) [26, 27] while Table 1.2 summarizes the core elements of the LTE architecture.

1.2.1 Communications Perspective Challenges in LTE Networks

Though LTE has proven to be a promising technology, it is a complex network and there are some challenges that need to be carefully addressed for optimum functionality.

1.2.1.1 Signalling System

In LTE networks, one of the major issues is to avoid or limit signalling overhead and overlapping in the control part of the network. A large number of connections between nodes and network fragmentation causes rapid increase in signalling traffic. Any failure in signalling system will drag operators towards increased system latency and

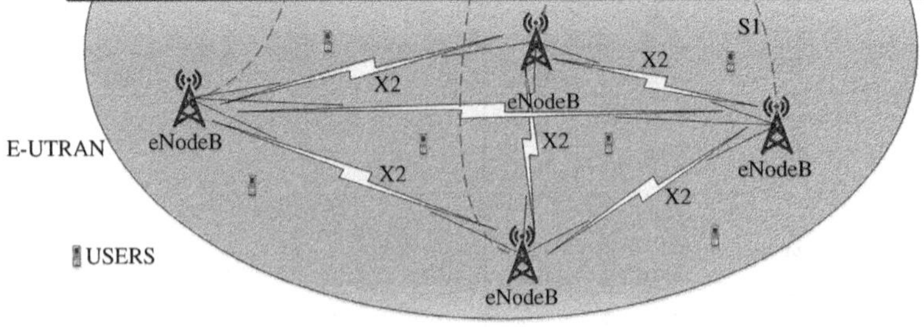

Figure 1.1 LTE architecture.

Table 1.2 LTE network components.

Components	Description
Evolved Packet System (EPS)	Provides IP connectivity using E-UTRAN.
Mobility Management Entity (MME)	Responsible for authorization, security, handover, roaming and mobility of users.
S1 Interface	It connects EPC with BSs (base stations).
Serving Gateway (S.GW)	EPC terminates at this node. It is connected to E-UTRAN through S1 interfaces. Each user is allocated unique S.GW, which is responsible for handover, packet routing and forwarding functions.
Packet data network gateway (PDN-GW)	PDN-GW provides UEs with access to packet data network by allocating IP addresses. It is also responsible for secure connection with untrusted devices from non-4G networks.
HnodeB	Femtocells that are employed to improve seamless connectivity in coverage holes.
eNodeB	Also known as BS that serves the UEs.
HnodeB GW	Provides connection to the core network.
X2 Interface	Provides communication between two BSs.

outages resulting in to loss of revenues [28, 29]. Increased signalling traffic also leads to increased energy consumption and definitely needs to be looked into carefully.

1.2.1.2 Backward Compatibility

LTE is usually compatible with all other relevant major standards. The combination of devices, network interfaces and equipment to support other standards complicates end-to-end functionality testing and interoperability testing (IOT) [30, 31].

1.2.1.3 BS Efficiency

Due to the employment of OFDMA in LTE, signals have high amplitude variability known as Peak-to-Average Power Ratio (PAPR), which reduces transmitter efficiency. Furthermore, the BS provides high data rate at the cost of high dynamic transmission power. Since high transmission power results in increased energy consumption and thereby increases Operational Expenditure, energy management has become a major challenge in LTE networks to stay profitable and also to reduce global warming [32].

1.2.2 LTE Radio Frame

The radio frame of LTE is defined as having a length of 10 ms as illustrated in Figure 1.2. It is divided equally into 10 sub-frames of duration $T_{FS} = 1\,ms$ per sub-frame. Each sub-frame is further divided into $N_{FS} = 2$ slots of length $T_S = \frac{T_{FS}}{2} = 0.5\,ms$. Each sub-frame contains $N_{sym} = 6$ or $N_{sym} = 7$ OFDM symbols on the length of the selected cyclic prefix. An extended cyclic prefix of 16.7 μs is allowed in LTE, which might be suitable in accommodating delay.

However, in femtocells, a normal length cyclic prefix ($T_{CP} = 5.2$ μs) might be enough due to its limited coverage area and short delay periods as compared with a Macrocell Base Station. More information about the frame structures can be found in [33].

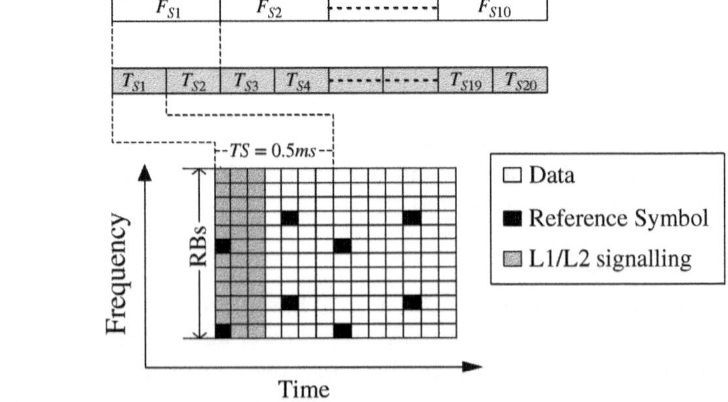

Figure 1.2 Downlink resource block and sub-frame structure in downlink LTE.

1.3 LTE Antennas

The antenna acts as a transducer between the guided electromagnetic wave travelling in a radio frequency circuit or transmission line and the unguided electromagnetic wave travelling in free space. It is the fundamental building block in the development of any wireless communications system.

The requirements for LTE antennas depend on the specific application or where it will be used since there is a need to meet the increased demand for a high data rate. Varying LTE applications consideration of a number of specific factors in antenna selection such as polarization, multi-frequency or multi-mode operation, multiple-input multiple-output (MIMO) structure, reconfigurability, directionality and certain specific absorption rate on top of common requirements of size, bandwidth, gain, radiation pattern and efficiency. The antenna can be put on mobile handheld terminal, laptop, BS, access points, high-speed trains or cars, aeroplanes and so on.

Antenna selection and design is a challenging task that necessitates the utmost care as a poorly chosen antenna can severely affect the cost and performance of the overall LTE system.

1.4 LTE Applications

LTE has become a global wireless foundation supporting continual enhancements. Its applications range from communications to health monitoring, surveillance to public safety and smart homes to entertainment.

1.4.1 Communications

The major application area of LTE technology is cellular communications. It carries inherent benefits of reduced latency and increased data rates offering peak downlink data rates of 300 Mbps, peak uplink rates of 75 Mbps and QoS measures allowing latency of less than 5 ms in the radio access networks. It can manage moving devices and supports multicast and broadcast streams. Both frequency division duplexing (FDD) and time-division duplexing (TDD) can be used in LTE. These advantages have made LTE the front-runner in mobile communications standards.

1.4.2 Public Safety

An important LTE application area is public safety. Initially, it was a broadband data service that eventually turned into mission-critical voice service.

Micro-location information from small cells allows emergency and health services to locate the emergency. The USA and the UK have developed authorities, namely the First Responder Network Authority (FirstNet) and Emergency Service Network, employing LTE for public safety. Use of LTE for this purpose has special requirements in terms of features, network deployment and device-level approaches that differ from general communication application.

1.4.3 Device-to-Device Communications

LTE supports autonomous discovery and communication of a device with nearby devices and services in a battery-efficient manner. A device can broadcast its needs and services and can also passively identify services without user intervention. In this application scenario, the LTE network performs configuration and authentication while communication can take place either via the network or directly between the devices. It is fast becoming popular for emergency scenarios and disaster management when the rest of the network is unavailable.

1.4.4 Video Streaming

LTE is widely used for video streaming that requires high data rates. An increasing number of video applications, such as Netflix and Skype, adapt their streaming rates based on available bandwidth enabling them to continue operation even when throughput rates drop. LTE also supports video streaming via multicast or broadcast functions.

1.4.5 Voice over LTE (VoLTE)

LTE offers a transition from circuit-switched voice (VoIP) to *Voice over LTE* (VoLTE). Using VoLTE, high-definition voice transfer is possible having improved clarity and intelligibility and reduced background noise using Multi-Rate Wideband voice codecs. Other advantages of VoLTE include ability to combine it with other services, such as video calling and presence and high voice spectral efficiency.

1.4.6 Internet of Things

LTE is one of the key enabling technologies from the Internet of Things (IoT). Though not fully implemented yet, early IoT applications do exist in the form of Machine-to-Machine (M2M communications) including vehicle infotainment, remote health, smart metering, security and home automation, construction and heavy equipment and industrial manufacturing. "Smart cities" initiatives are also supporting vast research and development activities. Although promising, the IoT market has to deal with numerous challenges such as varying communications requirements, long battery requirements, cost sensitivity and security concerns to name a few. Research is continuing to devise efficient methods addressing these issues.

1.4.7 Wearable Systems

One of the major application areas of the LTE systems is wearable systems for health monitoring, emergency services and entertainment. The user wears a body-worn LTE device, such as the smart phone, smart watch or health tracker. It gathers vital physiological parameters and transmits required information to the access point that relays the information to the relevant services such as hospitals or fire fighters for appropriate action. New Wearable Augmented Reality applications such as Google Goggles and Samsung Gear are also fast becoming available. These applications need micro-location information provided by the LTE femtocells. Apart from the location information, the user's interests, place and context can also be used in these applications to retrieve relevant information.

1.4.8 Cloud Computing

LTE is also being used in cloud computing where the delivery of computing services like servers, storage, databases, networking, software and analytics is made available over the Internet. Cloud computing eliminates the cost of buying site-specific hardware and software, offers high mobility, scalability and reliability through data backup, disaster recovery and business continuity. However, issues of security and privacy are restricting its universal acceptance up to now.

1.5 Book Organization

LTE technology has brought a revolution in the field of wireless communications. It has attracted huge attention due to its essential features of being an easily deployable network, offering high speeds and low latencies over long distances. Femtocells are envisioned as a step forward to smart homes and low-interference LTE systems. In this book, many challenging issues of LTE femtocells and LTE antennas are discussed giving solutions from a technology and application point of view.

The book is divided into two parts. Part I (Chapters 2–6) deals with femtocells and the topics of cognitive radio, interference mitigation and power management schemes for LTE femtocell systems. Part II (Chapters 7–11) discusses the design challenges, different approaches, performance enhancement and application case scenarios for LTE antennas. Chapter 12 presents the concluding remarks and future prospects for LTE femtocells.

Chapter 2 provides an introduction to the LTE communications in femtocells and the rationale for selecting this communication mode. Interference is one of the major hurdles in the deployment of an efficient, robust and reliable communications link. The ever-growing communication sector with an increasing number of devices and introduction of new technologies demands methods to mitigate it without altering the communication quality. This chapter also discusses various techniques for interference mitigation.

Chapter 3 discusses cognitive radio applications in LTE femtocells, which is considered as one of the key techniques to manage the increasingly important problem of spectrum shortage by allowing unlicensed users to utilize the licensed spectrum when the licensed user is not occupying it. This chapter introduces the concept of cognitive radio femtocells and deals with the issue of the interference by employing various mitigation strategies. A comparative analysis of these techniques is also presented to recommend an optimal approach.

Chapter 4 explains the fundamentals of coverage area based power control scheme and describes its usability in LTE femtocells to mitigate interference within a cell as well as across multiple cells using metrics of SINR, throughput and droppage.

Chapter 5 discusses importance of energy management in LTE femtocells that is one of the major constraints for wireless devices. Different energy saving schemes for the LTE femtocells are discussed presenting a comparative study to highlight advantages and disadvantages of these schemes while identifying the optimal solution.

Chapter 6 gives detailed discussion on working principles and operation of different sensing mechanisms employed in cognitive radio LTE femtocells as efficient sensing mechanisms are required to increase usability of the spectrum and minimize interference and collision of the secondary user with the primary user. It also identifies the strengths and weaknesses of these techniques through thorough comparative analysis.

Chapter 7 introduces antenna technology for LTE systems discussing fundamental parameters including bandwidth, gain, directivity, polarization, radiation pattern and efficiency. Complexity of the LTE antenna design, due to specific operational requirements on top of fundamental parameters such as form factor, SAR, working on various frequency bands and MIMO, is also highlighted.

Chapter 8 discusses the basics of the multiband antennas operating at multiple frequency bands and their importance in LTE systems to support various technologies. The design procedure and performance evaluation of three candidate antenna solutions for LTE femtocells are also described.

Chapter 9 deals with the fundamentals of reconfigurable antennas for multiple frequency LTE operation with a controlled switching mechanism to meet with the device size and form factor requirements. Different design approaches of reconfigurable antennas are also detailed along with the study of two candidate antenna solutions for LTE femtocell systems.

Chapter 10 covers the design challenges and proposes an effective solution for multi-mode antennas for LTE femtocells covering multi-frequency bands. The efficiency of multi-mode antennas in devising compact, cost effective, simple and highly efficient LTE devices is also discussed.

Chapter 11 investigates the performance of various LTE femtocell antennas for the human body effects considering different wearable scenarios as the human body is an integral part of practical portable LTE systems. Body-to-body communication is considered a special case of LTE communications for emergency services. The prospects of body-to-body communication between two LTE devices carried by two users are also discussed in this chapter in terms of antenna design.

Chapter 12 presents the applicability of the LTE femtocell systems in future communication devices. It also highlights potentials and discusses challenges that need to be addressed in order to maximize the abilities and benefits of the LTE Femtocell and LTE antennas.

References

1. A. F. Molisch, *Wireless Communications*, 2nd Edition, Wiley-IEEE, February 2011.
2. Ericsson. 2010, *Mobile data traffic surpasses voice, 23 March.* Stockholm, Sweden.
3. Cisco, "*Cisco visual networking index: Global mobile data traffic forecast update*, 2010/2015," Whitepaper, Feb. 2011.
4. R. Q. Hu, Y. Qian, S. Kota and G. Giambene, "Hetnets – a new paradigm for increasing cellular capacity and coverage [Guest Editorial]," in *IEEE Wireless Communications*, vol. 18, no. 3, June 2011, pp. 8–9.
5. Cellular Asset, "Femtocells – Benefits, Challenges and Deployment Scenarios" [Online]. Available: www.cellularasset.com (accessed December 2017).
6. V. Chandrasekhar, J. Andrews and A. Gatherer, "Femtocell networks: a survey," *IEEE Communications Magazine*, vol. 46, no. 9, September 2008, pp. 59–67.
7. H. Osman, H. Zhu, T. Alade and J. Wang, "Downlink transmission of distributed antenna systems in high building environments," *Communications (ICC)*, 2011 IEEE International Conference on, 5–9 June 2011, pp. 1–5.
8. L. Gatzoulis, A. Aragon, G. Povey, S. R. Saunders, "Performance analysis of in-building FDD deployment measured data," in Proc. 4th IEEE Intl. Conf. on 3G Mobile Communication Techn., June 2003, pp. 167–172.
9. K. J. Grandell, "Indoor antennas for WCDMA systems," in *Proc. 11th IEEE Intl. Conf. on Antennas and Propagation*, 2001, vol. 1, pp. 208–211.
10. M. S. Alouini and A. J. Goldsmith, "Area spectral efficiency of cellular mobile radio systems," *IEEE Trans. Vehic. Tech.*, vol. 48, no. 4, July 1999, pp. 1047–1066.

11 S. Shukla, V. Khare, S. Garg and P. Sharma, Comparative study of 1G, 2G, 3G and 4G. *Journal of Engineering, Computers & Applied Sciences,* vol. 2, no. 4, 2013, pp. 55–63.
12 A. R Mishra (2007). *Advanced Cellular Network Planning and Optimization.* Chichester: John Wiley & Sons, Ltd., pp. 1–51.
13 C. Cox (2012). *An Introduction to LTE: LTE, LTE-Advanced, SAE and 4G Mobile Communications.* Hoboken, NJ: John Wiley & Sons, Inc., pp. 1–7.
14 N. Takaharu, "LTE and LTE-advanced: Radio technology aspects for mobile communications," *General Assembly and Scientific Symposium,* 2011 30th URSI, Istanbul, 2011, pp. 1–4.
15 H. Lee, S. Vahid and K. Moessner, "A survey of radio resource management for spectrum aggregation in LTE-advanced," in *IEEE Communications Surveys & Tutorials,* vol. 16, no. 2, Second Quarter 2014, pp. 745–760.
16 N. Becker, A. Rizk and M. Fidler, "A measurement study on the application-level performance of LTE," *Networking Conference,* 2014 IFIP, Trondheim, 2014, pp. 1–9.
17 S. B. Manir, M. M. Rahman and T. Ahmed, "Comparison between FDD and TDD frame structure in SC-FDMA," *International Conference on Informatics, Electronics & Vision (ICIEV),* 2012, Dhaka, 2012, pp. 795–799.
18 L. Wan, M. Zhou and R. Wen, "Evolving LTE with flexible duplex," *2013 IEEE Globecom Workshops (GC Wkshps),* Atlanta, GA, 2013, pp. 49–54.
19 R. Ratasuk, A. Ghosh, W. Xiao, R. Love, R. Nory and B. Classon, "TDD design for UMTS Long-Term Evolution," *2008 IEEE 19th International Symposium on Personal,* Indoor and Mobile Radio Communications, Cannes, 2008, pp. 1–5.
20 R. Zheng, X. Zhang, X. Li, Q. Pan, Y. Fang and D. Yang, "Performance evaluation on the coexistence scenario of two 3GPP LTE systems," *Vehicular Technology Conference Fall (VTC 2009-Fall),* 2009 IEEE 70th, Anchorage, AK, 2009, pp. 1–6.
21 S. S. Prasad, C. K. Shukla and R. F. Chisab, "Performance analysis of OFDMA in LTE," *Computing Communication & Networking Technologies (ICCCNT),* 2012 Third International Conference on, Coimbatore, 2012, pp. 1–7.
22 L. A. M. R. de Temino, G. Berardinelli, S. Frattasi, K. Pajukoski and P. Mogensen, "Single-user MIMO for LTE-A uplink: Performance evaluation of OFDMA vs. SC-FDMA," *2009 IEEE Radio and Wireless Symposium,* San Diego, CA, 2009, pp. 304–307.

23 3GPP, "Evolved Universal Terrestrial Radio Access (E-UTRA) and Evolved Universal Terrestrial Radio Access Networks (E UTRAN): Overall description," V10.4.0., TS 36.300.

24 3GPP, "Technical Specification Group Radio Access Network; Further advancements for E-UTRA – LTE-Advanced feasibility studies in RAN WG4," V9.0.0., TR 36.815.

25 3GPP, "General Packet Radio Service (GPRS) enhancements for Evolved Universal Terrestrial Radio Access Network (E-UTRAN) access," 3rd Generation Partnership Project (3GPP), TS 23.401, Jun. 2011.

26 S. B. H. Said, M. R. Sama, K. Guillouard, et al., "New control plane in 3GPP LTE/EPC architecture for on-demand connectivity service," Cloud Networking (CloudNet), 2013 IEEE 2nd International Conference on, San Francisco, CA, 2013, pp. 205–209.

27 A. Cox, *in An Introduction to LTE: LTE, LTE-Advanced, SAE and 4G Mobile Communications*, Chichester, UK: John Wiley & Sons, Ltd, 2012, pp. 21–28.

28 K. P. Makhecha and K. H. Wandra, "4G wireless networks: Opportunities and challenges," *2009 Annual IEEE India Conference*, Gujarat, 2009, pp. 1–4.

29 A. Pande, V. Ahuja, R. Sivaraj, E. Baik and P. Mohapatra, "Video delivery challenges and opportunities in 4G networks," in *IEEE MultiMedia*, vol. **20**, no. 3, July-Sept. 2013, pp. 88–94.

30 M. Iwamura, K. Etemad, M. H. Fong, R. Nory and R. Love, "Carrier aggregation framework in 3GPP LTE-advanced [WiMAX/LTE Update]," in *IEEE Communications Magazine*, vol. **48**, no. 8, August 2010, pp. 60–67.

31 Y. Yuan, S. Wu, J. Yang, F. Bi, S. Xia and G. Li, "Relay backhaul subframe allocation in LTE-Advanced for TDD," *Communications and Networking in China (CHINACOM), 2010 5th International ICST Conference on*, Beijing, 2010, pp. 1–5.

32 S. Ahmadi, "*LTE-Advanced: A Practical Systems Approach to Understanding 3GPP LTE Release 10 and 11 Radio Access Technologies*," Waltham, MA, USA: Elsevier Inc., 2014, pp. 61–65.

33 A. Valcarce, G. De La Roche, A. Juttner, D. Lopez-Erez, and J. Zhang, "*Applying FDTD to the coverage prediction of WiMAX femtocells*", EURASIP Journal on Wireless Communications and Networking, Feb. 2009.

Part I

LTE Femtocells

2

LTE Femtocells

Ghazanfar Ali Safdar

School of Computer Science and Technology, University of Bedfordshire, Luton, UK

Recently, considerable attention has been devoted to the potential of the femtocell as a solution to poor indoor network coverage. It has been exploited to provide higher capacity and intelligent coverage with an improvement in quality of service (QoS) for future indoor services. Though the advantages of femtocells cannot be overemphasized, they introduce some challenges, mostly with respect to the possible inherent interference due to a co-channel deployment in two-tier architecture with the Macrocell Base Station (MBS) and User Equipments (UEs). In this chapter, the femtocell is analysed as a solution for indoor network coverage problems and local-convergence demands for indoor network applications. The chapter also focuses on the conventional methods/solutions used today to mitigate interference, which comprises spectrum allocation, power control and antenna approaches. These solutions can be combined with cognitive radio (CR) to introduce a higher degree of interference awareness with respect to the dynamic changes in the wireless environment.

2.1 Introduction

There will be a continuous increase in the demand for wireless spectrum in the foreseeable future with the introduction of Internet multimedia applications such as Facebook, YouTube and multimedia

LTE Communications and Networks: Femtocells and Antenna Design Challenges, First Edition. Edited by Masood Ur Rehman and Ghazanfar Ali Safdar.

networks to name a few. As a result, there is even more of an increase in demand for the spectrum compared with technological development, which aims at increasing spectrum efficiency [1]. However, the main concern mostly in an indoor environment is the poor indoor network coverage as opposed to the increased demand of real-time applications [2]. For these reasons, femtocells have gained attention to secure spectrum efficiency in near-future networks. One of the important benefits of femtocells is the elimination of coverage-area problems for indoor scenarios. Some other benefits of femtocells include increased average revenue per user, reduced capital expenditure (CAPEX) and operational expenditure (OPEX), deployment in operator-owned spectrum and reduced bandwidth load and power requirements [3].

However, one of the main problems is due to the inefficient policies in spectrum management rather than the real-time application on the available bandwidth [4]. Femtocells are typically designed to support simultaneous spectrum access of up to four mobile UEs in residential or small indoor environments. Predictions show that, in the near future, about 60% of voice traffic and about 90% of data traffic will originate from indoor environments such as home, office, airport and school [5]. Due to the limitations in spectrum availability, network operators prefer to deploy femtocells in a co-channel access mode with the MBS, which means same resources are shared simultaneously [6]. However, since it is relatively expensive for network operators to perform careful cell planning and optimization in femtocells, they are user deployed by subscribers, which introduces an inherent interference with the MBS. This is because the femtocell uses the same spectrum as the MBS it is currently sharing with. Since the importance of femtocells cannot be overemphasized, there is a need to introduce interference mitigation schemes to enable femtocells to co-exist alongside macrocells, as the former complements the latter in today's heterogeneous networks.

This chapter concentrates on elaborating the importance of femtocells in a helpful or harmful perspective. The chapter highlights possible interference scenarios in a two-tiered network with macrocells as well as probable solutions for mitigating interference.

2.1.1 Cross-Tier Interference

Cellular networks require dedicated spectrum to support high data rates. However, the current radio spectrum is very crowded and only leaves very limited space for future evolutions, which results in a compact arrangement of frequency bands between 4G/5G releases and other wireless systems. For instance, the 2400–2483.5 MHz ISM band is utilized by both Wireless Fidelity (Wi-Fi) and Bluetooth, while the operating band of the LTE-A Band 40 ranges 2300–2400 MHz. Due to the imperfect transceiver components, the compact arrangement of adjacent frequency bands introduces severe cross-tier interference, not only among wireless stations but also within a device with multiple radios such as femtocells, which causes immense interference to Macrocell User Equipments (MUEs) within its vicinity as depicted as I_3 in Figure 2.1. In [7], even usage of state-of-the-art radio frequency filters fails to provide mitigation to interference especially in adjacent channels.

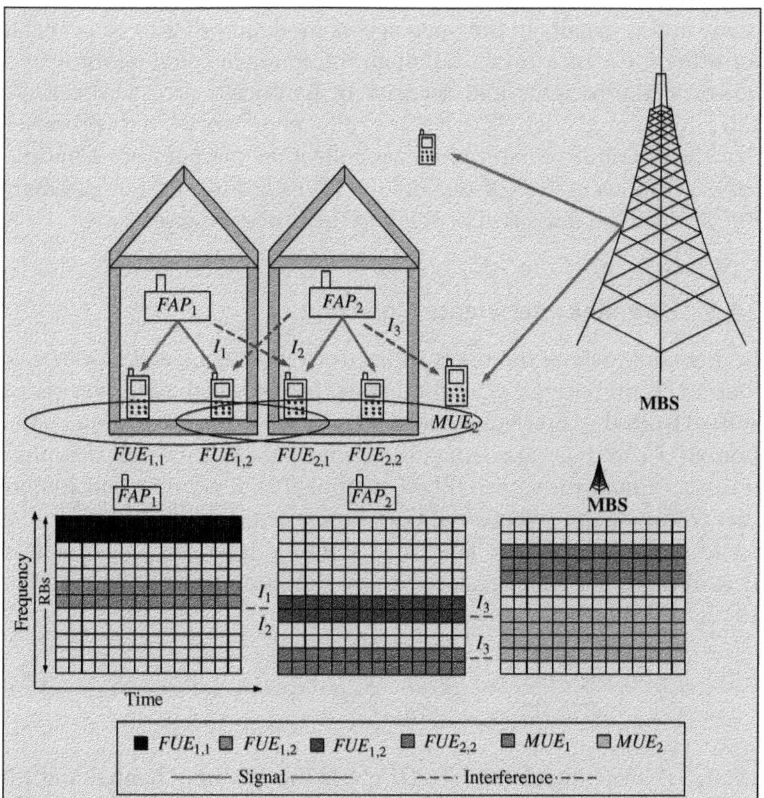

Figure 2.1 Femtocell and interference in a two-tiered network.

2.1.2 Co-Tier Interference

Due to the underlying deployment of femtocells without proper cell planning, severe interference may occur not only between a femtocell and a macrocell, but also among femtocells under highly dense deployments (Figure 2.1). There are different classes of interference in the HetNets, among those, the major challenge lies in the interference from a femtocell to a neighbouring femtocell (collocated) depicted as I_1 and I_2 in Figure 2.1. In the perspective of femtocells that are user deployed with no network planning, mitigating co-tier interference under the limitation or knowledge of the presence of another femtocell can be difficult. Without a centralized coordinator or fixed spectrum partition, one proposed solution to mitigate co-tier interference is to handover the victim UE to the nearby femtocell. In addition, a Femtocell Access point (FAP) cannot always be in the close access mode and is usually in the open access mode in order to be available for other UEs. As a result, although interference could be mitigated, however, the benefits and security of femtocells are sacrificed. To solve interference issues in future cellular networks, it is proposed that the distributive nature for data collection and parameter optimization empowers the CR technology [8] as a novel design paradigm [9]. This will be discussed in detail in the proceeding sections.

2.1.3 Downlink Interference Modelling

In downlink, where the users suffer from interference, it can be said that a certain user *UEx*, whose connected server (best server) is C_i, suffers from the interference of cell C_j, based on the following condition; If C_i and C_j are using the same sub-channel for downlink transmission at the same OFDM symbol [10]. It is important to note that UE_x could be a Femtocell User Equipment (FUE) or MUE and C_i could be a FAP or MBS. Therefore, the total interference suffered in Downlink (DL) by UE_x at slot $slot_{i,k,t}$ is the summation of the interferences coming from all neighbouring cells C_j.

$$I_{x,k,t}^{DL} = \sum_{j=0,j\neq i}^{N-1} \sum_{t=0}^{T-1} \left(P_{j,k} \cdot G_j \cdot L_j \cdot L_{pj,x} \cdot G_x \cdot L_x \right) \tag{2.1}$$

where, x is the interfered UE, UE_x; k is the kth sub-channel and t is the tth symbol; i is the best server, $C_{i,j}$ is an interfering cell, C_j; $P_{j,k}$ is the transmit power of cell C_j in a SC of the kth sub-channel; $Lp_{j,x}$ is the channel gain or path loss (PL) between C_j and UE_x; G_j and G_{xs} are the antenna gains in Cj and UE_x. Also, L_j and L_x stands for the equipment losses in cells, C_j and UE_x.

2.1.4 Uplink Interference Modelling

On the other hand, in UL, interference is suffered by the cells (MBS or FAPs in our scenario). The conditions are that if a certain cell C_i, serving user is UE_x, suffers from the interference of another UE UE_y, if UE_x and UE_y are using the same sub-channel for UL transmission at same OFDM symbol. Therefore, the total interference suffered in UL by cell C_i at slot $slot_{i,k,t}$ will be the summation of all the interferences emanating from all neighbouring UEs, UE_y.

$$I_{i,k,t}^{UL} = \sum_{y=1,y\neq x}^{M-1} \sum_{t=0}^{T-1} \left(P_{y,k} \cdot G_y \cdot L_y \cdot L_{py,i} \cdot G_i \cdot L_i\right) \tag{2.2}$$

where, i indicates cell suffering from interference, C_i; k is the kth sub-channel and t is the tth symbol; x is the user being served, UE_x; y is the user causing interference, UE_y; $P_{y,k}$ is the applied transmit power of UE_x in a subcarrier of the kth sub-channel; $Lp_{y,i}$ is the PL between user UE_y and cell C_i; G_y and G_i stand for the antenna gains for UE_y and C_i, respectively while L_y and L_i stand for the equipment losses in UE_y and C_i. Shadowing effects and multi-path fading should be taken into account; computing L_p. L_p can be deduced as:

$$L_p = L_{att} \cdot L_s \cdot L_{ff} \tag{2.3}$$

Where L_{att} is the attenuation, L_s is the shadow fading and L_{ff} is multi-path fading. Therefore, the SINR of each slot, $slot_{i,k,t}$ can be expressed as follows:

$$SINR = \frac{C}{I+\sigma} \tag{2.4}$$

where, C is the received power of the carrier and I the interfering signals. σ denotes the background noise. The received signal power C can expressed as:

$$C_{x,k}^{DL} = P_{i,k} \cdot G_i \cdot L_i \cdot L_{pi,x} \cdot G_x \cdot L_x \tag{2.5}$$

$$C_{i,k}^{uL} = P_{x,k} \cdot G_x \cdot L_x \cdot L_{px,i} \cdot G_i \cdot L_i \tag{2.6}$$

The background noise, σ, on the other hand, can be deduced by:

$$\sigma_n = n_o + nf_{eq} \tag{2.7}$$

$$n_o = -174\frac{dBm}{Hz}\cdot 10\log\left(F_{sam}\cdot\frac{SC_{used}}{SC_{total}}\right) \tag{2.8}$$

where, n_0 is noise, and nf_{eq} for the noise figure of the UE. Also, F_{sam} represents the sampling frequency, while SC_{used} and SC_{total} are the number of used and total sub-carriers, respectively.

Once the SINR of all slots allocated to a user are known, the effective SINR of the user is computed using the Mutual Information based Exponential SNR Mapping (MIESM) average [11].

2.2 Platform for Femtocell Deployment

In order to enable femtocells operate within a variety of networks a standard femtocell network architecture is required. This architecture enables a diversity of femtocells from different manufacturers to work in the networks of different operators. This section will cover the physical-layer details of Long Term Evolution (LTE), which comprises time slot structures and available data rates. The current evolutionary step in 3GPP roadmap for future wireless cellular systems was introduced in 3GPP Release 8 in December 2008 with minor improvements in Release 9 and Release 10, respectively [12].

This release is commonly known as the LTE and it introduces enhancements to previous specifications to achieve higher throughput, spectral bandwidth and more flexible spectrum management. The requirements for high peak transmission rates are 100 Mbps for downlink and 50 Mbps for uplink. The LTE specifications introduce a wide range of support for femtocells. The data rates achieved by LTE are higher than those provided by most network interfaces, which increases the advantages of femtocells based on this release. An overview of the main transmission schemes of the LTE radio interface is provided in the subsequent section.

2.3 LTE Architecture Overview

The radio frame of LTE is defined as having a length of 10 ms. This is divided equally into 10 subframes (SF) of duration $T_{FS} = 1ms$ per SF.

Each SF is further divided into $N_{FS} = 2$slots of length $T_S = T_{FS}/2 = 0.5\,ms$. Each SF contains $N_{sym} = 6$ or $N_{sym} = 7$ OFDM symbols on the length of the selected cyclic prefix (CP). An extended CP of 16.7 μs is allowed in LTE, which might be suitable in accommodating delay. However, in femtocells a normal length CP (T_{CP}= 5.2 μs) might be enough due to its limited coverage area and short delay periods as compared with a MBS. More information about the frame structures can be found in [13].

2.3.1 LTE Downlink Transmission

In LTE, the radio transmission in the downlink is OFDMA and it is defined by a subcarrier (SC) spacing of $\Delta f = 7.5\text{kHz}$ and $\Delta f = 15\text{kHz}$ for multicast and all other cases, respectively. A resource block (RB) in OFDMA is equivalent to $N_{sc} = 12$ adjacent SC; Therefore, the total number of SCs contained in 1 RB during a single time slot is $N_{sc}^{rb} = N_{sc} \times N_{sym} = 12.7 = 84$. LTE allows between 6 and 110 RBs based on the frequency that is between 1 and 20 MHz. In LTE, reference symbols (which are transmitted between the first and fifth OFDM symbols) are responsible for the modulation of certain SC in the OFDM grid. Also, the reference symbols are used for cell identification as well for channel sounding. LTE supports QPSK (Quadrature Phase Shift Keying), 16QAM and 64QAM (Quadrature Amplitude Modulation) as modulation schemes. Therefore, the minimum usable data rate of a RB with normal CP occurs for the case of QPSK ($N_{bit} = 2$ bits per symbol). Furthermore, LTE supports MIMO schemes that can accommodate up to four transmitting antennas [14].

2.3.2 LTE Uplink Transmission

In LTE, the radio transmission technology in the UL is known as Single Carrier FDMA (SC-FDMA). Reference signals differ in the UL but are also important as it allows the implementation of coherent demodulation in FAPs. Also, it provides useful insight into channel conditions. Reference signals are also transmitted in LTE for the purpose of channel sounding, which facilitates scheduling in the UL based on accurate channel knowledge. For each OFDM symbol in LTE, different RBs can be allocated to UEs. Most of the Physical (PHY) layer functionality in UL that includes but not restricted to channel coding, Hybrid Automatic Repeat Request (HARQ), Cyclic Redundancy check (CRC) insertion, inter leaving, scrambling and data modulation is similar to DL.

2.4 LTE Femtocell Interference Analysis

In this section, we provide a brief analysis of the advantages and disadvantages of femtocells in a two-tier interference scenario. Consider an OFDMA system where the femtocell and macrocell are deployed in a co-channel fashion. A tri-sector MBS is in the centre of the network and serves the randomly distributed 90 MUEs within its coverage area. Since there is more activity of MUEs in the downlink (DL) and less activity in the UL, the analysis is conducted on downlink interference. Sixty femtocells are randomly located within the coverage area of the MBS and accordingly there are number of MUEs within the coverage area of the femtocells. The simulation parameters are based on 3GPP LTE specifications [15], whereas four FUEs are attached to each FAP. The performance key indicator (PKI) used in the following analysis is SINR.

2.4.1 Scenario 1: Cross-Tier Interference Analysis

1) *Detriment of femtocells* – Figure 2.2a demonstrates the detrimental effects of femtocells. In a co-channel deployment, MUEs served by the MBS are interfered by the closely located FAPs, thereby resulting into reduced SINR (grey shaded curve). Whereas in the absence of femtocells, (i.e. FAPs), the SINR improves around 20 dB (black curve).
2) *Importance of femtocells* – In this analysis, the importance of femtocells in a network is demonstrated with the help of Figure 2.2b. This plot compares the output from a scenario where firstly all the UEs are located indoor and served by the respective MBS, thereby resulting in a seriously degraded SINR (grey curve). Later on, the same UEs become part of an FAP (i.e. FUEs) and subsequently are served by the respective femtocell access point, thereby improving the SINR (on average 11 dB, black curve).

2.4.2 Scenario 2: Effects of Femtocell Access Mode Deployment

The possible access modes of femtocells could be either an open or closed subscriber group (i.e. Open Subscriber Group - Open Access (OSG) and Close Subscriber Group - Close Access (CSG), respectively). Accordingly, the deployment mode can affect the SINR values of nearby

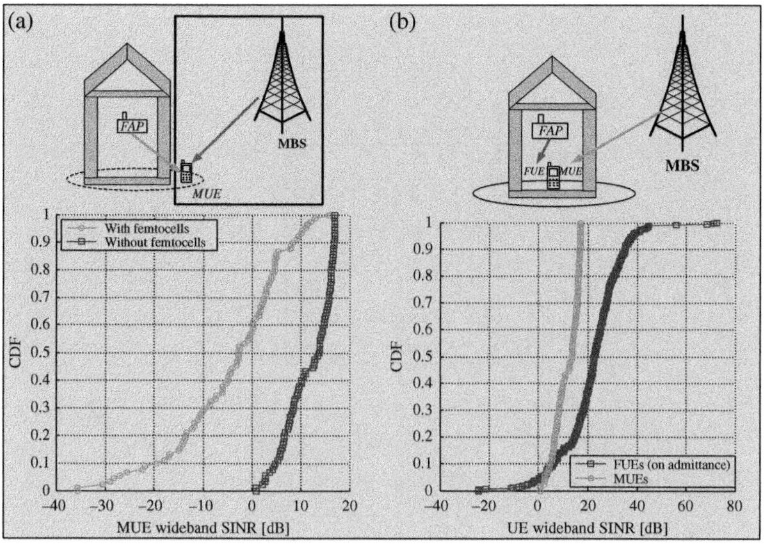

Figure 2.2 Cross-tier analysis with; (a) the detriment and (b) importance of femtocells in a heterogeneous network.

MUEs as shown in Figure 2.3. In CSG, the MUEs SINR suffer more degradation, that is, of around 8 dB compared to OSG due to the fact that OSG allows admittance of MUEs (to become FUEs) as opposed to CSG, which inhibits admittance thereby resulting into reduced SINR value.

2.4.3 Scenario 3: Co-Tier Interference Analysis

Figure 2.4 is a plot of two femtocell scenarios to include a standalone femtocell (STA FAP) and two collocated femtocells (COL FAP). SINR values show that the FUEs in standalone femtocells can reach average values of 29 dB, whereas in a collocated scenario FUEs can suffer an average SINR loss of 17 dB due to co-tier interference component.

2.4.4 Scenario 4: Effects of Varying FAP Transmit Power Levels on MUEs

It is fair to determine that an adaptive power level for FAPs presents a more viable solution to interference management than a fixed power control scheme. Figure 2.5 compares the plots for three power

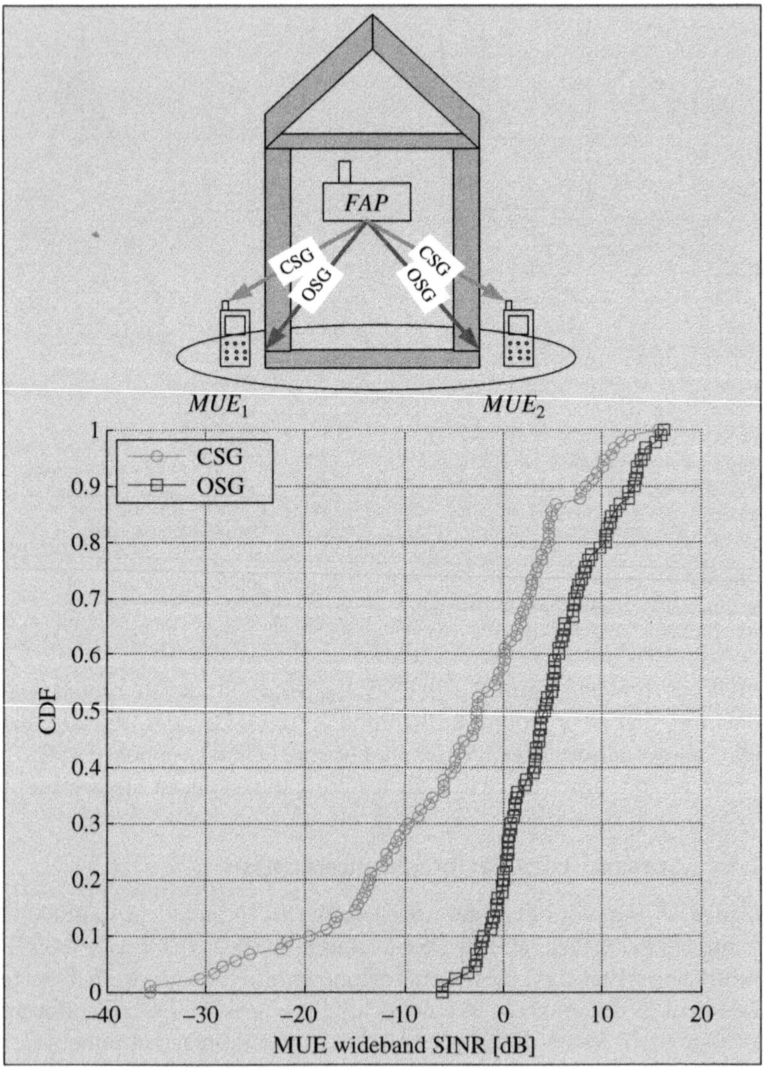

Figure 2.3 Effect of femtocell access mode on MUEs.

levels (0, 10 and 20 dBm) in terms of their effect on close by MUEs in a CSG deployed femtocell. At 20 and 10 dBm, MUEs attain average SINR values of –2 and 3 dB, respectively; however, an improved average SINR of 5 dB is experienced by MUEs when FAP transmit power is 0 dBm.

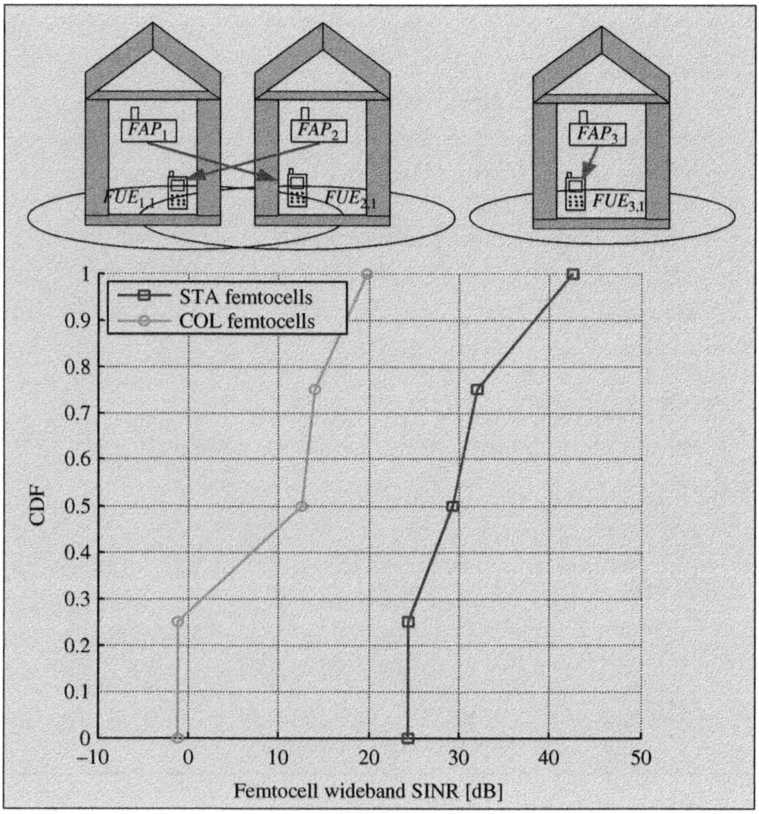

Figure 2.4 Co-tier interference in femtocells.

2.5 Interference Mitigation: Current State of the Art

The interference scenarios mentioned previously require interference mitigation schemes to curb the mutual interference in a heterogeneous network. Next are some of the conventional methods implemented in research today to mitigate interference.

2.5.1 Spectrum Access/Frequency Assignment

These schemes necessitate methods where a femtocell allocates its UE a spectrum with limited or no interference with collocated FAPs or

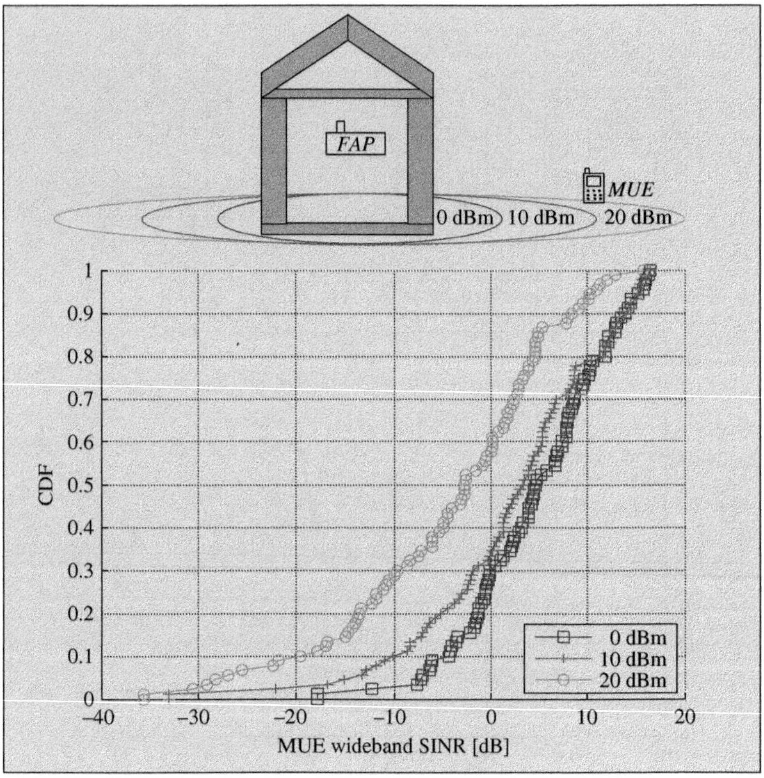

Figure 2.5 Varying FAP power levels and effect on MUEs.

FUE (co-tier) and MUEs or the MBS (cross-tier). The choice between dedicated or co-channel deployment can be employed with considerations such as the amount of vacant spectrum and density of femtocells in a specified area. A hybrid spectrum access schemes merge deployment modes where the spectrum is halved into different access modes with priority granted to the MBS. The drawback with hybrid schemes is that it needs the FAP to establish which region it falls into, thus introducing additional computation and complexity.

2.5.2 Power Control

The power transmission by FAPs comprises the FAPs traffic power and pilot power that determine the data and cell coverage area, respectively. The effect of interference on collocated FAPs and MBS is dependent on these power levels. A higher pilot power results in a larger coverage area, however, subsequently it has higher chances of causing interference. Power control is not only confined to the FAPs (centralized) as UEs in distributed schemes can also optimize their power levels or assist their FAPs to reduce interference to collocated FAPs and UEs.

2.5.3 Antenna Schemes

Beam directivity of the antennas in FAPs and FUE scan be utilized to avoid interference in heterogeneous networks. Conventional antenna schemes are employed that allow the FAP to direct their beams to Regions of Interest (ROI) while producing a null in other areas, thus mitigating interference.

In summary, the problems related to mitigating interference in femtocells are dependent on its two-tiered architecture in a co-channel mode of deployment. Also, since femtocells are randomly deployed, there is no central coordination between collocated femtocells and the MBS. If some information about the radio environment is known, such as characteristics of the interfering signals or sources, it can help in mitigating the interference. An ideal interference mitigation scheme for femtocells will be one that is aware of the interfering signals/ sources and takes into consideration the best deployment criteria to suit the needs of the subscribers while utilizing the network operators' resources efficiently.

2.6 Cognitive Femtocells: A Smart Solution to a Complex Problem

CR is the ability of a radio frequency (RF) to sense its surroundings and modify its features such as frequency, modulation, power and other operating parameters automatically to dynamically reuse spectrum [16]. It has been considered to be a technology capable of spectrum sensing, management and mobility in an opportunistic manner. Through spectrum sensing and management, it can detect an available and unused spectrum also known as white space or spectrum hole, vacated by a user known as the licensed or primary user (PU) and allocate to an unlicensed secondary user (SU).

Cognitive Femtocells (CF) are the integration of CR in femtocells that signifies a femtocell with CR capabilities [17]. Hence, CR enabled femtocells with the spectrum sensing capability of a CR can avoid interference (co-tier and cross-tier) by sensing the spectrum and assigning resources on a different spectrum to avoid interference. Also, the self optimization capabilities of CFs allow the radio environment to be sensed in a dispersed manner so as not to interfere with operational parameters. This entails a SU opportunistically occupying them in the absence of a PU and also vacating them for another possibly available white space as soon as a PU reclaims it. An accurate and efficient spectrum sensing technique is mostly influenced by two key metrics, sensing speed and accuracy.

The effect of CR on conventional interference mitigation schemes offers the advantages of CR with regards to spectrum management in a coordinated and interference void manner. The CFs have the ability to operate as normal femtocell but can also use an opportunistic spectrum access when a user requires higher QoS for certain services. As mentioned earlier, one of the conventional ways to avoid interference in a two-tier architecture will be in a dedicated approach where each UE is assigned a dedicated spectrum but that establishes a trade-off between interference and available spectrum. In this regard, a co-channel access is preferred at the detriment of interference. Unlike in conventional schemes that are restricted to licensed band only, CFs are able to allocate both licensed and unlicensed spectrum bands to UEs. As mentioned earlier, the random user deployment of FAPs is also a major problem as coordination between FAPs and MBSs is required to mitigate interference in most cases. Conventional schemes mostly employ a direct coordination between FAPs and MBS utilizing the backhaul that introduces overhead and delay as the cost.

In summary, to mitigate interference, the self optimization capabilities of CFs allow the radio environment to be sensed in a distributed manner to retrieve operating parameters. However, the operating parameters and features retrieved is usually specific to the radio technology but offers a wide range of options to include white spaces, SINR, Received Signal Strength Indicator (RSSI), noise, transmit power, channel statistics (channel gain and PL) and so on. By continuous sensing of the parameters, current as well as future interfering sources/signals can be deduced.

Although CR enabled schemes can be viable solution to effectively mitigate interference, there are various lessons that are to be learnt (Table 2.1) to identify the cost of using CR in femtocells.

Table 2.1 Cost of using CR in femtocells.

Tradeoffs	Description
Computational overhead	Cognition incurs increased communicational and computational overhead for FAPs/FUEs, thereby resulting in augmented losses such as delay and energy consumption.
Role specification	CR nodes necessitate role specification to perform spectrum sensing and spectrum sharing that require centralized or distributed coordination subsequently resulting in increased complexity.
Feedback	CR enabled femtocells require exchange and maintenance of channel lists through sophisticated communication mechanisms and feedback.
Signalling overhead	Coordination and feedback require dedicated or some other signalling channels leading to signalling overhead. This overhead can further point to increased delays, energy consumption and loss of bandwidth etc.
Customisation and operator software	Some CR enabled schemes are customized with the provision of guided operator software and hardware. This may well increase the unit cost of a FAP.
Security	Security is a big challenge during sensing, retrieval and feedback processes.

2.7 Summary

This chapter outlined the importance of deploying femtocells as a solution to poor indoor network coverage and introduced 4G as a suitable platform for deployment. However, the advantages of femtocell can be overshadowed with interference when deployed in a co-channel fashion with the MBS. The inherent interference scenarios in the uplink and downlink have been highlighted and a review into current avoidance or mitigation approaches mentioned. The question of whether femtocells are helpful or harmful in a co-channel deployment heavily depends on the effectiveness of an interference mitigating scheme. This chapter highlighted CR as a promising and smart solution that can be applied to conventional schemes to produce interference-aware schemes where present and future interfering sources can be detected and prevented. An ideal interference mitigation scheme for femtocells in a co-channel deployment can involve a CR where hybrid schemes are incorporated to complement each other, while keeping it easier to address both cross and co-tier interference.

References

1 Z. Du, Q. Wu, P. Yang, Y. Xu, J. Wang and Y-D. Yao, "Exploiting user demand diversity in heterogeneous wireless networks," in *Wireless Communications, IEEE Transactions on*, vol. 14, no. 8, Aug. 2015, pp. 4142–4155.

2 V. Chandrasekhar and J. Andrews, "Femtocell networks: A survey" *IEEE Commun. Mag.*, vol. 46, no. 9, Sept. 2008, pp. 59–67.

3 H. O. Kpojime and G. A. Safdar, "Interference mitigation in cognitive radio based femtocells". *IEEE Communications Surveys & Tutorials*, vol. 17, no. 3, Aug. 2015, pp. 1511–1534.

4 P. Demestichas, A. Saatsakis, and W. Koening, "An approach for realizing future Internet with cognitive technologies," *in Proc. 4th Int. Conf. Cognitive Radio-Oriented Wireless Networks and Communications*, Hannover, Germany, 2009, pp. 1–6.

5 M. Chowdhury, Y. Jang, and Z. Haas, "Network evolution and QoS provisioning for integrated femtocell/macrocell networks," *Int. J. Wireless Mobile Networks*, vol. 2, no. 3, 2010, pp. 1–16.

6 T. Zahir, K. Arshad, A. Nakata and K. Moessner, "Interference management in femtocells," *Communications Surveys & Tutorials, IEEE*, vol. 15, no. 1, February 2013, pp. 293–311.

7 CMCC, "RP-100671: Signalling and Procedure for In- Device Coexistence Interference Avoidance," 3GPP TSG RAN Meeting 48, June 2010.

8 S. Lien, K-C. Chen, Y-C. Liang and Y. Lin, "Cognitive radio resource management for future cellular networks," in *IEEE Wireless Communications*, vol. 21, no. 1, Feb. 2014pp. 70–79.

9 M. Danneberg, R. Datta, A. Festag and G. Fettweis, "Experimental testbed for 5G cognitive radio access in 4G LTE cellular systems," in *Sensor Array and Multichannel Signal Processing Workshop (SAM), 2014 IEEE 8th*, 22–25 June 2014, pp. 321–324.

10 J. Kim, A. Ashikhmin, A. van Wijngaarden, E. Soljanin and N. Gopalakrishnan, "On efficient link error prediction based on convex metrics", in *Proc. IEEE 60th Vehicular Technology Conference* (VTC2004-Fall), Sept. 2004, pp. 4190–4194.

11 L. Wan, S. Tsai and M. Almgren, "A fading-insensitive performance metric for a unified link quality model", in *Proc. IEEE Wireless Communications and Networking Conference* (WCNC2006. Apr. 2006, pp. 2110–2114.

12 Technical Specification Group RAN, E-UTRA; LTE physical layer – general description. 3GPP, Tech. Rep. TS 36.201 Version 8.3.0, March 2009.

13 A. Valcarce, G. De La Roche, A. Juttner, D. Lopez-Perez, and J. Zhang, "Applying FDTD to the coverage prediction of WiMAX femtocells", *EURASIP Journal on Wireless Communications and Networking*, Feb. 2009.

14 "Radio Resource Management", Aricent White Paper, January 2008.

15 3GPP, "Simulation assumptions and parameters for FDD HeNB RF requirements," R4–092042, TSG-RAN WG4, Meeting 51, 2009.

16 J. Mitola III and G. Q. Maguire Jr., "Cognitive radio: making software radios more personal," *IEEE Personal Communications*, vol. 6, no. 4, 1999, pp. 13–18.

17 G. Gur, S. Bayhan and F. Alagoz, "Cognitive femtocell networks: an overlay architecture for localized dynamic spectrum access [dynamic spectrum management]," *IEEE Wireless Communications*, vol. 17, no. 4, August 2010, pp. 62–70.

3

Interference Mitigation in Cognitive Radio-Based LTE Femtocells

Ghazanfar Ali Safdar

School of Computer Science and Technology, University of Bedfordshire, Luton, UK

Femtocell architecture involves the use of two separate layers – the macrocell and femtocell layers. In this architecture, the former is the conventional cellular network whereas the latter incorporates a range of shorter range cells. Femtocells are designed to co-exist alongside macrocells providing spatial frequency reuse, higher spectrum efficiency and cover areas where macrocells cannot. Femtocells positioned in the macrocell considerably improve the indoor coverage and provide better user experience. However, interference between the two layers is imminent; therefore, ways to manage it must be employed to efficiently avoid problems such as coverage holes in the macrocells. Essential limits of capacity and attainable data rates also mainly depend on the interference faced by a femtocell network. Recently, Cognitive Radio (CR), which has the ability to sense its environment and accordingly alter its characteristics, for example transmission parameters, has been merged with femtocells to exploit the capabilities of the former in the latter. CR enabled femtocells in a two-tier network can sense the environment and opportunistically allocate both licensed and unlicensed frequency bands to user equipment to avoid interference. This chapter examines interference mitigation in femtocells using CR and provides comprehensive survey of different CR enabled interference mitigation schemes. Presented schemes such as power control, spectrum

LTE Communications and Networks: Femtocells and Antenna Design Challenges, First Edition. Edited by Masood Ur Rehman and Ghazanfar Ali Safdar.

access, antenna and joint schemes are classified before they are compared for pros and cons. Finally, trade-offs and cost of using CR in femtocells are highlighted with some insight into future research issues and challenges.

3.1 Introduction

Wireless communication has involved relentless years of research and design and comprises cellular telephony, broadcast and satellite television, wireless networking to today's 3GPP and LTE technology. However, cellular telephony networks surpass the others in terms of usage [1]. Although cellular networks were designed to provide mobile voice services and low rate mobile data services, data services have excelled voice and findings show that global data traffic has grown by 280% since 2008 and is expected to double annually for the next 5 years [2]. Importantly, it has already exceeded those expectations in 2010 by nearly tripling and it is further predicted that by 2020 nearly 1 billion people will access the internet using a wireless mobile device [3]. The limited and in some cases under-utilized spectrum cannot accommodate this continuous increase in traffic, therefore network operators have to come up with ways of increasing spectrum efficiency.

The introduction of new or the upgrade of existing wireless standards such as the Institute of Electrical and Electronics Engineers (IEEE) Worldwide Interoperability for Microwave Access (WiMAX) and 3GPP's LTE have been developed to meet traffic and high data rates. Most of the methods to increase spectrum capacity in practice today are aligned towards; (1) improving the macro layer by upgrading the radio access, (2) densifying the macro layer by reducing the inter-site distances and (3) the use of low power nodes to complement the macro layer [4]. The macro layer deployment is the typical approach of deploying Base Stations (BS) in proximity to each other covering large distances with reduced handover frequency. Although it is the backbone of most wireless networks, it has proven to be inefficient as it does not guarantee a high-quality link in situations where the BS and Mobile Station (MS) are relatively far away. Moreover, a BS serving hundreds of contentious users of all vying for resources is old fashioned [5]. Researchers indicate that 50% of all voice calls and most of the data traffic, more than 70%, originate indoors [6]. However, indoor users may suffer from a reduced Received Signal Strength (RSS) due to low signal penetration through the walls or attenuation leading to total loss of signal in situations where the distance between transmitter and receiver is large. There is a need to provide solutions for poor indoor coverage to satisfy consumers. According to [5] the solutions to poor indoor coverage can be classified into two groups; DAS and Distributed Radios.

DAS comprise a group of Remote Antenna Units (RAU) spaced apart, providing not only enhanced indoor signal quality by significantly reducing transmission distance but also reducing transmit power (the power of the reference signal) [7]. Some of the challenges involved in deploying DAS are the choice of antennas and selecting a suitable location [8, 9].

Distributed radios involve the introduction of smaller cells to complement the deficiencies of the larger macrocell and the gains include an efficient spatial reuse of spectrum [10]. These small cells, which include picocells and microcells, are overlaid in the macrocell to provide voice and data services. Due to the two-tier nature of its architecture it is prone to interference, which may result to a low Signal to Interference plus Noise Ratio (SINR) and throughput, and in some cases a total disruption of service. As a result, there is a need to provide interference avoidance and mitigation schemes.

Lately, femtocell technology has emerged that promises to be a viable solution to indoor cellular communication. CR, which has the ability to sense its environment and accordingly alter its characteristics, has recently been merged with femtocells. CR enabled femtocells can sense the environment and opportunistically allocate both licensed and unlicensed frequency bands to user equipment to avoid interference. This chapter provides a thorough investigation and analysis of interference mitigation schemes in femtocells using CR. The chapter is organized as follows: Section 3.2 concisely introduces femtocells and different types of interference challenges experienced by them. Various deployment modes of femtocells and issues related to interference versus deployment are also presented in the same section before the section concludes by discussing typical interference mitigation schemes employed by femtocells. Section 3.3 offers the concept of CR and analysis of interference mitigation schemes using CR is presented. Section 3.4 provides some open research issues and challenges before the chapter is finally summarised in Section 3.5.

3.2 Femtocells

Femtocell provides the solution to poor indoor coverage in cellular communication, which has hugely attracted network operators and stakeholders. Femtocells are low powered, low cost and subscriber controlled units that provide a dedicated BS to indoor subscribers. The concept of femtocells, also known as home base stations, HNB or HeNB [11] and residential small cells [12] was first studied in 1999 by Bells lab of Alcatel-Lucent but it was in 2002 that Motorola announced the first 3G based home base station product [13]. Femtocell units, known as Femtocell Access Points (FAPs), connect standard mobile devices to the network of a mobile operator through residential DSL, optical fibres, cable broadband connections or wireless last-mile technologies as shown in Figure 3.1 [13–16]. By installing FAPs indoors the cell sites are reduced, thereby bringing the transmitter and receiver closer to each other. The use of the subscriber's broadband network to backhaul data offer improved indoor mobile phone coverage for both voice and data because of improved connectivity compared to the MBS. Femtocells are similar to Wi-Fi as both are connected to a wired backhaul but unlike Wi-Fi, femtocells make use of an existing cellular standard for their operation [15, 17].

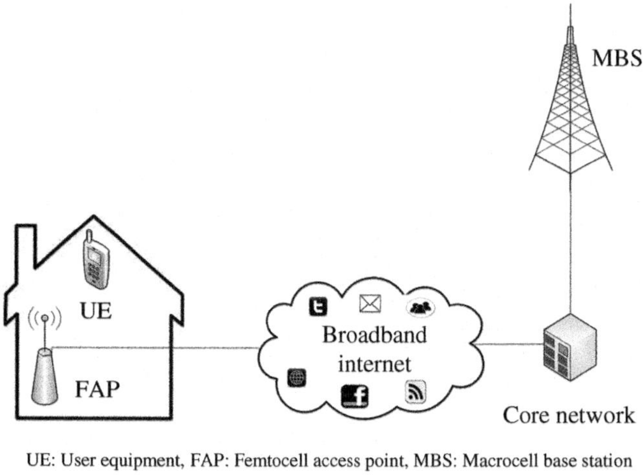

Figure 3.1 Femtocell deployment.

The benefits of femtocells now and in the long term cannot be overemphasized, as they have advantages for both network operators and subscribers. As far as network operators are concerned, the reduction in macrocells due to the deployment of femtocells will result in a huge saving in CAPEX (Capital Expenditure) of network operators and the reduction in traffic will also yield a saving in the OPEX (Operational Expenditure) through advanced self-management and optimization techniques. The subscribers also benefit as the close proximity of the transmitter and receiver offered by femtocells enables subscribers to have high speed services such as voice, video and multimedia. The close proximity greatly lowers transmission power and increases the battery life of mobile devices; with a dedicated FAP in their homes, it offers subscribers a single billing address for mobile phone, broadband and land line as they are all channelled through the same backhaul [13, 18]. Femtocells also act as a solution towards convergence of landline and mobile [19].

However, due to the two-tier architecture of femtocells and macrocells, interference is imminent. The cell sites covered by a number of FAPs (in some cases overlapping each other) is overlaid in the larger cell site of the macro base stations as shown in Figure 3.2. Interference

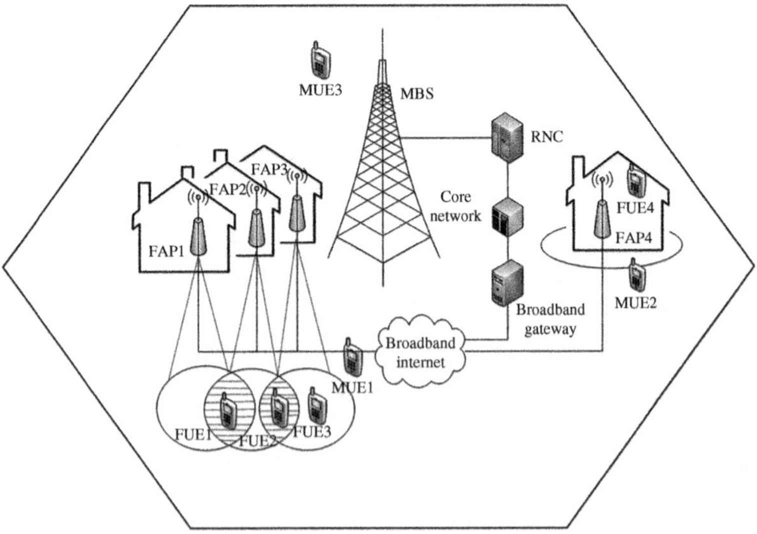

Figure 3.2 Femtocell scenario with co-tier and cross-tier interference.

here denotes the transmitted signals from the FAPs or MBS and their serving user equipment (UE) that appear as unwanted signals to each other.

Interference could be between femtocell and macrocell, which is known as cross-tier, or between neighbouring femtocells known as co-tier [20]. Interference can be further classified as Uplink (UL) or Downlink (DL) based on the sources that, besides the FAP and MBS, also includes the Femtocell User Equipment (FUE: a UE served by a FAP) and Macrocell User Equipment (MUE: a UE served by a MBS).

Uplink interference is caused by;

1) FUE interfering with the signals of a neighbouring FAP or the MBS.
2) MUE interfering with the signals of FAPs (while MUE is communicating with MBS).

While downlink interference is caused by;

1) A FAP interfering with the signals of UEs from a neighbouring FAP or MBS.
2) A MBS interfering with the signals of UEs (sent from FAP to UEs).

A femtocell scenario with four FAPs denoting four cell sites overlaid in a macrocell is shown in Figure 3.2 to describe co-tier and cross-tier interference. The cell sites covered by FAPs 1, 2 and 3 overlap each other thereby causing co-tier inference but they also suffer cross-tier interference as they are overlaid in the macrocell. FAP 4 on the other hand is a standalone FAP, therefore the only interference experienced is cross-tier from the MBS.

3.2.1 Femtocells – Interference versus Deployment

It is important to note that the scale at which interference affects a femtocell network is largely dependent on the deployment scenario. The 3GPP's technical specifications of the scenarios for deployment of femtocells are summarized in [18] and described as follows;

1) Spectrum Usage – Dedicated channel or co-channel deployment
2) Access Methods – Open access or Close Subscriber Group (CSG)
3) Transmit Power – Fixed Downlink (DL) transmit power or Adaptive DL transmit power.

Figure 3.3 summarizes different femtocell deployment scenarios followed by an explanation of interference versus deployment.

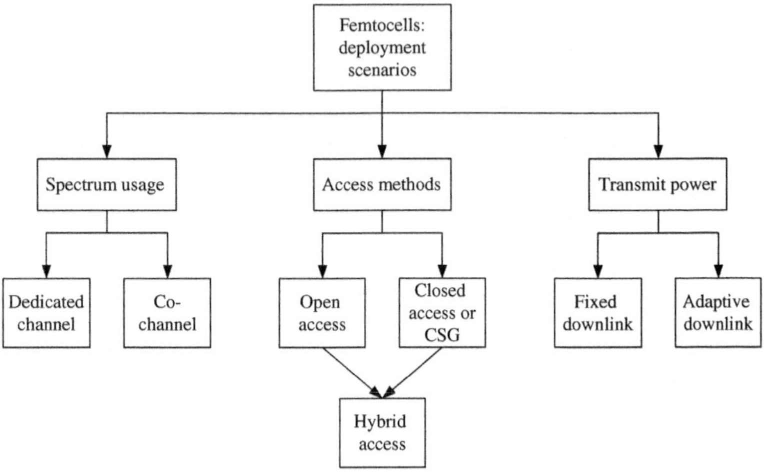

Figure 3.3 Femtocell deployment scenarios.

In a dedicated channel deployment, the licensed spectrum is split into different portions for each tiered network to operate in a dedicated manner whereas both tiers share the same licensed spectrum in co-channel deployment [21]. The choice of deploying any of the two requires a trade-off between spectrum availability and interference. In a dedicated channel, spectrum availability is limited as each portion is assigned a specific bandwidth to utilize. It still leads to a low cross-tier interference. On the other hand, spectrum is available for all users in a co-channel deployment but this result to high cross-tier interference. Network operators prefer a co-channel deployment due to the limited available bandwidth but will have to deal with the interference issues [22–25].

FAPs deployed in an open access allow connection for all users whereas in a CSG mode only the subscribed owners of the FAP have access. Cross-tier interference in open access is reduced due to the fact that users can connect to the nearest FAP with the strongest signal. In CSG, users experience strong UL and DL cross-tier interference. In UL, MUEs close to the FAP and far away from their BSs will have to increase their transmit power to gain a better throughput and SINR causing interference to the FAP. In DL, FAP interferes with the communication between an MUE and its serving BS. However, in CSG, higher SINR values are guaranteed for the served FUEs in contrast to an open access mode as FAP resources are restricted to subscribed users only as shown in Figure 3.4. A new access deployment is the hybrid access that combines open access and CSG by allowing only a limited amount of resources to all users [26]. The results presented in Figure 3.4 have been obtained through simulations where the scenario employs a carrier frequency of 2.14 GHz, bandwidth 20 MHz, MBS transmit power of 46 dBm, randomly distributed MUEs, FAPs, FUEs and a proportional fair scheduler.

In fixed mode, the DL transmit power is set to a predefined value, while in an adaptive mode the transmit power is controlled to avoid perceived interference in the environment, which implies reducing its power depending on its location to a BS such as at the cell edge and cell centre [27]. A very high transmit power will cause interference to neighbouring FAPs and MBSs while a very low transmit power on the other hand will limit the FAP coverage and in turn limit the QoS provided [28]. An adaptive transmit power is preferred over a fixed transmit power because of the inability of the FAP in a fixed mode to alter its transmit power when necessary to avoid interference [19].

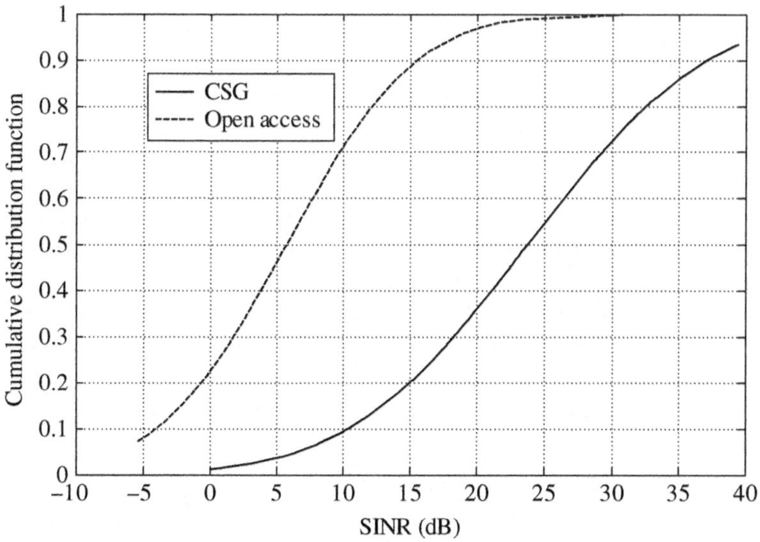

Figure 3.4 Open access versus CSG.

3.2.2 Femtocells – Typical Interference Mitigation Techniques

Although interference management can be classified under ways of avoiding, cancelling or distributed (randomized) approaches [20], the typical schemes to mitigate interference are aligned towards spectrum access/frequency assignment, power control and antenna schemes. Joint schemes consist of a combination of two or more schemes. A brief overview of mentioned schemes is given in the following section.

3.2.2.1 Spectrum Access/Frequency Assignment Schemes

Spectrum access schemes require methods where a FAP assigns its UE a spectrum with limited or no interference with neighbouring FAPs, FUEs, MUEs or the MBS [29–32]. The choice between a dedicated or co-channel deployment is implemented with considerations such as the amount of spectrum available and density of femtocells in a specified region. Hybrid spectrum access schemes combine both deployment modes where the spectrum is split into the two access modes with priority given to the MBS.

An example of a hybrid scheme involves a portion of the spectrum assigned for a dedicated channel mode specific to the MBS and its MUEs, whereas the remaining spectrum is accessed in co-channel fashion both by MBS, FAPs and their UEs [29]. Figure 3.5 illustrates an example of such an implementation with inner and outer regions denoting areas of dedicated and co-channel deployments, respectively. As shown in Figure 3.5, FAP1 overlaps in both inner and outer regions. This means FUEs falling into the region of intersection will have to access the spectrum based on that mode, whereas FAP2 and its UEs on the other hand access the spectrum only in a co-channel mode. The problem with hybrid schemes such as this is that it requires the FAP to ascertain which region it falls into, which introduces additional computation and complexity.

Frequency assignment schemes mitigate interference by exploiting the knowledge of the resources as well as the direct coordination between FAPs and MBS. These schemes thus prevent the use of Resource Blocks (RBs) of neighbouring FAPs and MBS such as in LTE systems [19].

3.2.2.2 Power Control (PC) Schemes

The radiated power transmitted by FAPs comprises the FAPs pilot power, which determines the cell coverage area and traffic power (consists of signalling and data) [33]. The effect of interference on

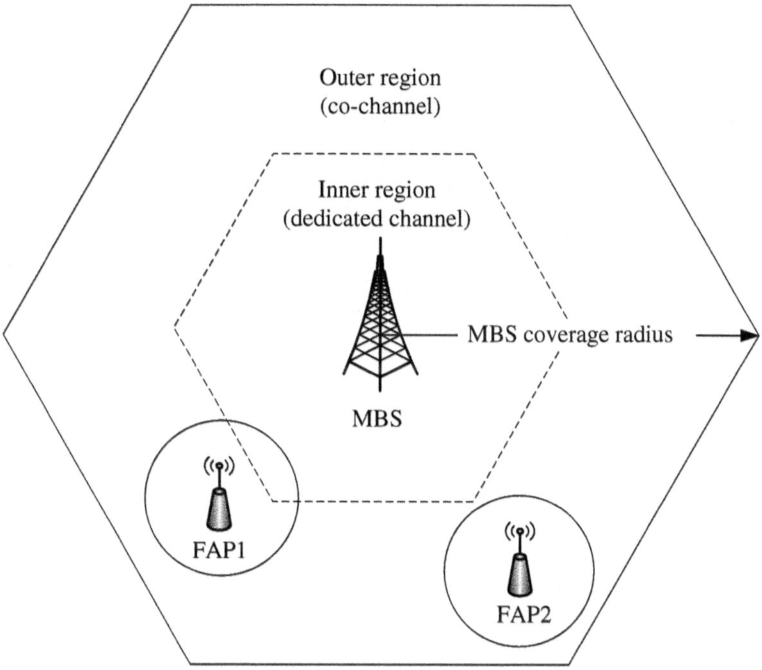

Figure 3.5 Interference mitigation – hybrid spectrum allocation.

other FAPs and the macrocell is dependent on these two power levels. A high pilot power will result into a large cell coverage area, which consequently has higher chances of causing interference. There is a need to optimize the transmit power in femtocells to avoid interference while maintaining a certain QoS.

Figure 3.6 illustrates a scenario where FAP1 and FAP2 are deployed in two flats, Flat A and Flat B, serving FUE1 and FUE2 respectively with their pilot power levels overlapping each other (solid circles in Figure 3.5). In this scenario, FAP1 employs a PC mechanism; that is, reduces its pilot power (i.e. coverage area, the dotted region in Figure 3.6) thus preventing co-tier interference to FUE2 and FAP2 and cross-tier interference with the MUE. PC, however, is not only restricted to the FAPs as UEs can also optimize their power levels or assist their FAPs to reduce interference to neighbouring FAPs and other UEs. Some of the typical PC schemes employed by femtocells are described in [33–40].

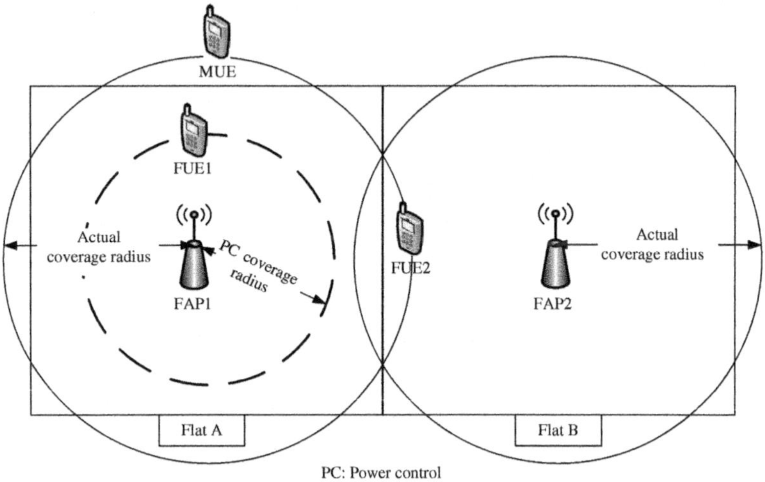

Figure 3.6 Interference mitigation – PC.

3.2.2.3 Antenna Schemes

Beam directivity of the antennas, both in FAP and FUEs, can be exploited to avoid interference in femtocell networks [41–44]. Typical antenna schemes are implemented that allow the FAP to direct their beams to specific UEs while creating a null in other zones, thus cancelling the interference to a greater extent. The scenario represented earlier in Figure 3.6 is replicated in Figure 3.7 to highlight how antenna schemes can generally be used to mitigate interference in femtocells.

To sum up, the problems associated with mitigating interference in femtocells are largely dependent on its tiered architecture and random deployment where there is no central coordination between its neighbouring femtocells and the macrocell network. Importantly, information about the radio environment, such as characteristics of all the interfering signals, if known, can be controlled to help mitigate interference in femtocells. An ideal femtocell interference mitigation technique is one that is aware of the interfering signals and takes into consideration the best deployment criteria to suit subscribers' needs while efficiently utilizing the network operators' resources. One example is the use of a co-channel deployment in a scarce spectrum environment and the use of CSG for subscribers who consider security a priority. In this regard, one of the most promising radio technologies that can sense its environment and allocate scarce radio resources

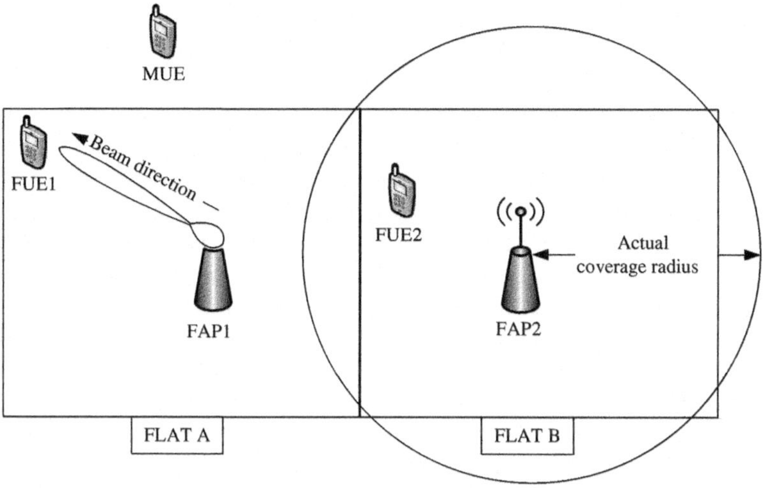

Figure 3.7 Interference mitigation – antenna schemes.

accordingly, known as CR, can be employed [45]. CR is briefly introduced in the next section before the focus of the chapter is made towards its core contribution, which is interference mitigation in femtocells using CR.

3.3 Interference Mitigation in Femtocells using Cognitive Radio

The Federal Communications Commission (FCC) and the European Telecommunications Standards Institute (ETSI) acknowledged the scarcity of the available spectrum thus the idea of CR came as a solution to the limited wireless spectrum where most of the frequency bands are already assigned or in some cases under-utilized [46]. CR is the ability of Radio Frequency (RF) to sense its environment and automatically alter its characteristics such as frequency, modulation, power and other operating parameters to dynamically reuse whatever spectrum is available [47]. Unlicensed spectrum bands such as the Industrial, Scientific and Medical (ISM) [1], which is set aside to encourage innovation, is either too congested or heavily under-utilized. The FCC released a 'notice of proposed rule-making' for the use of technological capabilities in the exploration of unused bands such as television broadcast bands [48].

CR has been regarded as a technology capable of achieving this with its opportunistic ability of spectrum sensing, management and mobility [28, 49–52]. Through spectrum sensing and management CR can detect the availability of an unused licensed spectrum also known as spectrum hole or white space, vacated by the user who is the licensed or primary user (PU) and assign it to an unlicensed secondary user (SU). Also, CR detects and vacates the spectrum into another spectrum hole as soon as a PU re-emerges while maintaining its seamless connection, which defines its mobility capability. A three-dimensional view of white spaces and spectrum occupancy is shown in Figure 3.8.

CR, through its ability to manage the spectrum and handoff when appropriate, allows a SU to choose the best frequency band among the available to support QoS requirements [53]. In situations where the FAPs overlap each other as in Figure 3.2, a dynamic spectrum access technique (DSA) is required to allow SUs (i.e. FUEs) share spectrum resources between themselves and the PUs while avoiding interference. There are mainly three spectra sharing concepts in CR, named underlay, overlay and interweave, which can be applied to mitigate interference while also ensuring that maximum capacity is maintained [54].

Cognitive Femtocells: The integration of CR in femtocells is known as Cognitive Femtocells (CFs) [55], or the Cognitive Femtocell Base Station (CFBS) [56], which signifies a femtocell with CR capabilities. Femtocells having the spectrum sensing capabilities of a CR can avoid

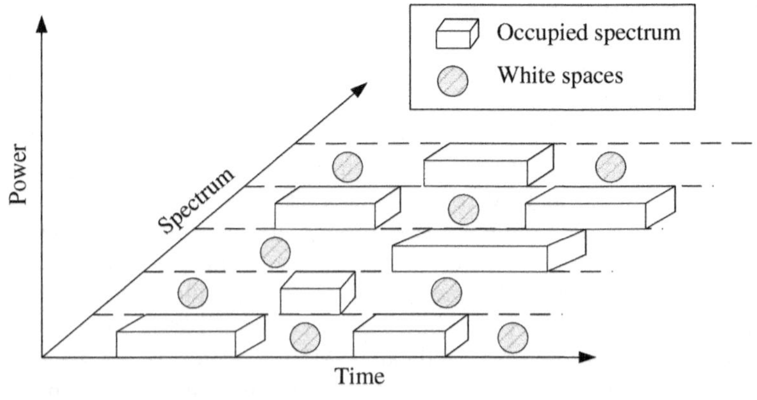

Figure 3.8 White spaces denoting availability of spectrum.

co-tier and cross-tier interference by sensing the spectrum and allocating resources on a different spectrum to avoid interference. In this regard, The FUEs are able to sense the environment for white spaces and opportunistically occupy them in the absence of a PU and also vacate them for another possibly available white space as soon as the PU reclaims it. The self-optimization capabilities of CFs allow the radio environment to be sensed in a dispersed manner to salvage operational parameters in order to manage interference. Typical spectrum sensing techniques employed by CFs are explained next.

3.3.1 Cognitive Interference Mitigation

Figure 3.9 provides an overview of the cognitive interference mitigation schemes discussed in this chapter. Presented schemes are classified based on their functionality and conferred in greater detail before salient features offered by each scheme are highlighted.

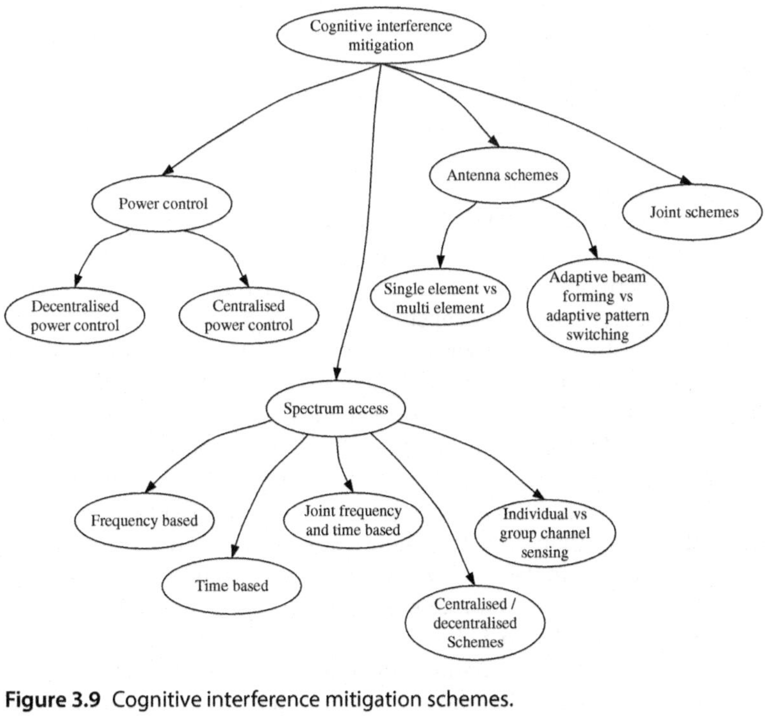

Figure 3.9 Cognitive interference mitigation schemes.

3.3.1.1 Cognitive Interference Mitigation – PC

PC schemes can either be decentralised or centralised, as presented in the following subsections.

3.3.1.1.1 Decentralised PC

A scheme that requires the exploitation of a MBS control channel by an FAP and FUEs in a network to mitigate interference is presented in [57]. By using CR, a FAP and its UEs are able to decode control information such as number, location and power of each active MUE in a spectrum in order to adjust their transmit power. Two rules are implemented after the acquisition of the necessary control information to mitigate interference. In the first rule, the MBS and MUEs are given top transmission priorities in what is known as the forbidden radius, which is the area of an active MUE or a MBS. FAPs within the range of this forbidden radius are restricted to transmit. In the second rule, FAPs and FUEs are tasked with controlling their transmit power such that the interference temperature received at the MBS or MUE does not exceed a set value. A MUE assisted PC scheme is adopted in [58] where MUE measures the received power from its serving MBS and forwards the information to all FAPs in its vicinity. Each FAP subsequently calculates its path loss from the MUE utilizing CR to optimize its power level to avoid interference with the MUE.

3.3.1.1.2 Centralised PC

As mentioned earlier, FAPs can be deployed in an open, closed and hybrid access that largely depend on the subscriber's choice. A novel approach with the ability of FAPs to switch between these access modes based on cognitive sensing and PC is presented in [59]. This self-configurability approach requires each FAP to sense the radio environment and identify white spaces or slots with less interference. For each slot identified, a SINR threshold is set based on channel conditions and a PC algorithm updates the transmit power to define its coverage range. This predefined SINR threshold determines the change in access mode. For example, if a received SINR for a particular slot is less than this threshold it means it can accommodate unregistered UEs in its vicinity and therefore switches to open/hybrid access so long as it does not affect the FUEs being served by the FAP, otherwise it switches to a closed/hybrid access to limit the UEs. This scheme highlights the dynamic capabilities of the femtocell; however, in reality, most subscribers would prefer a single CSG mode as added security becomes a factor in open or hybrid modes. Moreover, femtocells are paid for by the subscribers who would not want to share resources with unknown users.

A PC scheme using Q-learning that enables FAPs to allocate power optimally in a cognitive underlay approach is proposed in [60] to mitigate cross-tier interference in the downlink. A FAP carries out a distributed learning technique by sensing the radio environment to observe its state and takes an action to determine its consequences that can be assessed as a reward (low interference and high MUE capacity) or penalty (high interference and low MUE capacity). By repeating this process, it analyses the entire radio environment and is able to find an optimal power allocation policy to mitigate interference while maintaining MUE capacity. The problem with this scheme is that by trying to determine a suitable policy to mitigate interference, it accumulates a lot of signalling overhead that also leads to delay.

Another PC scheme utilizing communication in the uplink (UL) and composed of three phases (channel sensing, channel training and data transmission) is described in [61]. During channel/spectrum sensing, a FAP senses the radio environment by employing any of the well-known techniques to find unoccupied spectrum. A hypothesis test is conducted to make a decision as to whether a spectrum is occupied (MUE(s) present) or null (absent). The channel coefficient between the FAP and the FUE is estimated in channel training by the FUE sending a signal known as training signal to its FAP. The cognitive FAP thus optimizes the rate at which power is transmitted by allowing its FUE transmit at a reduced or maximum power when a MUE is present or absent respectively to avoid interference during data transmission.

The technique proposed in [62] is based on the overlay approach where each FAP in the network periodically sense and deduce the macrocell path loss, denoted *PLM*, between itself and the MBS. Spectrum sensing in this case is divided into two stages, the uplink and downlink sensing. In the uplink stage, FAP deduces the *PLM* of MUE by measuring a parameter such as Reference Signal Received Power (RSRP). In downlink sensing, based on the *PLM* deduced, the spectrum sensing threshold, denoted the γth, is calculated. Using this information, a FAP is able to identify a channel as unoccupied if the received signal power, denoted *PMC*, on that channel does not exceed γth. On each sensed channel, FAP is able to allocate a Lagrangian based transmit power function on its FUE for data transmission to mitigate interference.

In a nutshell, most PC schemes utilize an adaptive PC mode where the pilot power levels of an FAP is controlled effectively not only to mitigate interference but to reduce the need for handover of close-by UEs in open access mode. Although UEs can also employ PC techniques, distributed PC schemes assisted by UEs of a FAP or MBS may have a detrimental effect on the UEs as it increases their overhead, which may result to increased battery drain. On the other hand, a centralised PC scheme will require a continuous update of the information about all its UEs in real time, which makes it computationally complex. In our opinion, since most UEs are mobile devices it is better to leave the optimization of power to FAPs and MBS (centralised) who have a dedicated power supply and make the computation of the algorithms less complex.

3.3.1.2 Cognitive Interference Mitigation – Spectrum Access

Most spectrum allocation schemes involve the MUE as the PU and the cognitive FAP and its FUE as SUs. This paradigm differentiates and defines how spectrum is allocated in a co-channel mode with top priorities to the MBS and MUEs. Cognitive spectrum access schemes are largely based on the opportunities that exist in various domains such as frequency and time. In the following subsections, the spectrum access schemes are grouped based on these opportunities as well as how they are implemented (distributed/centralised) and how the channels are sensed (individual/group sensing).

3.3.1.2.1 Frequency Based

Unlike the conventional PU and SU analogy, a scheme that recognizes a femtocell in the same regard as macrocell is proposed in [63] with the argument that FAPs could be densely populated with large amount of data and traffic requiring high priorities like MBS. The scheme employed in LTE-A Macro-Femto Orthogonal Frequency Division Multiple Access (OFDMA) systems utilizes the concept of cross cognition and graph colouring concept technique to mitigate cross-tier and co-tier interference, respectively. In cross cognition, all the parties perform cognition in the system (FAPs FUEs, MBSs and MUEs) and the spectrum is divided into licensed and unlicensed parts with the MBSs and FAPs having access to both parts. FUEs and MUEs utilize the licensed spectrum offered by the corresponding serving FAP and

MBS but opportunistically utilize the unlicensed spectrum when the licensed spectrum is exhausted.

A scheme to reduce the macro uplink interference (interference from a MUE to a FAP) in heterogeneous networks combining channel sensing and resource scheduling is proposed in [64]. FAPs sense channel occupation by analysing the energy in the sub-channels and subsequently assign those with the lowest interference to its users. The scheme employs the concept of Fractional Frequency Reuse (FFR) where the spectrum is divided into Frequency Assignments (FA) comprising various sub-channels. Since the FAP cannot effectively sense the weak DL from the MBS, it capitalizes on the strong UL transmit power of the macro user as it tries to reach its MBS to find out available FAs. It subsequently employs a hypothesis that determines the presence and absence of a macro user signal [65]. If a macro user is present it immediately vacates the FA to protect the macro users UL and DL signal transmission. Under this hypothesis, the number of sub-channels for each FA, false alarm and detection probability are deduced that helps the FAP determine a FA not occupied (idle) by a macro user. It identifies this FA as the operating FA and in a situation where there is no idle FA, identifies the one with the lowest signal energy as the operating FA. To allocate resources, the FAP estimates the DL interference signature of the sub-channels in the operating FA using an algorithm and allocates it to its users for signal transmission. However, a DL approach is preferred in [66] for WiMAX femtocells by capitalizing on the downlink activity of MUEs by arguing that spectrum sensing accuracy is more achievable since there is more activity of MUEs in the DL and less activity in the UL as proposed in [62]. The scheme that utilizes cyclostationary feature detection for spectrum sensing allocates resources to FUEs only if the activity of MUEs is not significant.

A scheme based on the interweave paradigm with Gale–Shapely spectrum sharing (GSOIA) is presented in [67] where the CFs opportunistically and orthogonally assign PU channels based on a one-to-one matching policy to avoid interference. In GSOIA, the utility of each channel is calculated by each FAP and subsequently sets a back-off timer for each channel. A channel is deemed available for communication if an FAP detects no busy tone at the expiration of the timer, otherwise it abandons it and waits for the expiration of the next back-off timer until each FAP is able to communicate on an available channel.

To limit the downlink cross-tier interference of a FAP to a nearby MUE, a novel spectrum access scheme is proposed in [68]. In this scheme an MUE joins a nearby open access FAP while freeing up its allocated sub-channel. A FAP capitalizes and takes control of this available sub-channel from the MBS while adding the MUE on the list of its UEs. Utilizing CR, the FAP assigns its FUEs and the MUE sub-channels from its list of available spectra including the freed-up channel. Although open access schemes appear to conveniently tackle cross-tier interference, some issues need to be addressed such as the criteria for joining a FAP. Such a choice is suggested for an FAP with the strongest interference, which, although is feasible in standalone FAPs, will prove to be difficult in a highly populated FAP scenario. This is because close-by MUEs will opt for the same FAP while ignoring other FAPs that prompt for coordination between FAPs in closed access.

A cognitive empowered FUE method is proposed in [69] to assist the FAP in allocating resources. FAP continuously senses the radio environment for available spectrum in what is regarded as the proactive sensing phase. It shares the channel statistics with each of its FUEs who in turn perform a standalone spectrum sensing to verify if the sub-channel are actually 'available'. This is because the 'available' may have been concurrently sensed with a neighbouring FAP who might have occupied it. The FUE subsequently sends an acknowledgement (ACK) message to its FAP to confirm if the sub-channel is still available or otherwise to enable the FAP update its list of available sub-channels. This serves as a measure to tackle co-tier interference. By exploiting the delivered and undelivered packets through the Automatic Repeat request (ARQ) feedback between MBS and MUE, a cognitive FAP is able to mitigate interference as proposed in [70]. A FAP senses the initial DL transmission to decode information about the (re)reception and (re)transmission sub-channels. In the event of an unsuccessful packet the FAP capitalizes on the time between retransmissions to communicate with its FUE, thereby avoiding interference.

Cognitive Radio Resource Management (CRRM) and a Strategic Game-based Radio Resource Management (SGRRM) are combined in [48] to manage cross-tier and co-tier interference, respectively. In CRRM, each FAP cognitively senses the spectrum periodically to identify which RBs are unoccupied. A FAP senses the received interference power for each of the RBs in a frame and compares it against a threshold thereby identifying a RB as occupied if it exceeds this threshold and vice versa. Resultantly, the RBs found unoccupied are assigned to the UEs. SGRRM is developed to avoid co-tier interference due to the selfish nature of FAPs as they view the same resources as unoccupied after cognition and try to assign their UEs to the available spectrum. In SGRRM, overlapped (collocated) FAPs coordinate to autonomously ascertain the total number of RBs that is available and randomize its use to avoid co-tier interference.

3.3.1.2.2 Time Based

The exploitation of available bands in GSM networks can also be used for efficient spectrum access as proposed in [71]. To mitigate interference, it requires FAPs and FUEs to sense a spectrum based on per-time slots rather than the conventional method of continuous sensing of all the available bands. This scheme is based on the possibility of analysing the channels of GSM bands in consecutive frames based on per-time-slots of a common sequence number. First, each FUE carry out a fast standalone cognitive sensing using the current status of the bands where FUE senses an individual spectrum and identify idle slots for transmission as illustrated in Figure 3.10. An FUE performs a fast sensing for a time T_s, which is a fraction of the time slot T_{slot}. A request

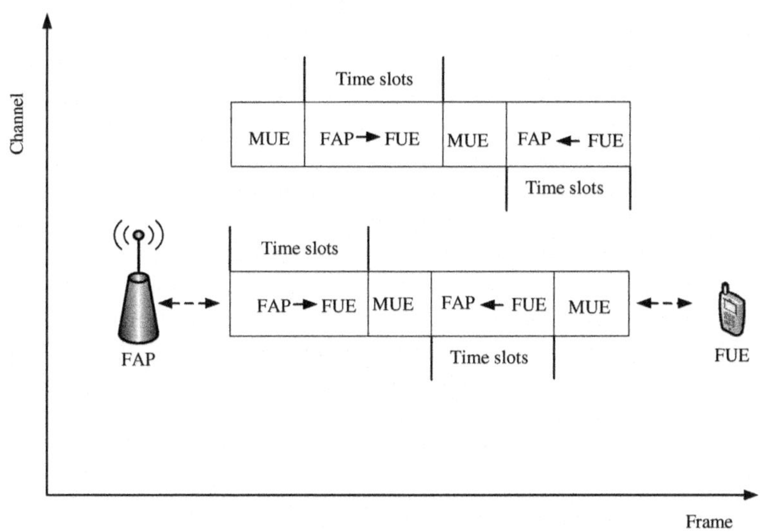

Figure 3.10 Interference mitigation on a per-time slot basis.

message is then sent to the FAP if any idle slot is identified and FAP assigns one or more channels to a FUE for transmission from the available time slots.

To tackle the problem of coordinating CFs in CF networks (CFNs), the use of dedicated signalling channels in LTE networks is proposed in [72]. In the network, each FAP analyses its radio environment as well as the activity of other FAPs and capitalizes on the uplink (UL) communication between a MUE and the MBS (Time Division Duplex mode) to measure the level of interference it has on a MUE. This information is shared between the femtocell nodes through the signalling channel and a dynamic spectrum allocation method is implemented to mitigate interference with the macrocell.

3.3.1.2.3 Joint Frequency and Time Based

A cognitive spectrum selection scheme for LTE-A femtocells based on a distributed carrier selection process is presented in [73]. This scheme employs carrier aggregation in LTE-A systems where spectrum is made available by combining component carriers (CC) on to the physical layer allowing UE to be assigned single or multiple CCs for transmission. To mitigate interference, each FAP must transmit on a different CC and select an anchor CC known as Primary CC (PCC), which acts as a control for the other CCs as illustrated in Figure 3.11. To select a PCC, an FAP senses its environment and randomly picks a CC if no other FAPs are on site. In a situation where other FAPs are detected, it analyses which CCs are occupied by its neighbours and then chooses the farthest. A secondary CC (SCC) selection is conducted if a PCC fails to satisfy the requirements of a FUE by cognitively measuring the path loss between FUE and neighbouring FAPs.

A co-tier and cross-tier downlink cognitive interference management scheme for heterogeneous femto/macro cell networks in CSG mode is proposed in [74]. This scheme allocates resources both in time and frequency in terms of transmission opportunities and RBs. It requires feedback from UEs and coordination between neighbouring FAPs and MBSs to ascertain the level of interference caused to each other's UEs. The scheme involves identifying the victims of interference by each UE in the network measuring the RSRP of other FAPs and MBS and reporting it back to its serving FAP/MBS. Each FAP and MBS in turn classifies the UEs in a safe/victim table that it shares with its neighbours similar to a routing table scenario (Figure 3.12, Table 3.I). A UE is considered to be safe if it is free from an FAP or

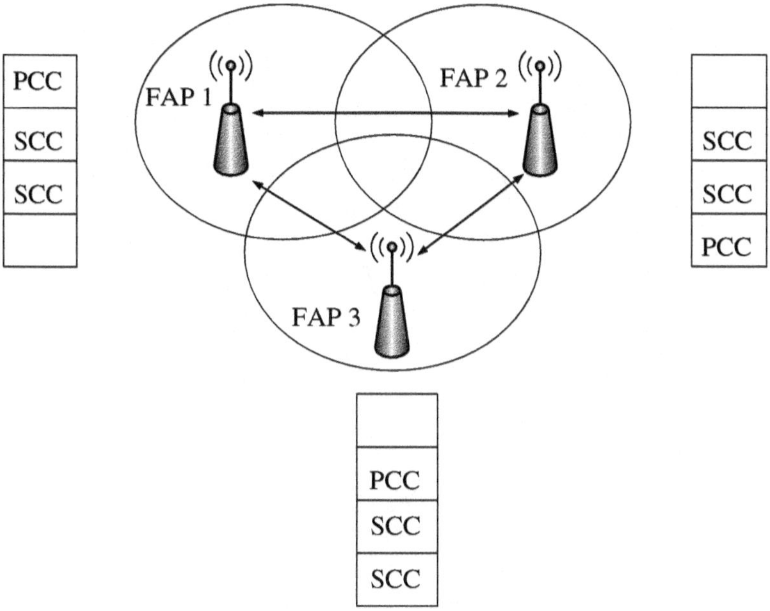

Figure 3.11 Distributed carrier selection process – PCC and SCC.

MBS interference and likewise considered a victim if it experiences a significant level of interference from a FAP or MBS.

Four different spectrum access schemes, Slotted Aloha, Opportunistic Aloha, Distance Sense Multiple Access (DSMA) and Opportunistic DSMA, are proposed in [75]. In Slotted Aloha, each FAP can only access a random subset of the spectrum sub-channels with equal probability, if it gets a head in a coin toss thereby preventing co-tier interference. In opportunistic Aloha a predefined threshold is set and an FAP is only allocated a sub-channel if its channel gain exceeds threshold. In DSMA, each FAP creates a guard zone which is given a radius to protect itself from interference. It notifies interferers of its guard zone by generating and exchanging real numbers between (0, 1) with the smallest number identifying its location. Lastly, Opportunistic DSMA combines Opportunistic Aloha and DSMA, the FAP with the highest gain in the guard zone is given preference in this scheme. A cognitive hybrid division duplex (CHDD) suitable for heterogeneous networks is proposed in [76] where both frequency division duplex (FDD) and time division duplex (TDD) are simultaneously performed

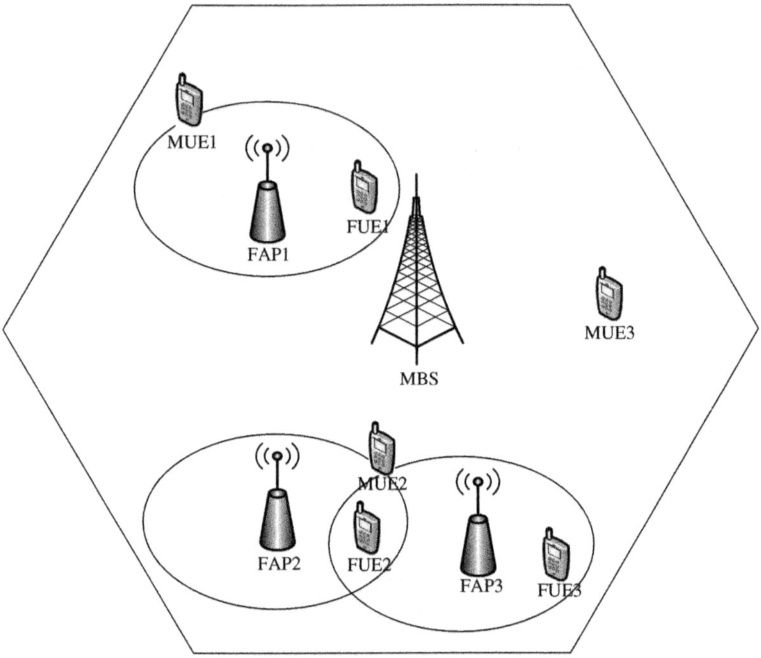

Figure 3.12 Interference mitigation scenario to illustrate safe/victim UE.

Table 3.1 Classification of safe/victim UE.

	Safe/Victim			
MUE/FUE	**MBS**	**FAP1**	**FAP2**	**FAP3**
MUE1	Safe	Victim	Safe	Safe
MUE2	Safe	Safe	Victim	Victim
MUE3	Safe	Safe	Safe	Safe
FUE1	Victim	Safe	Safe	Safe
FUE2	Victim	Safe	Safe	Victim
FUE3	Victim	Safe	Safe	Safe

by CF underlaid in the macrocell. In this scheme FDD is implemented at the macrocell tier to prevent cross-tier interference whereas TDD is implemented at the femtocell tier to prevent co-tier interference.

3.3.1.2.4 Centralised/Decentralised Schemes

A group of CFs can form a Cognitive Femto Network (CFN) based on a criterion such as location to enhance the sensing capabilities and share channel conditions to effectively mitigate interference. Group sensing in CFNs greatly improves the capabilities of sensing and the power levels of each node can be maintained or reduced accordingly based on the shared information. A scheme built around this concept can be implemented in a centralised or decentralised approach to assign spectrum in a CFN [77]. However, coordination on information gathering and spectrum allocation besides specification of the roles of each sensing node is a major issue in these schemes.

A game theory approach presented in [78] makes use of correlated equilibrium policy [79] to mitigate co-tier interference among cognitive FAPs for the downlink OFDMA LTE networks. Correlated equilibrium is used in preference to the traditional Nash equilibrium policy as it is a decentralised and adaptive algorithm that allows the agents to directly coordinate their strategies achieving better performance. In this case, the decision to allocate spectrum is assisted by the global and local FAP utility functions. The global utility function is an analysis of the entire network and provides fairness by giving spectrum access priority to all FAPs and it is dependent on factors such as the demand levels of each FAP. On the other hand, the local FAP utility function is dependent on each FAP to make the relevant decision to maximize its estimate of the global utility function.

The Logit Equilibrium (LE) is another decentralised scheme presented in [80] where FAPs experiment all the available options after a distributed learning algorithm to mitigate interference. The interference scheme that incorporates game and learning theory with a stochastic approximation relies on sensing of the SINR of MUEs acquired through communication with its serving MBS. FAP utilizes the constant SINR information to dynamically configure the probability distribution of the spectrum, which includes information about the spectrum and transmit power. Interference is mitigated by guaranteeing a minimum time-average SINR for the MUEs in the network at the equilibrium.

An adaptive spectrum access scheme is proposed in [81] that involves a central FAP with a two-fold CR management. The two-fold management comprises the Cognitive Manager for Spectrum Management (CM-SM), which is tasked with providing the spectrum opportunities by sensing the radio environment and accessing all the necessary information such as geo-location databases, whereas Cognitive Manager for Resource Management (CM-RM) manages the allocation of spectrum to all FUEs. The combination of the two processes provides the information for active channels (used by MUEs), operating channels (used by the allocated FUEs) and the reserve channels, which are the channels that are reserved for FUEs to relocate in case PU returns. In this scheme, the FUE also have CR capabilities that make it possible for them to report to the FAP for any interference. Upon receiving feedback from FUEs, the FAPs also request for interference measurements from neighbouring FAPs and subsequently tabulates two tables, A and B, based on all the statistics with table A containing a list of interference free resources and B containing resources that are restricted or shared. The level of allowable transmit power for resources in both tables varies with table A allowed a maximum transmit power and table B a dedicated transmit power derived from an algorithm. Accordingly based on this information, FAP allocates resources for FUEs to avoid interference.

A sensing and scheduling information based scheme to avoid cross-tier interference, both UL/DL, for OFDMA femtocell and macrocell wireless networks is introduced in [82]. It encompasses a FAP that assigns RBs to its FUEs not in use by a MUE or in a spectrum reuse approach the RBs of far away MUEs to avoid the interference of closely located MUEs. To accomplish this, a FAP obtains the scheduling information of MUEs from the MBS using two methods: the first method includes the backhaul connection and the second method involves the air interface. The second approach which makes use of air interface implies the FAP connecting to the MBS as a MUE and retrieving the information. The FAP then senses the spectrum to find the occupied RBs of nearby MUE through methods such as energy detection and compares it with the received scheduling algorithm and assigns a RB to an FUE from the available opportunities (Figure 3.13).

3.3.1.2.5 *Individual versus Group Channel Sensing*

Some schemes prefer individual channel sensing compared to group channel sensing. An FAP retrieves the SINR and received power of

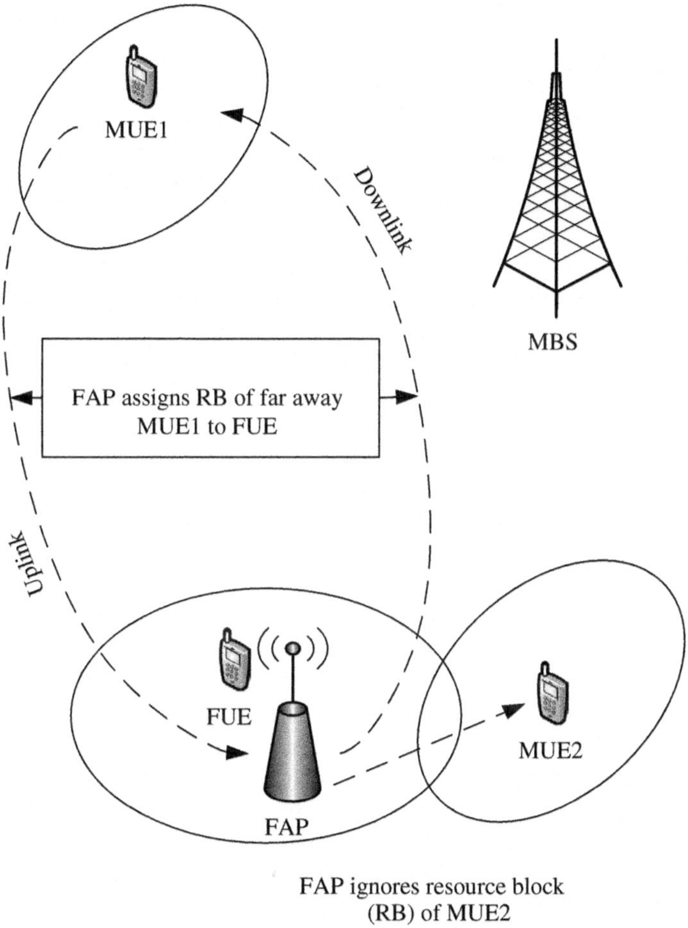

Figure 3.13 Spectrum reuse – FAP assigns RBs of a far away MUE to FUE.

each individual channel of the spectrum utilizing the UL communication of nearby MUEs. By sensing individual channels, the femtocell is not only able to calculate the level of interference but also the style of interference [83]. In such a scheme, each femtocell is classified into three categories of channels:

1) MUE occupied channels (*Mc*) – MUEs the source of interference to FAPs and FUEs

2) FAP interfered channels (*Fc*) – FAPs the source of interference to neighbouring FAPs and FUEs
3) Qualified Channels (*Qc*) – Channels that are free from both FAP and MUE interference.

Both co-tier and cross-tier interference can be mitigated by using the aforementioned channel classification. To mitigate cross-tier interference, FUEs are allocated orthogonal resources with *Mc* while an algorithm that consists of a predefined threshold, γ, with $\text{SINR} \geq \gamma = Qc$ or $\text{SINR} < \gamma = Fc$ to distinguish *Qc* from *Fc* is implemented to mitigate co-tier interference. Individual spectrum sensing can decrease the probability of false detection but it introduces sensing overhead and the idea of group of FAPs sensing an individual channel concurrently may degrade the whole process of sensing. Cooperative sensing performed over the entire range of available channels help to improve probability of misdetection and solve the hidden node problem. A hybrid channel sensing is proposed in [84] where cooperative sensing of the available channels is conducted by all the CF to mitigate cross and co-tier interference.

Most spectrum access schemes are based on the analogy of the two-tier system, where the MUE is considered a PU and cognitive FAP and its FUEs are considered SUs. Spectrum access in cognitive-enabled femtocells is not only restricted to licensed resources but spectrum in TV white spaces (TVWS) can also be utilized. Spectrum access schemes mostly employ channel sensing and since spectrum access approaches could also be either centralised or distributed, a distributed spectrum access approach that requires no or little coordination is more feasible due to the fact that FAPs are usually randomly and densely deployed. Since there is more MUEs activity in the downlink, spectrum access schemes that involve spectrum sensing in the downlink are more promising compared to the uplink schemes. An ideal cognitive spectrum access scheme is the one that overlays, underlays or interweaves its signal into the licensed spectrum undetected without causing any harm to PU or satisfying all the PU constraints.

3.3.1.3 Cognitive Interference Mitigation – Antenna Schemes

The CR enabled antenna schemes are sub-divided in single versus multi-element antenna schemes and adaptive beam-forming versus adaptive pattern switching schemes as described next.

3.3.1.3.1 Single-Element versus Multi-Element

A low cost multi-element antenna and UE assisted cognition has been employed in [85] to optimize coverage and to mitigate interference. In this scheme, FUEs are employed to sense and retrieve pilot power signals from the MBS and report to their serving FAP at places, described as 'crucial places' (CP), where power leaks are prone to occur such as windows and doors. This involves a subscriber walking around CPs and sending signals to its FAP by pressing a specific button on their device that has guided operator software. The sensed information is processed using the conventional Direction of Arrival (DoA) estimation algorithm on a six-element antenna that adjusts the antenna gains; this method thus optimizes the coverage area of the FAP, thereby mitigating interference. However, the problem with this scheme is that it would require special and compatible software installed on each handset. Also, FAPs are required to accommodate multi-antenna elements for all antenna patterns and this is difficult as FAPs are size constrained.

To tackle the problem of a multi-antenna system, cognitive FAPs equipped with omni-directional antennas form a network known as Smart Cognitive-Femto Network (SCFN) [86]. In this scheme, the cooperation of a group of cognitive FAPs who operate like a BS with multiple antennas dispersed in a large network is exploited. Although each of the cognitive FAPs is equipped with omni-directional antennas, the SCFN acts like a multi-antenna network system. Each FAP can sense the environment to mitigate interference by directing the main only towards a desired FUE and creating a null towards others that may be causing interference (Figure 3.14).

3.3.1.3.2 Adaptive Beam-Forming versus Adaptive Pattern Switching

Two antenna schemes, an adaptive pattern switching and adaptive beam-forming antennas, are designed and independently employed in [87] for cognitive LTE femtocells to exploit the available white spaces in TVWS. In the adaptive pattern switching, four patch antennas in a single FAP is proposed with each element at an angle of 0°, 90°, 180° and 270°, respectively. The FAP measures the Received Signal Strength Indication (RSSI) and the uplink Sounding Reference Signal (SRS), which is a parameter to analyse the quality of the uplink channel in what is known as the training process. One or more of the elements in the antenna can be switched ON or OFF based on this information to a specific target to mitigate interference. The adaptive beam-forming

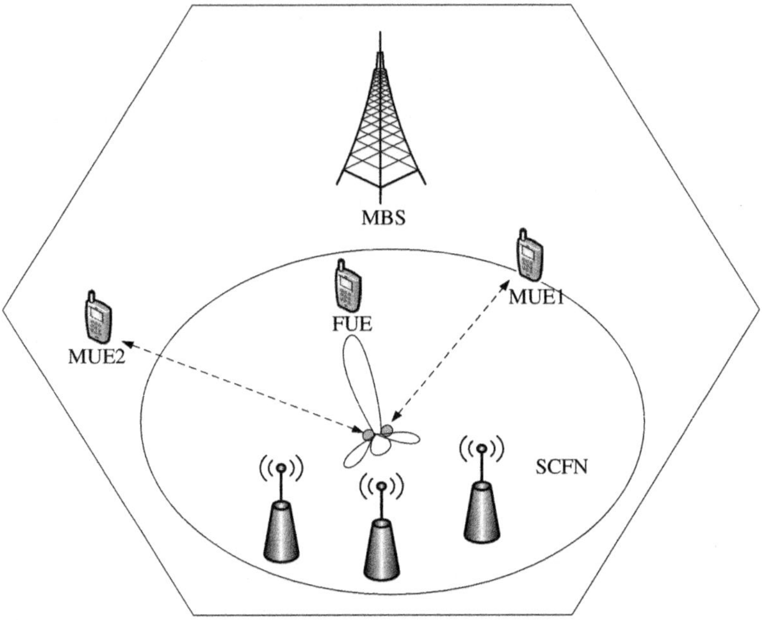

Figure 3.14 SCFN: Interference mitigation with main beam direction.

antenna scheme is similar to the adaptive pattern switching with an antenna array consisting of four elements in each FAP, whereas the difference lies in the fact that it is able to perform the Eigen-based beam-forming introduced in [84].

3.3.1.4 Cognitive Interference Mitigation – Joint Schemes

Joint or hybrid schemes introduce the combination of two more schemes with the aim of complementing and improving the overall performance of a single scheme. Some hybrid schemes are presented in the section that follows.

A Joint PC and cognitive spectrum access is proposed in [56] to mitigate the downlink cross-tier interference. The scheme comprises of FAPs with cognitive capabilities (CFBS) in cognitive femtocell network architecture. However, in this scheme, cognition is not restricted just to the CFBS; the femtocell and macrocell UEs have cognitive abilities too, accordingly known as cognitive femtocell users (CFUs) and cognitive macrocell users (CoMUs), respectively. The common macrocell users (CMUs) are defined to represent macrocell users with no cognitive capabilities. The CFBS that are randomly deployed in the MBS periodically senses the radio environment within its defined range known as cognition range (although it may increase its transmission power to exceed this range) to gather statistics such as SINR, unoccupied spectrum bands and mobility information from all the cognitive users in the environment. To mitigate downlink interference in the network, based on the received statistics, the CFBS controls its power or enhances its coverage area for nearby CoMUs and CMUs. In this case it lowers its transmission power to avoid its interference to a closely located CMU, whereas for the CoMUs it may apply either of the two. For instance, it will increase its cognition area to provide coverage for CoMUs whose SINR fall below a defined threshold or reduce its power to avoid interference on a CoMU with a high SINR from its serving MBS.

A novel spectrum allocation scheme to mitigate interference in LTE-femtocell networks is proposed in [88] that involves the sensing of available unlicensed TVWS. In this scheme, all MUEs in the network sense the transmission power of nearby femtocells and compare it with a predefined threshold and tabulate two lists:

1) Interfering Femtocell List – FAPs with transmission power above threshold
2) Non-Interfering Femtocell List – FAPs with transmission power below threshold

The interfering femtocell list is sent back to the MBS and based on this feedback it allocates RBs and TVWS to MUEs and FAPs, respectively, in the non-interfering and interfering femtocell lists, respectively.

Another similar scheme called Fractional Frequency Donation (FFD) mitigates cross-tier interference in co-channel deployment by utilizing TVWS and cellular bands [89]. In this scheme, each FAP with a good throughput is required to donate bands to poor ones and this is achieved by the FAP first finding suitable donors and subsequently bands to donate. To find a suitable donor a FAP measures a FUE's noise power plus interference power and estimates its average channel gain. After spectrum sensing, a band is regarded as a polluted or polluting if the donor FAP receives strong interference from, or if it causes strong interference to other frequency bands, respectively. As a result, FAP donors perform what is known as Selfish Frequency Access (SFA) without utilizing the polluted or polluting bands. Poor FAPs mitigate cross-tier interference by allocating more power on the donor bands and less on the undonated bands.

A PC algorithm that merges the concept of DSA with fractional PC (FPC) is described in [90]. FPC is used to define the SNR targets of both cell centre and edge users based on a Game-based Resource Allocation in Cognitive Environments (GRACE). GRACE is a fully distributed algorithm that evaluates a marginal utility function that trades off capacity and outgoing interference between CFs and neighbouring CFs and macrocells, respectively. Importantly, cell centre and cell edge are terms used to describe the proximity of a UE to its serving BS. Cell centre users experience a better SNR as they are well within the coverage range of the BS whereas cell edge users are prone to poor SNR due to their increased distance from the serving BS.FPC allows the transmission of high data rate of both cell centre and edge users by allocating power to each channel defining the transmission power and path loss, L, in dB. The path loss in UL is specific for UEs but FPC capitalizes on the advantages of keeping a constant power spectral density in DL, which includes an efficient spectrum sensing by other CFs. It achieves this by setting the DL path loss, L, to its highest among all the UEs served by the FAP which subsequently provides enough transmit power even to the weakest link. In GRACE, which performs a cognitive cycle of sensing, analysis, decision and action, a CF makes decisions to update its channel allocation so as to coordinate the inter-cell interference without a need for inter-cell communication. Figure 3.15 illustrates functioning of FPC scheme with cell centre user FUE_1 and cell edge users FUE_2 and FUE_3, respectively.

A cognitive spectrum reuse and power allocation scheme for 3G femtocells is introduced in [91] where a FAP recognizes the interference signature from its environment and avoids interference in the network by cognitively allocating spectrum to its UE with emphasis on reusing the spectrum in a proper way. In this scheme, the FAP periodically senses the network environment to retrieve the interference signatures of its neighbours. Upon acquiring the interference signatures, an opportunistic channel scheduler is invoked that allocates the best spectrum for reuse, where the best spectrum in this context is the

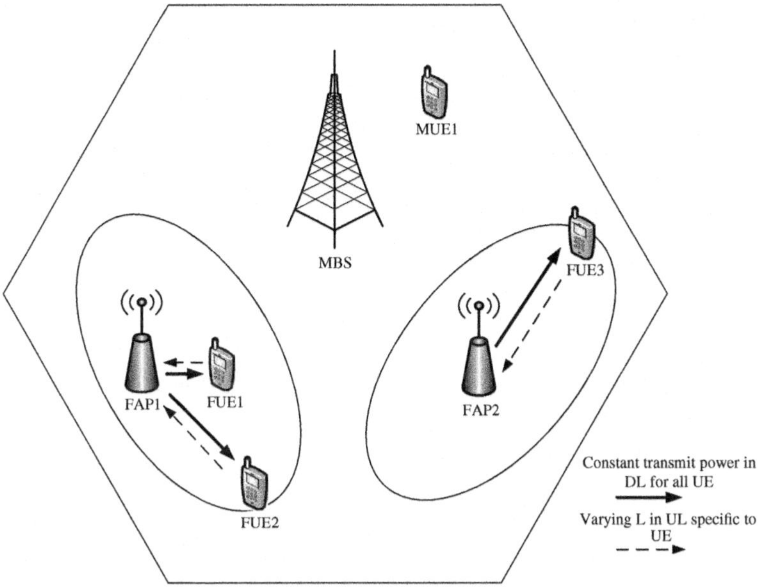

Figure 3.15 Interference mitigation through FPC.

spectrum that causes the lowest interference. The FAP then allocates power to FUE to reuse the spectrum only if the spectrum is found to satisfy QoS, otherwise a handoff is triggered to allocate another spectrum. The same scheme is further studied in [92] and proposed with some improvements to mitigate uplink interference in 4G cellular femtocells.

A channel assignment and power allocation scheme is proposed in [93] for cognitive LTE-femtocell network to minimize downlink co-tier and cross-tier interference. CF utilize available licensed spectrum of the macrocell network as well as TV broadcast systems and avoid co-tier interference based on some safety distance achieved through clustering. Neighbouring femtocells are grouped into Physical Clusters (PCs) based on their physical location and the number of available finite channel so that each femtocell in the cluster uses a different channel to avoid interfering with each other. Femtocells in different PCs but sharing same channel are grouped in Virtual Clusters (VCs). A spatial correlation, however, is defined to avoid the interference that may occur between two PCs that are closely located.

A downlink subcarrier and power allocation scheme is implemented to reduce the imminent aggregate interference due on the subcarriers. In line with the scheme presented in [68], an uplink subcarrier and power allocation scheme that use disruptive solutions against optimal solutions are implemented for OFDMA femtocells in [94] to help mitigate interference.

The pros and cons of CR enabled PC, spectrum access, antenna and joint/hybrid schemes are shown in Table 3.2 whereas Table 3.3 gives a quick synopsis of cognitive interference mitigation schemes discussed in this chapter.

3.3.2 Cognitive Interference Mitigation versus Conventional Interference Mitigation

The effect of CR on conventional interference mitigation schemes cannot be overemphasized as it offers the advantages of CR with regards to spectrum management in a coordinated and interference void manner. The CF have the capability to operate as normal femtocell as well but can also use opportunistic spectrum access when a user requires higher QoS for certain services. As mentioned earlier, one of the ways to mitigate interference in a two-tier network will be in a dedicated manner where each UE is assigned a dedicated spectrum but that introduces a trade-off between interference and available spectrum, thus co-channel is preferred at the detriment of interference. However, CR enabled femtocells in a two-tier network are able to opportunistically allocate both licensed and unlicensed spectrum bands to UEs unlike the conventional schemes that are restricted to licensed bands only. The random user deployment of FAPs is also a major problem as coordination between FAPs and MBSs is required to mitigate interference. Conventional or traditional schemes in most cases employ a direct coordination between FAPs and MBS through the backhaul, which introduces overhead and delay. The self-optimization capabilities of cognitive enabled FAPs and FUEs allows the radio environment to be sensed in a distributed manner to retrieve operating parameters in order to mitigate interference. The operating parameters retrieved is specific to the radio technology but offers a wide range of options to include transmit power, channel statistics (channel gain and path loss), available spectrum (white

Table 3.2 Pros and cons of cognitive interference mitigation schemes.

Scheme	*Pros*	*Cons*
PC	• Power can be optimized in an adaptive manner to mitigate interference • Flexible PC reduces the need for handover in open access as it restricts its coverage area	• UE assisted power optimization schemes introduce signalling overhead, which results in battery drain • PC schemes based on the pilot power of a FAP, which determines its coverage area becomes less significant if a FUE is located close to a neighbouring FAP
Spectrum Access	• Distributed access to spectrum with more spectrum opportunities for femtocells • Access to licensed systems not restricted to MBS but also TVWS • Universal frequency reuse can be achieved in co-channel deployment	• Spectrum inefficiently utilized in dedicated channel deployment • Added security concern introduced in open/hybrid access modes • MUE is given more priority over FUE even when FUE incurs more voice and data traffic
Antenna	• FAP targets a specific FUE while creating a null towards other users • Complex beam-forming schemes can be replaced with antenna switching systems, which reduce signalling overhead	• Multi-antenna elements may be required to accommodate antenna patterns in a FAP that is size constrained • Diversified-antenna elements incur costs that will increase the unit price of a FAP
Joint/ Hybrid	• Combines the advantages of two or more schemes	• Introduces complexity and signalling overhead

spaces), SINR, RSSI, SRS, noise and so on. By continuous sensing of the parameters, current as well as future intentions of interfering sources can be deduced, and in our opinion thus CR brings improved interference mitigation over the conventional schemes due to the previously mentioned facts (Figure 3.16).

Table 3.3 Synopsis of CR enabled interference mitigation schemes.

					Deployment Scenarios					
		Interference Mitigation Schemes			Spectrum Usage		Access Methods		Transmit Power	
[Refs]	*Cognitive Femtocell*	Spectrum Access	Power Control	Antenna Techniques	Dedicated Channel	Co-Channel	Open Access	Closed Access	Fixed Downlink	Adaptive Downlink
57	✓		✓			✓				✓
58	✓		✓			✓				✓
59	✓		✓		✓		✓	✓		✓
60	✓		✓			✓				✓
61	✓		✓			✓				✓
62	✓		✓			✓				✓
63	✓	✓			✓	✓		✓		
64	✓	✓				✓				
66	✓	✓				✓				
67	✓	✓				✓				
68	✓	✓			✓	✓	✓			
69	✓	✓				✓				
70	✓	✓				✓				
71	✓	✓				✓				
72	✓	✓				✓				

73	✓	✓			✓				
74	✓	✓			✓		✓		
75	✓	✓			✓		✓		
76	✓	✓			✓		✓		
77	✓	✓			✓				
78	✓	✓			✓				
79	✓	✓			✓	✓			
80	✓	✓			✓				
81	✓	✓			✓				
82	✓	✓			✓		✓		
83	✓	✓			✓		✓		
85	✓			✓	✓				
86	✓			✓	✓	✓			
87	✓			✓	✓				
88	✓	✓	✓		✓				
89	✓	✓	✓		✓				
90	✓	✓	✓		✓		✓		✓
91	✓	✓	✓		✓		✓	✓	
92	✓	✓	✓		✓				✓
93	✓	✓	✓		✓				
94	✓	✓	✓		✓				

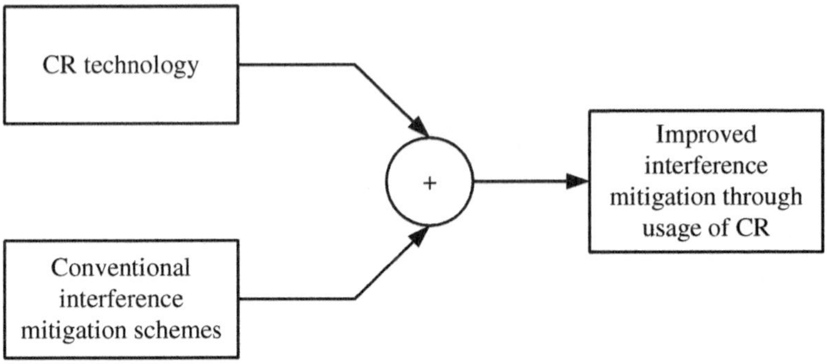

Figure 3.16 CR enabled interference mitigation.

3.4 Summary

Networks nowadays require higher data capacity, improved connectivity and a QoS guarantee. Soon it will not be possible for macrocell based networks to provide the required QoS to all users, thus femtocells play a vital part. Femtocells are considered to be the solution to meet the future needs for high data rates and capacity in the cellular networks. Even though femtocells are discovering an important role, the issue of interference mitigation needs to be addressed. CR has the ability to sense its environment and automatically alter its characteristics such as frequency, modulation, power and other operating parameters to dynamically reuse whatever spectrum is available. The integration of CR in femtocells is known as CF or CFBS, which signifies a femtocell with CR capabilities. Femtocells with the spectrum sensing capabilities of a CR can escape co-tier and cross-tier interference by sensing the spectrum and allocating resources on a different spectrum to avoid interference. Unlike the traditional femtocells, CR enabled femtocells are able to opportunistically allocate both licensed and unlicensed spectrum bands to user equipment. This chapter presented a comprehensive review of CR enabled interference mitigation schemes. The mentioned schemes in the chapter were compared in order to highlight their pros and cons. Tradeoffs and cost of having CR in femtocells was discussed before some of the important research issues were presented. Despite some of the constraints of using CR in femtocells, the conclusion is that CR enabled femtocells can provide improved interference mitigation over traditional femtocells because of the very nature of cognitive radio.

References

1 A.F. Molisch, *Wireless Communications*, 2nd Edition, Wiley-IEEE, February 2011.
2 Ericsson, "Mobile data traffic surpasses voice", March 2010. [Online]. Available: www.ericsson.com/thecompany/press/releases/2010/03/1396928.
3 Cisco, "Cisco visual networking index: Global mobile data traffic forecast update, 20102015," Whitepaper, Feb. 2011.
4 S. Landström, A. Furuskär, K. Johansson, L. Falconetti and F. Kronestedt "Heterogeneous Networks (HetNets) – an approach to increasing cellular capacity and coverage," Feb 11, 2011.
5 Cellular Asset, "Femtocells – Benefits, Challenges and Deployment Scenarios" [Online]. Available: www.cellularasset.com (accessed December 2017).
6 V. Chandrasekhar, J. Andrews and A. Gatherer, "Femtocell networks: a survey," *IEEE Communications Magazine*, vol. 46, no. 9, September 2008, pp. 59–67.
7 H. Osman, H. Zhu, T. Alade and J. Wang, "Downlink transmission of distributed antenna systems in high building environments," *Communications (ICC), 2011 IEEE International Conference on*, 5–9 June 2011, pp. 1–5.
8 L. Gatzoulis, A. Aragon, G. Povey and S.R. Saunders, "Performance analysis of in-building FDD deployment measured data," *in Proc. 4th IEEE Intl. Conf. on 3G Mobile Communicaton Techn.*, June 2003, pp. 167–172.
9 K.J. Grandell, "Indoor antennas for WCDMA systems," *in Proc. 11th IEEE Intl. Conf. on Antennas and Propagation*, vol. 1, 2001, pp. 208–211.
10 M.S. Alouini and A.J. Goldsmith, "Area spectral efficiency of cellular mobile radio systems," *IEEE Trans. Vehic. Tech.*, vol. 48, no. 4, July 1999, pp. 1047–1066.
11 3GPP. *A Global Initiative*, Website, available at: www.3gpp.org (accessed December 2017).
12 Small Cell Forum. Website, available at: www.smallcellforum.org (accessed December 2017).
13 J. Zhang and G. Roche, *Femtocells: Technologies and Deployment*, Chichester, UK: John Wiley & Sons, Ltd, 2010.

14 F. Liu, E. Bala, E. Erkip and R. Yang;, "A framework for femtocells to access both licensed and unlicensed bands," *International Symposium on Modelling and Optimization in Mobile, Ad Hoc and Wireless Networks (WiOpt)*, 9–13 May 2011, pp. 407–411.

15 J.G Andrews, H. Claussen, M. Dohler, S. Rangan and M.C Reed, "Femtocells: Past, present, and future," *IEEE Journal on Selected Areas in Communications*, vol. 30, no. 3, April 2012, pp. 497–508.

16 J. Fildes, "Home cells signal mobile change," 2007. [Online]. Available: http://news.bbc.co.uk/1/hi/technology/6916125.stm (accessed December 2017).

17 S.F. Hasan, N.H. Siddique and S. Chakraborty, "Femtocell versus Wi-Fi – A survey and comparison of architecture and performance," *1st International Conference on Wireless Communication, Vehicular Technology, Information Theory and Aerospace & Electronic Systems Technology*, May 2009, pp. 916–920.

18 H.A. Mahmoud and I. Guvenc, "A comparative study of different deployment modes for femtocell networks," *IEEE 20th International Symposium on Personal, Indoor and Mobile Radio Communications*, Sept. 2009, pp. 1–5, 13–16.

19 N. Saquib, E. Hossain, L.B. Le and D.I. Kim, "Interference management in OFDMA femtocell networks: issues and approaches," *Wireless Communications, IEEE*, vol. 19, no. 3, June 2012, pp. 86–95.

20 T. Zahir, K. Arshad, A. Nakata and K. Moessner, "Interference management in femtocells," *IEEE Communications Surveys & Tutorials*, vol. 15, no. 1, February 2013, pp. 293–311.

21 Y. Bai, J. Zhou and L. Chen, "Hybrid spectrum usage for overlaying LTE macrocell and femtocell," *IEEE Global Telecommunications Conference, GLOBECOM*, Nov. 2009, pp. 1–6.

22 D.L. Perez, A. Valcarce, G.D.L. Roche, E. Liu, and J. Zhang, "Access methods to WiMAX femtocells: A downlink system-level case study," in *Proc. IEEE Int. Conf. Commun. Syst. (ICCS)*, Guangzhou, China, Nov. 2008, pp. 1657–1662.

23 H. Claussen, "Performance of macro- and co-channel femtocells in a hierarchical cell structure," in *Proc. IEEE Int. Symp. Personal, Indoor, Mobile Radio Commun. (PIMRC)*, Athens, Greece, Sep. *2007*, pp. 1–5.

24 L.T.W. Ho and H. Claussen, "Effects of user-deployed, co-channel femtocells on the call drop probability in a residential scenario," in *Proc. IEEE Int. Symp. Personal, Indoor, Mobile Radio Commun. (PIMRC)*, Athens, Greece, Sep. 2007, pp. 1–5.

25 V. Chandrasekhar and J.G. Andrews, "Uplink capacity and interference avoidance for two-tier cellular networks," in *Proc. IEEE Global Telecommunication Conference, November* 2007, pp. 3322–3326.
26 G. de la Roche, A. Valcarce, D. Lopez-Perez and J. Zhang, "Access control mechanisms for femtocells," *IEEE Communication Magazine*, vol. 48, no. 1, Jan. 2010, pp. 33–39.
27 J. Chen, P. Rauber, D. Singh, C. Sundarraman, P. Tinnakornsrisuphap and M. Yavuz "Femtocells – Architecture & Network Aspects", 2007. [Online]. Available: http://www.qualcomm.com/media/documents/files/femtocells-architecture-network-aspects.pdf (accessed December 2017).
28 V.U. Sankar and V. Sharma, "Subchannel allocation and power control in femtocells to provide quality of service," *National Conference on Communications (NCC)*, 3–5 Feb. 2012, pp. 1–5.
29 I. Guvenc, M. Jeong and F. Watanabe, "A hybrid frequency assignment for femtocells and coverage area analysis for co-channel operation", *IEEE Communication Letters*, no. 12, no. 12, Dec. 2008, pp. 880–882.
30 V. Chandrasekhar and J. Andrews, "Spectrum allocation in tiered cellular networks," *IEEE Transactions on Communications*, vol. 57, no. 10, October 2009, pp. 3059–3068.
31 H. Li, X. Xu, D. Hu, X. Qu, X. Tao and P. Zhang, "Graph method based clustering strategy for femtocell interference management and spectrum efficiency improvement," *6th International Conference on Wireless Communications Networking and Mobile Computing (WiCOM)*, September 2010, pp. 1–5.
32 H. Su, L. Kuang and J. Lu, "Interference avoidance in OFDMA based femtocell network," *Information, Computing and Telecommunication, IEEE Youth Conference on (YC-ICT)*, 20–21 Sept. 2009, pp. 126–129.
33 Y. Zhang, Y.Y. Li, E.S. Sousa and Q. Zhang, "Pilot power minimization in HSDPA femtocells," *IEEE Global Telecommunications Conference (GLOBECOM 2010)*, Dec. 2010, pp. 1–5.
34 Z. Wang, W. Xiong, C. Dong, J. Wang and S. Li, "A novel downlink power control scheme in LTE heterogeneous network," *International Conference on Computational Problem-Solving (ICCP)*, 21–23 Oct. 2011, pp. 241–245.
35 P. Mach and Z. Becvar, "Dynamic power control mechanism for femtocells based on the frame utilization," *Wireless and Mobile*

Communications (ICWMC), 2010 6th International Conference on, Sept. 2010, pp. 498–503.
36 D. Roche, G. Ladányi, D. López-Pérez, D.C. Chong and J. Zhang;, "Self-organization for LTE enterprise femtocells," *IEEE GLOBECOM Workshops (GC Wkshps),* Dec. 2010, pp. 674–678.
37 L. Mohjazi, M. Al-Qutayri, H. Barada and K. Poon, "Femtocell coverage optimization using genetic algorithm," *Technical Symposium at ITUTelecom World (ITU WT),* 24–27 Oct. 2011, pp. 159–164.
38 Y.Y. Li and E.S. Sousa, "Base station pilot management for user-deployed cellular networks," *Communications, IEEE International Conference on (ICC '09),* June 2009, pp. 1–5.
39 B. Choi, E.S. Cho, M.Y. Chung, K. Cheon and A. Park, "A femtocell power control scheme to mitigate interference using listening TDD frame," *Information Networking (ICOIN), 2011 International Conference on,* Jan. 2011, pp. 241–244.
40 J. Han-Shin, M. Cheol, M. June and Y. Jong-Gwan, "Interference mitigation using uplink power control for two tier femtocell networks," *Wireless Communications, IEEE Transactions on,* vol. 8, no. 10, Oct. 2009, pp. 4906–4910.
41 S. Park, W. Seo, S. Choi, and D. Hong, "A beamforming codebook restriction for cross-tier interference coordination in two-tier femtocell networks," *Vehicular Technology, IEEE Transactions on,* vol. 60, no. 4, May 2011, pp. 1651–1663.
42 M. Husso, J. Hämäläinen, R. Jäntti, J. Li, E. Mutafungwa, R. Wichman, et al. "Interference mitigation by practical transmit beamforming methods in closed femtocells' *EURASIP Journal on Wireless Communications and Networking,* May 2010.
43 V. Chandrasekhar, M. Kountouris and J.G Andrews, "Coverage in multi-antenna two-tier networks," *IEEE Transactions on Wireless Communications,* vol. 8, no. 10, Oct. 2009, pp. 5314–5327.
44 M.R. Sanatkar and B. Natarajan, "Power selection for maximizing SINR in femtocells with sectorized antennas," *IEEE Consumer Communications and Networking Conference (CCNC),* Jan. 2012, pp. 690–692.
45 S. Cheng, S. Lien, F. Chu and K. Chen, "On exploiting cognitive radio to mitigate interference in macro/femto heterogeneous networks," *Wireless Communications, IEEE,* vol. 18, no. 3, June 2011, pp. 40–47.

46 A. Goldsmith, S.A. Jafar, I. Maric and S. Srinivasa, "Breaking spectrum gridlock with cognitive radios: an information theoretic perspective," *Proceedings of the IEEE*, vol. 97, no. 5, May 2009, pp. 894–914.

47 J. Mitola III and G. Q. Maguire Jr., "Cognitive radio: making software radios more personal," *IEEE Personal Communications*, vol. 6, no. 4, 1999, pp. 13–18.

48 S.Y. Lien, Y. Lin and K. Chen "Cognitive and game-theoretical radio resource management for autonomous femtocells with QoS guarantees" *IEEE Transactions on Wireless Communications*, vol. 10, no. 7, Jul. 2011, pp. 2196–2206.

49 Federal Communications Commission. FCC Website available at: http://www.fcc.gov (accessed December 2017).

50 F. Akyildiz, W. Lee, M. C. Vuran and S. Mohanty "NeXt generation/dynamic spectrum access/cognitive radio wireless networks: A survey" *The International Journal of Computer and Telecommunications Networking*, vol. 50, no. 13, 15 Sept. 2006, pp. 2127–2159.

51 J. Palicot, J. Mitola, Z. Lei, and F.K. Jondral "Special Issue on 10 years of cognitive radio: State of the art and perspectives" *EURASIP Journal on Wireless Communications and Networking*, 2012.

52 A. Jovicic and P. Viswanath, "Cognitive radio: an information-theoretic perspective," *IEEE Transactions on Information Theory*, vol. 55, no. 9, Sept. 2009, pp. 3945–3958.

53 K.A. Yau, F.G. Tan, P. Komisarczuk and P.D. Teal, "Exploring new and emerging applications of Cognitive Radio systems: Preliminary insights and framework," *IEEE Colloquium on Humanities, Science and Engineering (CHUSER)*, Dec. 2011, pp. 153–157.

54 B. Wang and K.J.R Liu, "Advances in cognitive radio networks: A survey," *IEEE Journal of Selected Topics in Signal Processing*, vol. 5, no.1, Feb. 2011, pp. 5–23.

55 G. Gur, S. Bayhan, and F. Alagoz, "Cognitive femtocell networks: an overlay architecture for localized dynamic spectrum access [dynamic spectrum management]," *IEEE Wireless Communications*, vol. 17, no. 4, August 2010, pp. 62–70.

56 Q. Li, Z. Feng, W. Li, Y. Liu and P. Zhang "Joint access and power control in cognitive femtocell networks," *International Conference on Wireless Communications and Signal Processing (WCSP)*, Nov. 2011, pp. 1–5.

57 A. Adhikary, V. Ntranos and G. Caire, "Cognitive femtocells: Breaking the spatial reuse barrier of cellular systems," *Information Theory and Applications Workshop (ITA)*, 6–11 Feb. 2011, pp. 1–10.

58 J.E. Hakegard, A. Lie and T.A. Myrvoll, "Power control in HetNets and cognitive networks", *Communications and Information Technologies (ISCIT), 2012 International Symposium on*, 2–5 Oct. 2012, pp. 568–573.

59 F.H. Khan and Y. Choi, "Towards introducing self-configurability in cognitive femtocell networks," *9th Annual IEEE Communications Society Conference on Sensor, Mesh and Ad Hoc Communications and Networks (SECON)*, 18–21 June 2012, pp. 37–39.

60 H. Saad, A. Mohamed and T. ElBatt, "Distributed cooperative Q-learning for power allocation in cognitive femtocell networks," *Vehicular Technology Conference (VTC Fall), 2012 IEEE*, 3–6 Sept. 2012, pp. 1–5.

61 S.E. Nai and T.Q.S. Quek, "Coexistence in two-tier femtocell networks: Cognition and optimization," *International Conference on Computing, Networking and Communications (ICNC)*, Jan. 2 2012, pp. 655–659.

62 D. Sun, X. Zhu, Z. Zeng and S. Wan, "Downlink power control in cognitive femtocell networks," *International Conference on Wireless Communications and Signal Processing (WCSP)*, Nov. 2011, pp. 1–5.

63 P. Hu, J. Ye, F. Zhang, S. Deng, C. Wang and W. Wang, "Downlink resource management based on cross-cognition and graph coloring in cognitive radio femtocell networks," *Vehicular Technology Conference (VTC Fall), 2012 IEEE*, Sept. 2012, pp. 1–5.

64 D. Oh, H. Lee, and Y. Lee, "Cognitive radio based femtocell resource allocation," *International Conference on Information and Communication Technology Convergence (ICTC)*, Nov. 2010, pp. 274–279.

65 Y. Saleem, A. Bashir, E. Ahmed, J. Qadir and A. Baig, "Spectrum-aware dynamic channel assignment in cognitive radio networks," *Emerging Technologies (ICET), 2012 International Conference on*, Oct. 2012, pp. 1–6.

66 S. Bose and B. Natarajan, "Reliable spectrum sensing for resource allocation of cognitive radio based WiMAX femtocells," *IEEE Consumer Communications and Networking Conference (CCNC)*, Jan. 2012, pp. 889–893.

67 L. Huang, G. Zhu and X. Du, "Cognitive femtocell networks: an opportunistic spectrum access for future indoor wireless

coverage," *Wireless Communications, IEEE*, vol. 20, no. 2, Apr. 2013, pp. 44–51.

68 L. Li, C. Xu and M. Tao, "Resource allocation in open access OFDMA femtocell networks," *Wireless Communications Letters, IEEE*, vol. 1, no. 6, Dec. 2012, pp. 625–628.

69 X.Y. Wang, P.-H. Ho and K.-C. Chen, "Interference analysis and mitigation for cognitive-empowered femtocells through stochastic dual control," *IEEE Transactions on Wireless Communications*, vol. 11, no. 6, June 2012, pp. 2065–2075.

70 S. Cheng, W.C. Ao and K. Chen, "Efficiency of a cognitive radio link with opportunistic interference mitigation," *IEEE Transactions on Wireless Communications*, vol. 10, no. 6, June 2011, pp. 1715–1720.

71 K.A. Meerja, P.H. Ho and B. Wu, "A novel approach for co-channel interference mitigation in femtocell networks," *IEEE Global Telecommunications Conference (GLOBECOM 2011)*, Dec. 2011, pp. 1–6.

72 A. Attar, V. Krishnamurthy and O.N. Gharehshiran, "Interference management using cognitive base-stations for UMTS LTE," *Communications Magazine, IEEE*, vol. 49, no. 8, Aug. 2011, pp. 152–159.

73 L. Zhang, L. Yang and T. Yang, "Cognitive interference management for LTE-A femtocells with distributed carrier selection," *IEEE 72nd Vehicular Technology Conference Fall (VTC 2010-Fall)*, 6–9 Sept. 2010, pp. 1–5.

74 S. Kaimaletu, R. Krishnan, S. Kalyani, N. Akhtar and B. Ramamurthi, "Cognitive interference management in heterogeneous femto-macro cell networks," *IEEE International Conference on Communications (ICC)*, June 2011, pp. 1–6.

75 S. Cheng, W.C. Ao, F. Tseng and K. Chen, "Design and analysis of downlink spectrum sharing in two-tier cognitive femto networks," *IEEE Transactions on Vehicular Technology*, vol. 61, no. 5, June 2012, pp. 2194–2207.

76 Y.S. Soh, T.Q.S. Quek, M. Kountouris and G. Caire, "Cognitive hybrid division duplex for two-tier femtocell networks," *IEEE Transactions on Wireless Communications*, vol. 12, no. 10, Oct. 2013, pp. 4852–4865.

77 3GPP RAN1, "RAN1–090133 Huawei Beamforming enhancement in LTE-Advanced," Jan. 2009.

78 J.W. Huang and V. Krishnamurthy, "Cognitive base stations in LTE/3GPP femtocells: A correlated equilibrium game-theoretic approach," *IEEE Transactions on Communications*, vol. 59, no. 12, Dec. 2011, pp. 3485–3493.

79 R.J. Aumann, "Correlated equilibrium as an expression of Bayesian rationality," *Econometrica*, vol. 55, no. 1, Jan 1987, pp. 1–18.
80 M. Bennis and S.M. Perlaza, "Decentralized cross-tier interference mitigation in cognitive femtocell networks," *IEEE International Conference on Communications (ICC)*, June 2011, pp. 1–5.
81 S. Leveil, Le Martret, J. Christophe, H. Anouar, K. Arshad, T. Zahir, et al., "Resource management of centrally controlled cognitive radio networks," *Future Network & Mobile Summit (FutureNetw)*, July 2012, pp. 1–9.
82 M.E. Sahin, I. Guvenc, M. Jeong and H. Arslan, "Handling CCI and ICI in OFDMA femtocell networks through frequency scheduling," *Consumer Electronics, IEEE Transactions on*, vol. 55, no. 4, Nov. 2009, pp. 1936–1944.
83 Y. Ma, T. Lv, J. Zhang, H. Gao and Y. Lu, "Cognitive interference mitigation in heterogeneous femto-macro cell networks," *Personal Indoor and Mobile Radio Communications (PIMRC), 2012 IEEE 23rd International Symposium on*, Sept. 2012, pp. 2131–2136.
84 E. Mugume, W. Prawatmuang and D.K.C. So, "Cooperative spectrum sensing for green cognitive femtocell network," *IEEE 24th International Symposium on Personal Indoor and Mobile Radio Communications (PIMRC), 2013*, Sept. 2013, pp. 2368–2372.
85 L. Yizhe, F. Zhiyong, Z. Qixun, T. Li, and T. Fang, "Cognitive optimization scheme of coverage for femtocell using multi-element antenna," *IEEE 72nd VTC*, Sept 2010, pp. 1–5.
86 M. Z. Shakir, R. Atat and M. Alouini, "On the interference suppression capabilities of cognitive enabled femto cellular networks," *International Conference on Communications and Information Technology (ICCIT)*, June 2012, pp. 402–407.
87 Z. Zhao, M. Schellmann, H. Boulaaba and E. Schulz, "Interference study for cognitive LTE-femtocell in TV white spaces," *Telecom World (ITU WT), 2011 Technical Symposium at ITU*, Oct. 2011, pp. 153–158.
88 F. Peng, Y. Gao, Y. Chen, K.K. Chai and L. Cuthbert, "Using TV white space for interference mitigation in LTE femtocell networks," *IET International Conference on Communication Technology and Application (ICCTA 2011)*, Oct. 2011, pp. 5–9.
89 G. Zhao, C. Yang, G.Y. Li and G. Sun, "Fractional frequency donation for cognitive interference management among femtocells," *Global Telecommunications Conference (GLOBECOM 2011), 2011 IEEE*, Dec. 2011, pp. 1–6.

90 G.W.O. da Costa, A.F. Cattoni, V.A. Roig and P.E. Mogensen, "Interference mitigation in cognitive femtocells," *Proc. IEEE GLOBECOM Workshops*, Dec. 2010, pp. 721–725.

91 Y. Li, M. Macuha, E.S. Sousa, T. Sato and M. Nanri, "Cognitive interference management in 3G femtocells," *20th International Symposium on Personal Indoor and Mobile Radio Communications (PIMRC)*, Sept. 2009, pp. 1118–1122.

92 Y. Li and E.S. Sousa, "Cognitive uplink interference management in 4G cellular femtocells," *IEEE 21st International Symposium on Personal Indoor and Mobile Radio Communications (PIMRC)*, Sept. 2010, pp. 1567–1571.

93 X. Tao, Z. Zhao, R. Li, J. Palicot and H. Zhang, "Downlink interference minimization in cognitive LTE-femtocell networks," *Communications in China (ICCC), 2013 IEEE/CIC International Conference on*, Aug. 2013, pp. 124–129.

94 C.C. Chai, "Distributed subcarrier and power allocation for OFDMA-based cognitive femtocell radio uplink," *IEEE 24th International Symposium on Personal Indoor and Mobile Radio Communications (PIMRC)*, 2013, Sept. 2013, pp. 2876–2880.

4

Coverage Area-Based Power Control for Interference Management in LTE Femtocells

Ghazanfar Ali Safdar

School of Computer Science and Technology, University of Bedfordshire, Luton, UK

An important issue affecting cellular networks is to make services available to regions of bad or no reception. Femtocells are considered to be the solution to meet the future needs for better indoor coverage, high data rates and capacity in the cellular networks. Since FAPs are usually user deployed, blind placement of FAPs is inevitable leading to problems of power spillage causing severe co- and cross-tier interference and subsequent performance degradation. Even though femtocells are discovering an important role, the issue of interference as a result of blindly placed FAPs needs to be addressed. This chapter presents performance analysis of a coverage radius based power control scheme to circumvent the problems caused by blind placement of FAPs. Our co scheme does not require FAPs to be relocated to optimal positions for interference mitigation; rather a self-update algorithm is implemented by FAPs to reduce their cell radius by adaptive adjustment of power values for interference management. Using system level simulations, the performance of the scheme has been analysed for densely deployed single and multicell femtocell scenarios. Compared to existing schemes, our scheme provides improved interference mitigation and throughput results. The results reveal that coverage radius of FAPs has considerable effect on both co- and cross-tier SINR values. It is also shown that there are coverage radius bounds that provide balanced co- and cross-tier SINR values. These

LTE Communications and Networks: Femtocells and Antenna Design Challenges, First Edition. Edited by Masood Ur Rehman and Ghazanfar Ali Safdar.

bounds provide valuable information for effective distribution of FUE to avoid performance degradation in blindly placed densely deployed femtocells.

4.1 Introduction

Indoor cellular usage, which accounts for 50% of all voice calls and 70% of data traffic, is mostly faced with poor reception due to low signal penetration through walls and attenuation, which may lead to total loss of signal [1]. A MBS, which motivates existing wireless strategies, is supported by the deployment of smaller cells to address these growing issues. Most of the benefits of smaller cells overlaid in the MBS arise from the fact that a significantly reduced distance between transmitter and receiver results in a better-quality link. The femtocell is a new indoor solution to cellular communication, which aims to provide better indoor network coverage and has attracted network operators and stakeholders. Femtocells are low powered, low cost and subscriber controlled units that provide a dedicated Base Station (BS) to subscribers indoors.

The concept of femtocells, also known as home BSs, Home NodeB (HNB) or Home eNodeB (HeNB) [2] and residential small cells [3] was first studied in 1999 by Bell Labs of Alcatel-Lucent, but it was in 2002 that Motorola announced the first 3G based home BS product [4]. Femtocell units, known as FAPs, connect standard mobile devices to the network of a mobile operator through residential Digital Subscriber Line (DSL), optical fibres, cable broadband connections or wireless last-mile technologies [4–6]. By installing FAPs indoors, the cell sites are reduced, thereby bringing the transmitter and receiver closer to each other. The use of the subscriber's broadband network to backhaul data offer improved mobile phone coverage indoors for both voice and data compared to the MBS. The close proximity greatly lowers transmission power and increases the battery life of mobile devices and with a dedicated FAP it offers subscribers a single billing address for mobile phone, broadband and land line as they are all channelled through the same backhaul [7]. Additionally, it acts as a solution to the convergence of landline and mobile systems [8].

Due to the two-tier architecture of femtocells and macrocells, interference is imminent. The cell sites covered by a number of FAPs (in some cases overlapping each other) are overlaid in the larger cell site of the macro BSs. Interference could be between a femtocell and macrocell, which is known as cross-tier interference or between neighbouring femtocells known as co-tier interference [9]. Interference can be further classified as uplink (UL) or downlink (DL) based on the sources, which besides the FAP and MBS, also include FUE – a UE served by a FAP) and Macrocell User Equipment (MUE – a UE served by a MBS). It is important to note that the scale at which interference affects a femtocell network is largely dependent on the deployment scenario. In a *dedicated channel* deployment, the licensed spectrum is split into different portions for each tiered network to operate in a dedicated manner whereas both tiers share the same licensed spectrum in *co-channel deployment* [10]. Network operators prefer a co-channel deployment due to the limited available bandwidth but will have to deal with the interference issues [11–13]. FAPs deployed in an *open access* allow connection for all users, whereas in a *close subscriber group* (CSG) mode only the subscribed owners of the FAP have access. A new access deployment is *hybrid access*, which combines open access and the CSG by allowing only a limited amount of resources to all users [14]. The power transmitted by a FAP, which comprises of pilot and traffic power values, also has a detrimental effect in a network. For instance, a high pilot power will result into a large cell coverage area, but consequently has higher chances of causing increased interference. Thus, there is a need to optimize the power levels in femtocells to avoid interference while maintaining a certain QoS. Some of the typical power control schemes employed by femtocells for interference mitigation are described in [15–17].

Nonetheless, a FAP is user deployed and is usually blindly placed in an indoor environment such as near walls and windows. In the absence of any antennas' beam directivity and so on, the FAP power could spill out in the surrounding regions, thereby causing considerable co- and cross-tier interference (Figure 4.1). As opposed to blind placement, research work carried out in [18–20] investigates optimal positions to place a FAP in an indoor environment to improve the throughput and mean capacity. However, it might not always be possible or necessary to find the optimal positions and move the FAPs to effectively mitigate the interference caused. Thus, there is a serious need for development of interference mitigation schemes for blindly placed LTE femtocells. In this chapter, the problem of interference due to blind placement of FAPs is addressed by analysis of a coverage

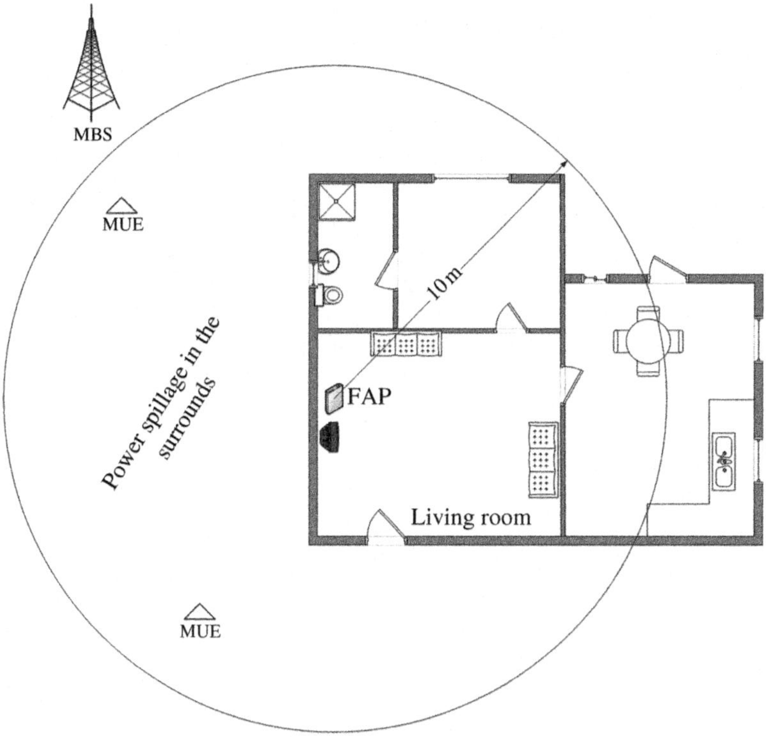

Figure 4.1 Blind placement of a FAP and power spillage.

radius based power control scheme, which adaptively varies the pilot power of FAP based on its distance from the farthest served FUE. The scheme investigated in the chapter does not require FAPs to be located or relocated at optimal positions; rather, interference is mitigated by coverage radius based power control scheme. Based on balanced co- and cross-tier SINR values, the important research findings are coverage radius bounds that provide valuable information towards radial placement of FUE in densely deployed blindly placed femtocells.

The rest of the chapter is organized as follows: the coverage radius based power control scheme is described in Section 4.2 before Section 4.3 presents the system model and simulation parameters. Section 4.4 details the performance analysis, results and discussion whereas finally the chapter is summarized in Section 4.5.

4.2 Coverage Radius Based Power Control Scheme (PS)

The coverage radius based PS is described next with the help of equation 4.1 and illustrated in Figure 4.2.

$$P_t = min\left[P_m + G(\theta) - L_m(d) + L_f(r),\ P_{max}\right] \tag{4.1}$$

Where P_t is the FAP power value, P_m is the MBS power value, $G(\theta)$ is the MBS antenna gain in the direction of the FAP, $L_m(d)$ is the MBS path loss relative to a femtocell distance *d*, $L_f(r)$ is the FAP path loss relative to the target radius *r*, and P_{max} is an upper limit power value of P_t set at 20 dBm. The varying value of *r* that is configured at each FAP

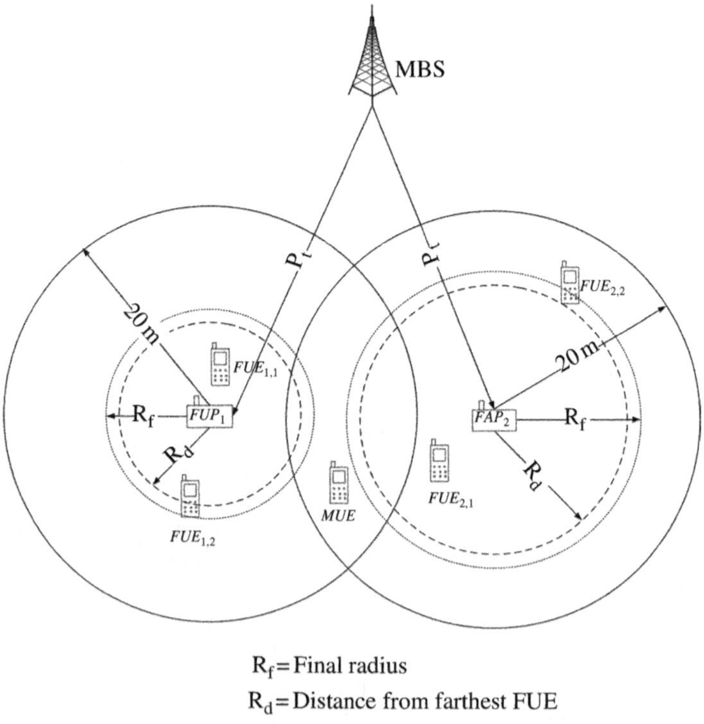

Figure 4.2 Coverage radius based PS.

is determined to be the distance between the FAP and its farthest served UE with constraints as described next.

4.2.1 Radius Limit Setting

Each FAP sets a coverage radius upper limit, R_u and lower limit R_l to be 10 and 3 m, respectively. The reason is to provide adequate coverage radius for user premises while effectively mitigating the interference caused. As already suggested in most research work, FAPs are not always installed at the centre of premises rather blindly placed, therefore in reality, even at a 10-m radius, the pilot power of a CSG FAP could spill out of the premises and affect non-subscribed users in the vicinity, such as MUE [21].

4.2.2 Initial Coverage Radius

The initial coverage radius of FAP is set to be R_u assuming no FUE are present. With the presence of FUE, the FAP measures the distance between itself and the farthest FUE, which is denoted R_d. FAP employs RSSI value of an FUE to deduce distance (i.e. R_d) between itself and the FUE. In the event where only one FUE is served, R_d becomes the distance between the FAP and that FUE.

4.2.3 Self-Update

Since FUE is usually not static, in order to account for mobility of FUE, the FAP puts on a radius cap of *2m* to R_d to make the final radius $R_f = R_d + 2m$, thus ensuring seamless coverage for mobile FUE and further avoiding any handover to MBS. The FAP conducts a self-update per unit of time (e.g. 1 s chosen in our simulations) to determine a new coverage radius. The choice of 1 s and threshold of 2 m is based on the fact that the indoor user velocity is usually between 0–3.5 m/s [22].

4.2.4 Final Radius

The final radius R_f is always compared against R_u and R_l to maintain a coverage radius such that $R_l \leq R_f \leq R_u$.

The proposed scheme can be represented with the help of an algorithm as follows.

Coverage Radius Based Power Control Scheme

Initialization;
Set $R_u = 10\text{m}$, $R_l = 3m$,
1: if $(FUE > 1)$
2: for 1:n (n = number of all active FUEs) do
3: calculate each FUE: compare R_u and $R_l$to deduce $E = R_u : R_l$
4: sort $E = \{1,2,\&,K\}$
5: set $R_d = \text{max\_value from } E$
6: set $R_f = R_d + 2\text{m}$: final FAP coverage radius
7: end if
8: Continue loop next TTI = 1s
9: end

4.3 System Model

This section describes the system model and simulation parameters based on 3GPP LTE specifications [23]. The simulation tool used in our analysis is the well-known LTE system level simulator v1.7r1119, which is offered for free under an academic, non-commercial use licence [24]. However, the simulator has some limitations because it only supports a fixed transmit and pilot power value algorithm implementation for femtocells. Eventually, its functionality was enhanced by the implementation of existing adaptive and proposed coverage radius based adaptive power control algorithms. The main simulation parameters are given in Table 4.1. The considered network topology consists of two simulation scenarios; a single tri-sector and a densely deployed urban scenario that consists of seven tri-sector hexagonal MBS cells with an inter-site distance of 500 m. Ten FAPs are blindly distributed in each sector with four pieces of FUE attached to each FAP in a CSG fashion. Additionally, 30 MUE are randomly deployed in each sector of the MBS to investigate effects of cross-tier interference. WINNERII+ channel modelling [25] is used and traffic load is uniformly distributed among all the users.

Table 4.1 Simulation parameters.

Parameter	Value
Carrier Frequency	2.14 GHz
Bandwidth	20 MHz
MBS Inter-Site Distance	500 m
MBS/FAP Tx Power	46/Variable dBm
Scheduler	Proportional Fair
UE Receiver Noise Figure	9 dB
UE Thermal Noise	−174 dBm/Hz
Single cell	
No. of MBS/FAPs	1/30
MUE	90
FUE per FAP/Total FUE	4/120
Multicell	
No. of MBS/FAPs	7/210
MUE	630
FUE per FAP/Total FUE	4/840

To account for the macrocell propagation model, our simulations employ the macroscopic path loss model as proposed in [26] in an urban environment and defined in equation 4.2.

$$PL(dB) = 40 \cdot \left(1 - 4 \cdot 10^{-3} \times Dhb\right) \cdot \log_{10}(R) - 18 \cdot \log_{10}(Dhb) + 21 \cdot \log_{10}(f) + 80 \tag{4.2}$$

Where R is the distance between BS and UE in km, Dhb is the height of the BS antenna above ground in metres and f is the carrier frequency in MHz. The path loss model implemented at the femtocell is the dual slope path loss for urban deployment while ignoring shadowing and penetration losses as defined in equation 4.3.

$$PL(dB) = 38.45 + 20log_{10}(r) + 0.7d_{2D,indoor} \tag{4.3}$$

where $d_{2D,indoor}$ in this context is the indoor distance between an FAP and its serving FUE.

The azimuth antenna gain pattern used is proposed in TS36.942 [23] and given as follows in equation 4.4.

$$A(\theta) = -\min\left[12\left(\frac{\theta}{\theta_{3dB}}\right)^2, A_m\right] \tag{4.4}$$

where $\theta_{3dB} = 70°$ is the gain pattern angle and $A_m = 20\,\text{dB}$ is the side lobe gain.

4.4 Performance Analysis

The performance analysis carried out is to investigate and compare our coverage radius based scheme with other power control schemes in both single- and multicell scenarios. However, the main purpose is to investigate the coverage radius bounds and subsequently their impact on SNIR for both these scenarios. Simulations have thus been performed for all the possible values of R_f between 10 and 3 m for our coverage radius based scheme and compared with three existing power control schemes. The first is a baseline scheme where all FAPs are assigned a fixed value of 20 dBm. It is important to note that this scheme is used for simulation and comparison analysis and not for implementation because a fixed maximum power is not an ideal solution for mitigating interference in femtocells. The second scheme assigns FAP power value based on the power it receives from its closest MBS while maintaining a target femtocell radius of 10 m [27]. The third scheme is a distance based power control scheme proposed in [28] that intends to limit the impact a FAP has on the aggregate macrocell downlink throughput. In the distance based scheme the MBS is divided into three regions with power values assigned for FAPs in each region as defined in equation 4.5. d_{FAP} here denotes the distance between a FAP and the closest MBS.

$$P_{FAP}(d_{FAP}) = \begin{cases} 20dBm & 46m \le d_{FAP} < 85m, \\ 10dBm & 85m \le d_{FAP} < 156m \\ 0dBm & 156m \le d_{FAP} < 289m \end{cases} \tag{4.5}$$

A graphical comparison of the schemes is presented in Figure 4.3 where sector 1 illustrates the distance based power control algorithm

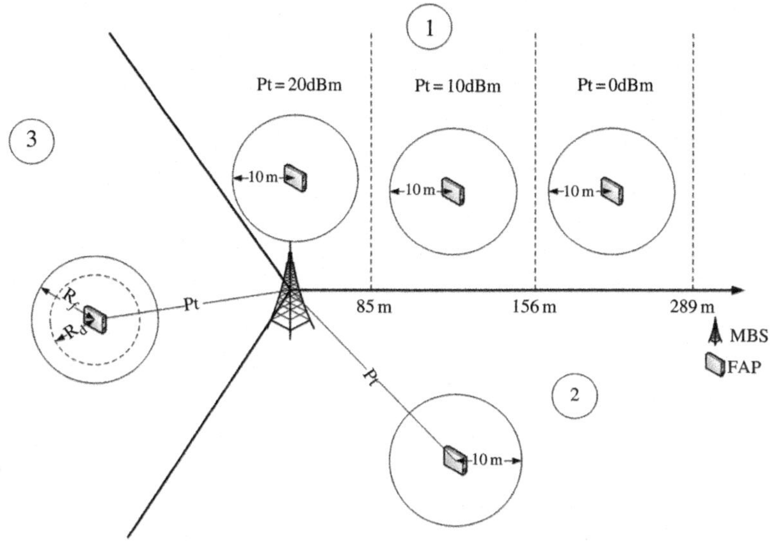

Figure 4.3 Comparison of PS with other schemes.

as defined in equation 4.5. Sector 2 illustrates the constant radius scheme and sector 3 illustrates our coverage radius based PS. The value of P_t in sectors 2 and 3 is as defined in equation 4.1.

The result of the fixed power value of 20 dBm for all FAPs is denoted 'FP' while the distance based power scheme is denoted 'DB'. The constant radius power scheme is denoted 'CR' and for our coverage radius based adaptive PS, the results are shown for the values of R_f at 7, 6 and 5 m denoted PS-7 and PS-6 and PS-5, respectively. The results provided in the chapter are from simulations performed for single- and multicell scenarios, whereas the single cell scenario is composed of single tri-sector hexagonal MBS cell compared to a multicell scenario that consists of seven tri-sector hexagonal MBS cells with an inter-site distance of 500 m. The reason single and multicell scenarios are chosen is to investigate the effect of coverage radius bounds on the variations of SINR in each scenario as described next.

4.4.1 Results and Discussion

4.4.1.1 SINR Cross-Tier (Single Cell)

Figure 4.4 shows the cumulative distribution function (CDF) plot of SINR value for all MUE. The proposed scheme takes the cross-tier

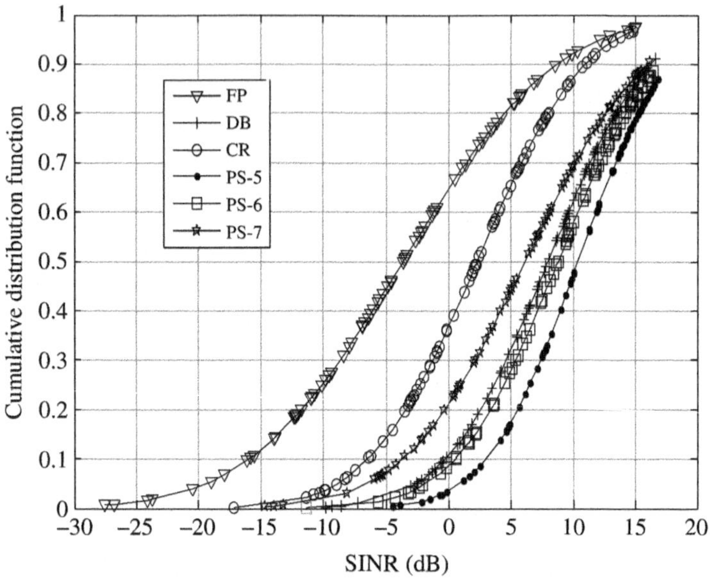

Figure 4.4 SINR cross-tier (single cell).

interference impact it has on MUE fully into consideration. The transmit power value in proposed scheme is directly proportional to the coverage radius. Due to its low transmit power (P_t)value for smaller FAP coverage area, our scheme at PS-5 with a mean SINR value of 10.35 dB performs far better compared to other schemes. With a slightly increased coverage area the mean SINR values of PS-6 and PS-7 are 8.75 and 6.13 Db, respectively. On the other hand, due to maximum$P_t$in case of FP, MUE experiences heavy cross-tier interference with a low mean SINR value (−3.60 dB). The mean SINR values for DB and CR are 7.95 and 2.26 dB, respectively. As a whole, our proposed scheme at PS-5 improves SINR by 13.90 dB while compared to FP, 8.09 dB compared to CR and 2.04 dB compared to DB schemes.

4.4.1.2 SINR Co-Tier (Single Cell)

Figure 4.5 shows the SINR results for all the FUE. FP performs better compared to our proposed scheme because some of the femtocells in the simulated scenario are standalone with a maximum fixed value of P_t,thus resulting in better co-tier SINR. This improvement is due to the fact that in FP, FAPs satisfy their serving FUE, however, completely

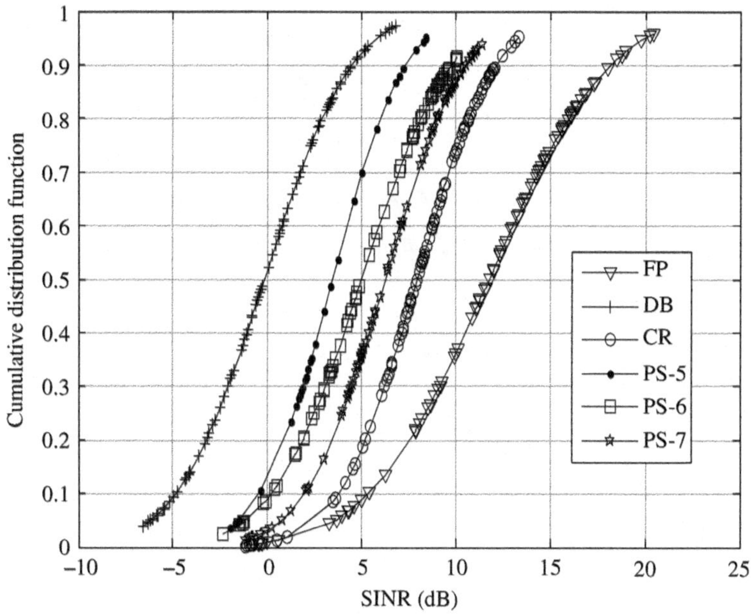

Figure 4.5 SINR co-tier (single cell).

disregarding neighbouring FAPs and MUE (causing serious cross-tier interference as already shown in Figure 4.4).

The mean SINR values of DB, CR, PS-5, PS-6 and PS-7 are −0.16, 7.93, 3.45, 4.99 and 6.24 dB, respectively. The slightly lower SINR values in our proposed scheme as compared to CR are attributed to lower values of P_t because of smaller coverage radius.

4.4.1.3 Downlink Throughput (Single Cell)

Figure 4.6 shows the CDF plot for downlink throughput over all FUE. In accordance with SINR results, the baseline scenario FP performs better due to its high P_t value with a mean throughput of 12.06 Mbps. With mean throughput values of 6.72, 7.64 and 8.42 Mbps for PS-5, PS-6 and PS-7, respectively, our scheme shows that a significantly high throughput can still be achieved with a varying coverage radius as compared to CR (8.64 Mbps) with fixed coverage radius. DB with a mean throughput value of 3.99 Mbps performs lower than the other schemes.

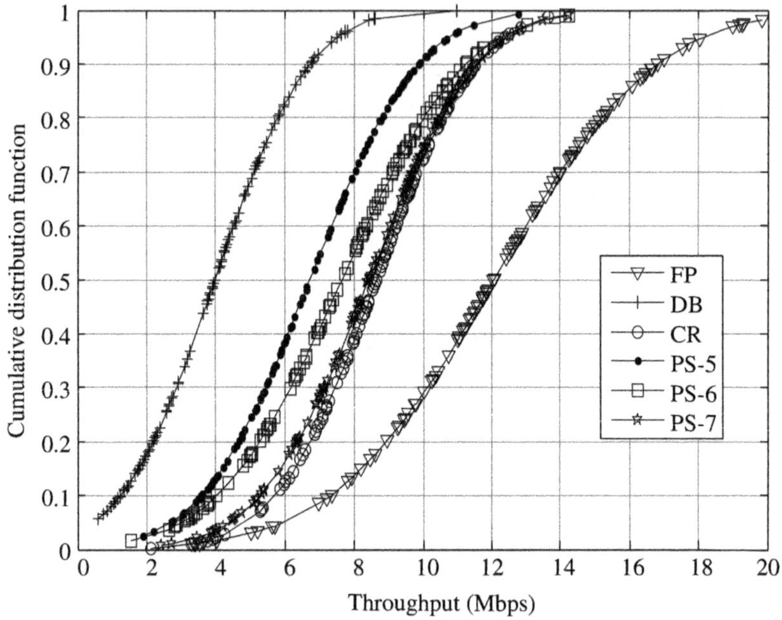

Figure 4.6 Downlink throughput (single cell).

4.4.1.4 Co- and Cross-Tier SINR (Single Cell versus Multicell)

Figures 4.7 and 4.8 show how co- and cross-tier SINR, respectively, vary for all the mentioned schemes compared against single- to multicell scenarios. Compared to a single-cell scenario, clearly for all the schemes both co- and cross-tier SINR values dropped in a multicell scenario because of enhanced interference experienced by any cell due to increased number of blindly placed FAPs. Figure 4.7 further reveals that no matter what the scenario is (single-cell or multicell), the DB scheme resulted into extremely poor co-tier SINR values. Compared to other schemes, even though FP scheme exhibited better co-tier SINR values for both scenarios (Figure 4.7), however, as expected, FP due to its very nature resulted into very bad cross-tier SINR values (Figure 4.8). Despite the fact that CR scheme compared to our PS resulted into better co-tier SINR values (Figure 4.7), however, it is important to note that CR scheme suffered serious degradation for its cross-tier SINR value (Figure 4.8). Lastly, even though all schemes resulted into droppage in SINR values compared

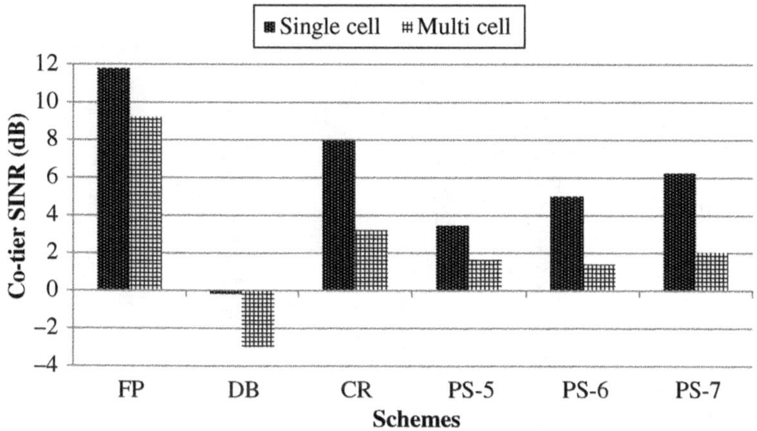

Figure 4.7 Co-tier SINR comparison (single versus multicell).

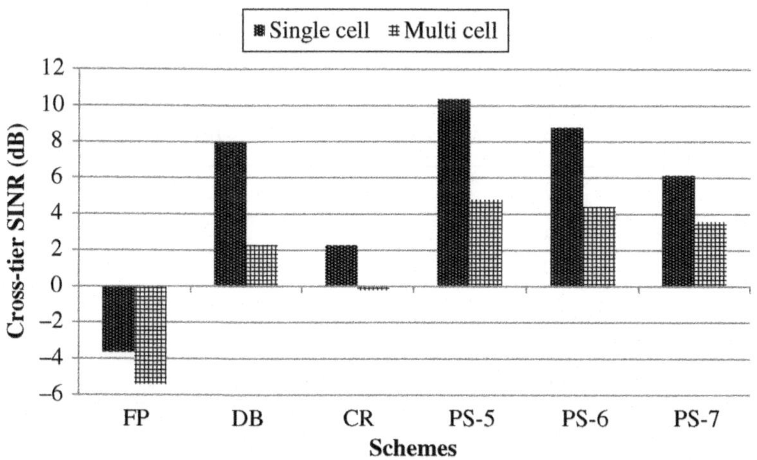

Figure 4.8 Cross-tier SINR comparison (single versus multicell).

against single and multicell scenarios, our proposed scheme (PS) clearly proven to be the best scheme in terms of SINR values.

4.4.1.5 Droppage in SINR (Single Cell versus Multicell)

Figure 4.9 compares mentioned schemes for a percentage drop in SINR values for single- and multicell scenarios. Table 4.2 shows how

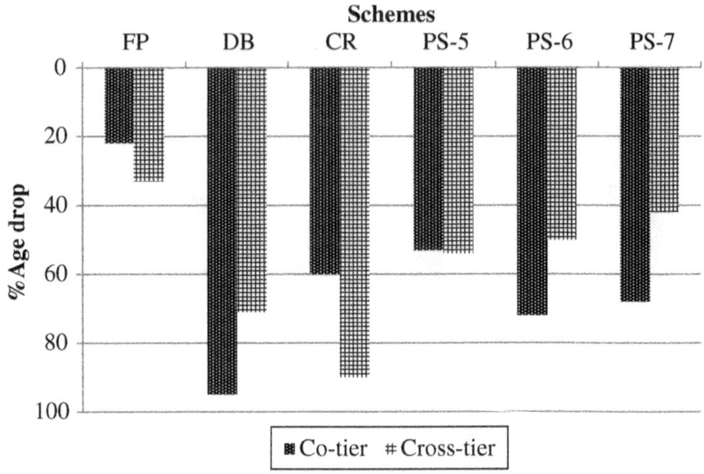

Figure 4.9 Percentage droppage in SINR (single- versus multicell).

Table 4.2 Percentage drop due to change in scenario (single to multicell).

Scheme	*Co-Tier SINR (%age)*	*Cross-Tier SINR (%age)*
FP	22	33
DB	95	71
CR	60	90
PS-5	53	54
PS-6	72	50
PS-7	68	42

SINR values have reacted to the change in scenario (single to multicell).

The shaded fields in Table 4.2 provide very important information about the schemes and how change of scenario has affected their SINR values. It might appear that compared to other schemes FP scheme suffered the lowest droppage in its SINR values, however, careful consideration would reveal that irrespective of the scenarios, cross-tier SINR for FP has never been of any importance because it has always stayed negative (−3.63 dB for single-cell and −5.425 dB for multicell, Figure 4.8). It further proves that a fixed power scheme such as FP is not at all a suitable scheme for cross-tier interference mitigation in densely deployed blindly placed femtocells.

On the same lines, DB scheme behaved very poorly for co-tier SINR values (highlighted in Table 4.2, it suffered a drop of 95% with the change of scenario). Co-tier SINR values for DB scheme stayed at −0.160 dB for single-cell and −2.977 dB for multicell scenarios (Figure 4.7). Lastly, even though the CR scheme showed some promise in terms of co-tier SINR values, it suffered droppage of 90% for its cross-tier SINR values (highlighted in Table 4.2). Despite the droppage due to change in scenario, our proposed scheme has always been promising both for co and cross-tier SINR values. The bounds of coverage radius and its impact on SINR values for our PS are further described in Section 4.6.

4.4.1.6 Coverage Area Bounds and Impact on SINR (Single Cell versus Multicell)

Figure 4.10 plots how co- and cross-tier SINR values responded to the change in coverage radius and also the change in the scenario for our proposed scheme.

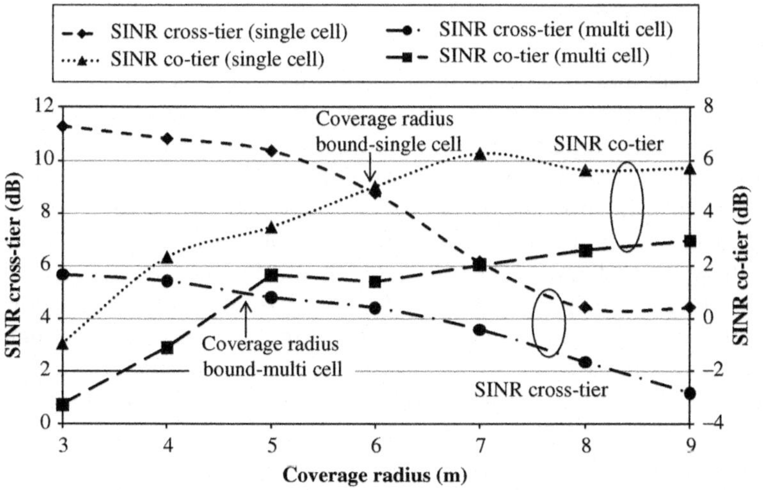

Figure 4.10 Coverage radius bounds and effect on SINR (single- versus multicell).

Irrespective of the scenario (single- or multicell), lower coverage radius gave higher cross-tier SINR values, whereas increase in coverage radius resulted into better co-tier SINR and vice versa. This further validates the results already shown in Figures 4.4, 4.5, 4.7 and 4.8. Higher cross-tier SINR values are obtained because of the fact that with reduced coverage radius more MUE is left out, thereby resulting in lower interference and better cross-tier SINR. On the other hand, since coverage radius is directly proportional to the transmit power, lower coverage radius resulted into lower signal power per FUE compared to increased interference due to densely deployed blindly placed femtocells, thus resulting in reduced co-tier SINR. It is fair to conclude that in our proposed interference mitigation scheme, for both scenarios (single- and multicell) a lower coverage radius favours cross-tier SINR whereas higher coverage radius values favour co-tier SINR. However, from our simulations it is shown that there are coverage radius bounds (i.e. ~6 m for single-cell and ~4.7 m for multicell), which resulted into balanced (optimum) value for both co- and cross-tier SINR values. These findings are very important in blindly placed densely deployed femtocells for the possible distribution and radial mobility of FUE to avoid significant performance degradation.

4.5 Summary

Femtocells aim to improve poor indoor network coverage in cellular communication, which has attracted network operators and stakeholders. Even though femtocells are discovering an important role, the issue of interference as a result of blindly placed FAPs needs to be addressed. In this chapter, a coverage radius based adaptive power control scheme to mitigate interference for blindly placed LTE femtocells was investigated. The proposed scheme does not require FAPs to be relocated on optimal locations for effective interference mitigation, rather it implements a self-update algorithm for FAPs to reduce their cell radius and adjust power values in an adaptive manner. The performance of the scheme was analysed using system level simulations for single- and multicell scenarios. The results have shown that our proposed scheme has an improved value of cross-tier SINR, throughput and lower co-tier SINR compared to baseline and existing adaptive interference mitigation schemes. Further, the results have proven that, irrespective of the scheme, the change of scenario from single to multicell, affected and resulted in lower co- and cross-tier SINR values for multicell compared to single-cell values. It was also found that our proposed adaptive power control scheme contributed towards coverage radius bounds, which provide balanced co- and cross-tier SINR values. In terms of densely deployed blindly placed LTE femtocells, coverage radius bounds are a very important finding because they can be helpful in the effective distribution of FUE to achieve balanced co and cross-tier SINR values while maintaining other performance parameters too; for example, throughput.

References

1 V. Chandrasekhar, J. Andrews and A. Gatherer, "Femtocell networks: a survey", *Communications Magazine, IEEE*, vol. **46**, no. 9, September 2008, pp. 59–67.

2 3GPP. *A Global Initiative*, Website, available at: www.3gpp.org (accessed December 2017).

3 Small Cell Forum. Website, available at: www.smallcellforum.org (accessed December 2017).

4 J. Zhang and G. Roche, *Femtocells: Technologies and Deployment*, John Wiley & Sons, Ltd, 2010.

5 F. Liu, E. Bala, E. Erkip and R. Yang, "A framework for femtocells to access both licensed and unlicensed bands", *International Symposium on Modeling and Optimization in Mobile, Ad Hoc and Wireless Networks (WiOpt)*, **9–13** May 2011, pp. 407–411.

6 J.G. Andrews, H. Claussen, M. Dohler S. Rangan, and M.C. Reed, "Femtocells: Past, present, and future", *IEEE Journal on Selected Areas in Communications*, vol. **30**, no. 3, April 2012, pp. 497–508.

7 H.A. Mahmoud and I. Guvenc, "A comparative study of different deployment modes for femtocell networks", *IEEE 20th International Symposium on Personal, Indoor and Mobile Radio Communications*, **13–16** Sept. 2009, pp. 1–5.

8 N. Saquib, E. Hossain, L.B. Le and D. Kim, "'Interference management in OFDMA femtocell networks: issues and approaches", *Wireless Communications, IEEE*, vol. **19**, no. 3, 2012, pp. 86–95.

9 T. Zahir, K. Arshad, A. Nakata and K. Moessner, "Interference management in femtocells", *Communications Surveys & Tutorials, IEEE*, vol. **15**, no. 1, Feb. 2013, pp. 293–311.

10 Y. Bai, J. Zhou and L. Chen; "Hybrid spectrum usage for overlaying LTE macrocell and femtocell", *IEEE Global Telecommunications Conference, GLOBECOM*, Nov. 2009, pp. 1–6.
11 D.L. Perez, A. Valcarce, G.D.L. Roche, E. Liu, and J. Zhang, "Access methods to WiMAX femtocells: A downlink system-level case study", in *Proc. IEEE Int. Conf. Commun. Syst. (ICCS)*, Guangzhou, China, Nov. 2008, pp. 1657–1662.
12 L.T.W. Ho and H. Claussen, "Effects of user-deployed, co-channel femtocells on the call drop probability in a residential scenario", in *Proc. IEEE Int. Symp. Personal, Indoor, Mobile Radio Commun. (PIMRC)*, Athens, Greece, Sep. 2007, pp. 1–5.
13 V. Chandrasekhar and J.G. Andrews, "Uplink capacity and interference avoidance for two-tier cellular networks", in *Proc. IEEE Global Telecommunication Conference*, Nov. 2007, pp. 3322–3326.
14 G. de la Roche, A. Valcarce, D. Lopez-Perez, and J. Zhang, "Access control mechanisms for femtocells", *IEEE Commun. Mag.*, vol. **48**, no. 1, Jan. 2010, pp. 33–39.
15 Z. Wang, W. Xiong, C. Dong, J. Wang and S. Li, "A novel downlink power control scheme in LTE heterogeneous network", *International Conference on Computational Problem-Solving (ICCP)*, **21–23** Oct, 2011, pp. 241–245.
16 P. Mach and Z. Becvar, "Dynamic power control mechanism for femtocells based on the frame utilization", *International Conference on Wireless and Mobile Communications (ICWMC)*, Sept. 2010, pp. 498–503.
17 D. Roche, G. Ladányi, D. López-Pérez, D.C. Chong and J. Zhang, "Self-organization for LTE enterprise femtocells", *IEEE GLOBECOM Workshops (GC Wkshps)*, Dec. 2010, pp. 674–678.
18 J. Liu, Q. Chen and H.D. Sherali, "Algorithm design for femtocell base station placement in commercial building environments", *INFOCOM, 2012 Proceedings IEEE*, 25–30 March 2012, pp. 2951–2955.
19 S. Wang, W. Guo and T. O'Farrell, "Optimising femtocell placement in an interference limited network: theory and simulation", *Vehicular Technology Conference (VTC Fall), 2012 IEEE*, 3–6 Sept. 2012, pp. 1–6.
20 W. Guo and S. Wang, "Interference-aware self-deploying femto-cell", *Wireless Communications Letters, IEEE*, vol. **1**, no. 6, Dec. 2012, pp. 609–612.

21 H.O. Kpojime, G.A. Safdar, "Efficacy of coverage radius-based power control scheme for interference mitigation in femtocells", *Electronics Letters*, vol. **50**, no. 8, April 10 2014, pp. 639–641.
22 G.A. Safdar, "Analysis of time correlated channel model for simulation of packet data networks", *Antennas and Propagation Conference (LAPC), 2011 Loughborough*, Nov. 2011, pp. 1, 4, 14–15.
23 Technical Specification Group RAN, "E-UTRA; LTE RF system scenarios", 3GPP, Tech. Rep. TS **36**.942, 2008–2009.
24 J.C. Ikuno, M. Wrulich, and M. Rupp, System level simulation of LTE networks, in *Proc. 2010 IEEE 71st Vehicular Technology Conference, Taipei, Taiwan, May 2010.* [Online] Available: http://publik.tuwien.ac.at/files/PubDat_184908.pdf (accessed December 2017).
25 L. Hentila, P. Kyosti, M. Kaske, M. Narandzic, and M. Alatossava, "MATLAB implementation of the WINNER Phase II Channel Model ver1.1", Dec. 2007.
26 ETSI TR 136 942 V10.3.0 "LTE; Evolved Universal Terrestrial Radio Access (E-UTRA); Radio Frequency (RF) system scenarios" July 2012.
27 H. Claussen, "Performance of macro- and co-channel femtocells in a hierarchical cell structure", in *Proc. IEEE Int. Symp. Personal, Indoor, Mobile Radio Commun. (PIMRC)*, Athens, Greece, Sep. 2007, pp. 1–5.
28 Small Cell Forum, "*Interference Management in OFDMA Femtocells*", Small Cell Forum, Mar. 2010.

5

Energy Management in LTE Femtocells

Kapil Kanwal[1]*, Ghazanfar Ali Safdar*[1]*, Masood Ur Rehman*[1]
and Xiaodong Yang[2]

[1] *School of Computer Science and Technology, University of Bedfordshire, Luton, UK*
[2] *School of Electronic Engineering, Xidian University, Xi'an, China*

Wireless cellular networks have seen dramatic growth in number of mobile users. As a result, data requirements and hence base-station (BS) power consumption has increased significantly. In turn, this adds to operational expenditures (OPEX) and also causes global warming. The BS power consumption in Long Term Evolution (LTE) means, therefore, it has become a major challenge for vendors to stay green and profitable in a competitive cellular industry. It necessitates novel methods to devise energy efficient communication in LTE. Importance of the topic has attracted huge research interests worldwide. Energy saving (ES) approaches proposed in the literature can be broadly classified in categories of energy efficient resource allocation, load balancing, carrier aggregation and bandwidth expansion. Each of these methods has its own pros and cons leading to a trade-off between ES and other performance metrics resulting in open research questions. This chapter discusses various ES techniques for the LTE systems and critically analyses their usability through a comprehensive comparative study.

LTE Communications and Networks: Femtocells and Antenna Design Challenges,
First Edition. Edited by Masood Ur Rehman and Ghazanfar Ali Safdar.

5.1 Introduction

Wireless communication has become one of the basic provisions of the modern world. Since the inception of first radio communication system by Marconi [1], wireless communication systems have seen a massive growth in the last few decades from having a couple of individuals to the majority of the world as their users [2, 3]. The concept of frequency reuse was first introduced in cellular radio communication systems by AT&T [4]. Further developments in radio communication introduced digital cellular systems, which pass through a long chain of evolution known as generations (G). We have seen usage of 1G, 2G, 3G and now 4G as the communication standard with each resulting in enhanced performance of cellular systems [5, 6]. Aiming for key achievements such as short transmission time, high throughput, low latency and security [7, 8], these systems generally consist of BS connected to core network. Each BS has designated coverage area, called cell and communicates directly with User Equipment (UE) within its coverage [9–11]. Whenever UE moves from serving cell to neighbour cell, its transfer of control is initiated through handover process [12, 13]. LTE is a 4G technology that transmits Digital Broadband Packets over Internet Protocol (IP) while offering a peak data rate of 100–300 Mbps [14–16]. This increased data rate in LTE is achieved by employing Orthogonal Frequency Division Multiple Access (OFDMA) based technology that promises low latency, high data rate and packet optimized radio access [17]. This enhanced performance of services compared to previous generations of the cellular networks has helped LTE systems to gain rapid popularity both commercially and academically.

5.2 Architecture of LTE Networks

Since the chapter is focused on energy management in LTE networks, this section presents a brief overview of the LTE architecture. LTE systems usually provide low latency, high data rate and packet optimized radio access. Compared to 3G, LTE additionally provides international roaming and compatibility with other legacy networks [18, 19]. The 4G systems make use of OFDMA and Single Channel Frequency Division Multiple Access (SC-FDMA) schemes to support flexible bandwidth [20–26]. LTE architecture is generally based on Evolved Packet Core

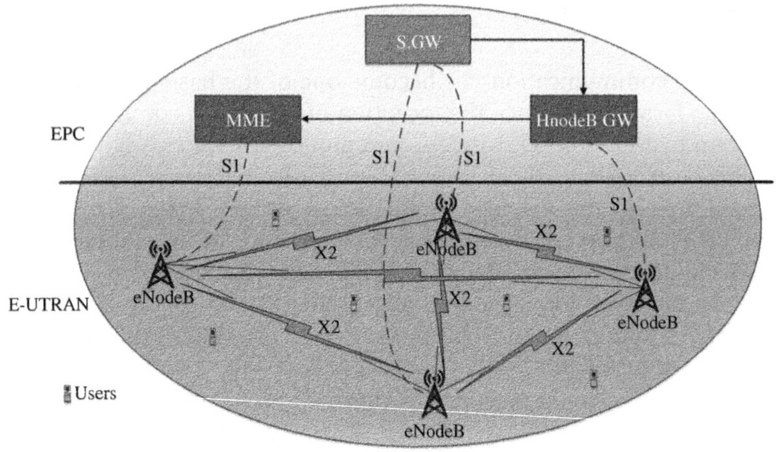

Figure 5.1 Architecture of LTE networks.

(EPC), Universal Terrestrial Radio Access (UTRA), and Universal Terrestrial Radio Access Network (UTRAN), each of which communicates with core network air interfaces and radio access network [27, 28].

Figure 5.1 illustrates the overall architecture of LTE networks showing both EPC and evolved UTRAN (E-UTRAN) [29–31] while Table 5.1 summarizes the core elements of LTE architecture.

5.2.1 Communications Perspective Challenges in LTE Networks

Though LTE has proven to be a promising technology, it is a complex network and there are some challenges that need to be carefully addressed for optimum functionality.

5.2.1.1 Signalling System

In LTE networks, one of the major issues is to avoid or limit signalling overhead and overlapping in control part of the network. A large number of connections between nodes and network fragmentation causes a rapid increase in signalling traffic. Any failure in signalling system will drag operators towards increased system latency and outages resulting in loss of revenue [32, 33]. Increased signalling traffic also leads to increased energy consumption and definitely needs to be looked in carefully.

Table 5.1 LTE network components.

Components	Description
Evolved Packet System (EPS)	Provides IP connectivity using E-UTRAN.
Mobility Management Entity (MME)	Responsible for authorization, security, handover, roaming and mobility of users.
S1 Interface	It connects EPC with BSs.
Serving Gateway (S.GW)	EPC terminates at this node. It is connected to E- UTRAN through S1 interfaces. Each user is allocated unique S.GW which is responsible for handover, packet routing and forwarding functions.
Packet data network gateway (PDN-GW)	PDN-GW provides UEs with access to packet data network by allocating IP addresses. It is also responsible for secure connection with untrusted devices from non-4G networks.
HnodeB	Femtocells that are employed to improve seamless connectivity in coverage holes.
eNodeB	Also known as BS that serves the UEs.
HnodeB GW	Provides connection to the core network.
X2 Interface	Provides communication between two BSs.

5.2.1.2 Backward Compatibility

LTE is usually compatible with all other relevant major standards. The combination of devices, network interfaces and equipment to support other standards complicates end-to-end functionality testing and interoperability testing (IOT) [34, 35].

5.2.1.3 BS Efficiency

Due to the employment of OFDMA in LTE, signals have high amplitude variability known as Peak-to-Average Power Ratio (PAPR), which reduces transmitter efficiency. Furthermore, the BS provides high data rate at the cost of high dynamic transmission power. Since, high transmission power results in increased energy consumption and thereby increases OPEX; energy management has become major challenge in LTE networks to stay profitable and also to reduce global warming [36].

5.2.2 Importance of Energy Management in LTE Networks

Since increased power consumption (using non-renewable energy sources) directly contributes in climate change, therefore it has become major obstacle for environmental and economic aspects [37, 38]. Vendors highlighted the raising trend of power consumption due to the increased data traffic. Number of users of the mobile networks has 10% annual increase across the world with an increase of 25–50% in each user's data requirements [39]. Therefore, provision of high data rate demanding services with minimum power consumption has become a major challenge for vendors to stay profitable [40, 41]. Information and Communication Technology (ICT) contributes approximately 10% of worldwide power consumption while it adds 2% to global warming [42, 43]. Moreover, global warming is growing swiftly due to the additional advance BSs being deployed to fulfil increased users demand thus resulting in a 15–20% increase per year and this increase almost becomes double every 5 years [44]. Noteworthy, ICT contribution in global warming will become 3% by 2020 [45]. Since, BS consumes major part of energy in LTE networks, reducing power consumption at BS could help cutting down OPEX [46]. Vendors choose to deploy automated networks to facilitate ES [47]. 3GPP has already introduced Self-Organized Networks (SON), which increase the level of automation achieved in operation and maintenance, thereby resulting in a decreased OPEX [48]. Apart from other functionalities, SON also provides opportunities for incorporation of enhanced ES techniques that can help achieve reduced OPEX values. Technologies based on the concept of SON (e.g. LTE), can enjoy a 19% reduced OPEX due to advanced ES techniques [49].

5.3 Classification of ES Schemes

The literature presents considerable amount of research work on energy efficiency in LTE systems. Each BS in cellular networks consists of Baseband Units (BBU) with one or more transceivers. Each transceiver contains Radio Frequency (RF) part, Power Amplifier (PA) and Antennas connected through cables [50–52]. All these components are located very close to each other in a unit called Radio

Resource Unit (RRU). PA is the main power-hungry element in this unit [53]. Since LTE employs OFDMA [54–56] and normally PA operates at a level that is 6–12 dB lower than the saturation point, this results in lower adjacent channel interference. Power consumption at BS can be categorized as static and dynamic power consumption [57]. Static power consumption belongs to hardware used in BS and remains nearly constant. Dynamic power (also known as communicational power) on the other hand, depends on traffic load between BS and UEs [58, 59]. The focus of this chapter is to investigate, classify and critically analyse existing ES techniques to control the dynamic power consumption.

5.3.1 Static Power Consumption

Static power is purely hardware based constant power consumption, which the BS needs to cater for necessary operations. The static power consumption can be improved by energy efficient hardware designs and subsequent intelligent deployments. However, in this chapter our main focus based on dynamic power consumption.

5.3.2 Dynamic Power Consumption

The dynamic power consumption depends on the BS's resources utilization and is directly affected by BS transmission operations. Therefore, it could be reduced by turning off of BS operations during idle states. Dynamic power management has attracted attention of researchers and could be classified in to three main categories, that is, energy efficient resources allocation, bandwidth expansion and load balancing as shown in Figure 5.2.

The dynamic (or communicational) power consumption can be reduced by appropriate activation and deactivation of BS's transceivers also known as Discontinuous Transmission (DTX) during off peak time periods. DTX based schemes allocate Multicast and Broadcast Single Frequency Network (MBSFN) subframes through traffic load consideration [60, 61]. The power aware algorithm analyses the traffic that cells need to serve, then calculates the amount of resources required and distributes them among the subframes (SFs) to minimize the power consumption. In lightly loaded conditions, there is a possibility of some frames being not utilized, which could help to achieve improved energy conservation by configuration and turning off of

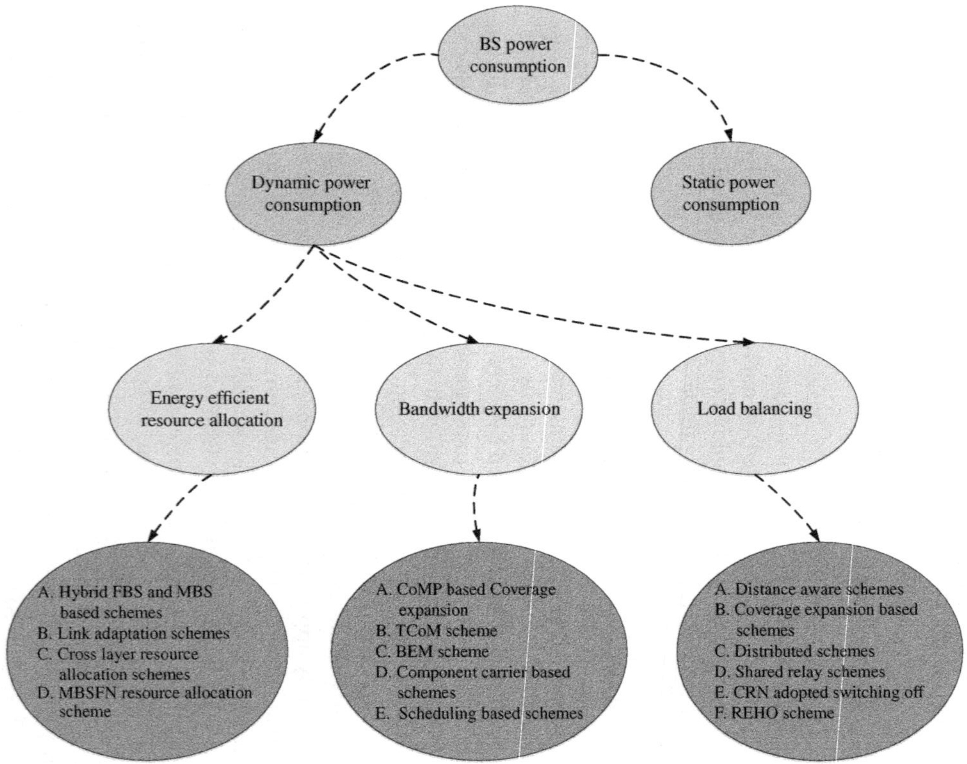

Figure 5.2 Classification of energy saving schemes.

idle frames. Importantly, micro cell DTX significantly reduces power consumption during low traffic rate; however, it does not work during high traffic loads because no empty SFs are left. Results indicate that average power consumption per cell without DTX is 170 W [60], while micro DTX enabled cell reduces this figure by 60%. Next to this, the work presented in [62–64] avail the concept of energy efficient resource allocation and significantly reduce overall BS power consumption. Aggregation of resource blocks through carrier aggregation algorithms can also help achieve better ES [65, 66]. This helps in reducing overheads, thereby increasing energy conservation. Along the same line, energy efficient carrier aggregation algorithms group together the component carriers (CC) to achieve greater ES [67]. Distance aware schemes, which involve switching off the BS that is greater distance from UEs, can also help to obtain better ES [68–70]. These schemes reduce energy consumption by appropriate activation/deactivation of the BS, based on information of varying distance and load. Another dynamic traffic-aware approach is introduced in [71], which uses time varying traffic information for energy conservation. Each BS divides its cell in different number of sectors, then switch off appropriate sector (with low traffic) providing power saving opportunities. Centralized and distributed schemes that engage UE migration also help achieve better ES [72, 73]. Centralized schemes select the highest loaded BS through analysing traffic information and determine if it could accommodate more UEs. Considering selected BSs, if available bandwidth is greater than the capacity required to serve neighbour cells, UEs with lowest load traffic will then be shifted towards heavily loaded BSs, resulting in switching off lightly loaded BSs for reduced energy consumption. Compared to centralized, distributed schemes in contrast, it is possible to select a pair of BSs and then determine the ES state of each BS. Initially, schemes activate ES on particular BS, which examines the neighbour cells list, and select one BS with the lowest load forming a pair. The BS preferring to keep powered ON is the one that can accommodate more UEs. On the same lines, work presented in [74–76] employs distributed scheduling and energy efficient power control approaches for reduced power consumption. Another study in [77] shares a relay between different operators, thus resulting in ES. In [78] authors introduced an energy efficient link adaptation scheme that combines the traditional link adaptation with power control, thereby resulting in improved energy efficiency at the BS. This scheme uses the BS's transmitted power as a new feedback parameter and predicts an optimal set of parameters in order to maximize the BS's energy efficiency and satisfy the Block Error Rate (BLER) constraint for the channel state. Another interesting scheme is presented in [79, 80], which suggests an energy efficient resource allocation scheme that operates in multi-cell OFDMA-based LTE networks. In the same context, [81, 82] introduce energy efficient resource allocation for reduced power consumption in LTE networks. This method combines dynamic Resource Block (RB) allocation with energy efficient power allocation and reduces overall BS power consumption. A bandwidth expansion scheme with load balancing is introduced in [83, 84], which employs the idea of moving UEs from overlapping area of lightly loaded cell to the heavily loaded cell. The Time Compression Mode (TCoM) is presented in [85], which saves power by reducing control signals overhead's transmission. RBs are compressed together in TCoM, either in time or frequency domain by usage of higher order modulation. ES is achieved through reduction in overhead signalling by appropriately turning off of the unused RBs [85]. On the same lines, optimized resource allocation could also lead to reduced energy consumption as discussed in [86]. Energy efficient BS deployment too has helped in improved energy conservation [87]. ES approaches for D2D communication in LTE networks resulting in reduced power consumption are presented in [88–90]. The most recent research work has established the idea of integrating Cognitive Radio Networks (CRNs) [91] with LTE infrastructure for improved ES. This predominantly lies in the fact of isolating users in two categories (i.e. PUs and SUs). During awake periods, BS transmits PUs data over the licensed spectrum while in contrast, SU data is sent over the unlicensed spectrum. BSs are switched in to sleep mode right after completion of data packets transmission thus resulting in opportunities for improved power conservation. ES can also be achieved through suitable cells coverage expansion and turning off of idle BSs [92]. Importantly, this scheme initially splits cells in two main categories; that is, cooperative cells and dormant cells. Where, cooperative cells serve associated users while dormant cells are turned off during low traffic time periods for ES. Next to this, intelligent resource allocation and power control [93, 94] can help reduce dynamic power consumption thus resulting in improved energy efficiency. It is worth noting that energy efficient schemes, while deployed at every BS, allocate lower transmit power to suitable resources in line with the associated Signal-to-Noise plus Interference (SINR) ratio. Among others, the

D2D communication based scheme presented in [95] uses energy efficient heterogeneous routing for enhanced energy conservation. A significant amount of research work has been carried out to develop different ES schemes, which help to reduce dynamic power consumption. However, the increasing trends of OPEX and global warming indicate that there is always a need to do more research work to achieve enhanced ES for future wireless systems. Based on the previous discussion, a broad classification of ES schemes is presented in Figure 5.2 while detailed insights into individual schemes are provided in the following sections.

5.4 Energy Efficient Resource Allocation

In order to transport UE data across wireless media, wireless cellular systems employee various control channels that segregate dissimilar types of data and transport them across Radio Access Network (RAN) in an orderly routine. LTEs consist of physical channels, transport channels and logical channels. Further, physical channels consist of Physical Broadcast Channel (PBCH), Master Information Block (MIB) and Physical Downlink Control Channel (PDCCH). Intelligent switching on and off of these control channels can result in increased ES. Some energy efficient schemes in energy efficient resource allocation category are explained next.

5.4.1 Hybrid FBS and MBS Based Schemes

The use of femto base stations (FBS) has proven to be promising technology to cover those areas where macro base stations (MBS) are limited. In the same context, work presented in [93] introduces power control based RBs allocation scheme in LTE network with MBS and FBS, which employ the concept of Almost Blank Subframe (ABS) and Reduced Power Blocks (RPBs) to allocate reduced transmission power to RBs, thereby resulting in to reduced downlink power consumption. This scheme is recommended for two tier heterogeneous networks with MBS and few FBSs as shown in Figure 5.3. The main idea lies in the fact that varying transmit power levels can be assigned to different resources thus resulting in reduced BS power consumption. The level of transmit power is measured through SINR, thus if users SINR is higher than the predefined threshold, then they are allocated with

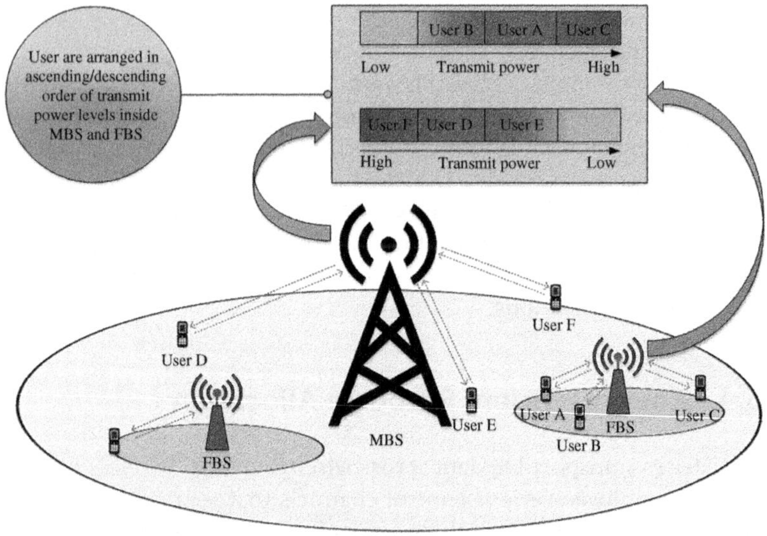

Figure 5.3 Hybrid FBS and MBS based ES scheme.

lower transmit power, while higher power is assign to users with lower SINR. Since SINR values changes rapidly, accordingly estimation of transmit power also changes continuously. Next to this 'Breathing', a technique is introduced for RBs allocation that divides users in two classes; that is, Inhale and Exhale. Users are arranged in ascending order in the Inhale class in relation to the required transmit power and are mapped with RBs in sequence. On the other hand, the Exhale class involves sorting users in descending order of their transmission power value [93].

5.4.2 Link Adaptation Schemes

LTE provides a high data rate through the effective resources utilization in available bandwidth. The Channel Quality Indicator (CQI), Precoding Matrix Indicator (PMI) and Rank Indicator (RI) parameters play key roles in efficient use of resources. PMI determines which precoding matrix should be used for downlink transmission while RI presents the number of layers that should be used for downlink transmission. CQI is reported from UEs to the BS that contains information about the supported Modulation and Coding Schemes (MCS).

CQI plays a major role in selection of MCS at downlink in BS. CQI values range from 0 to 15. A higher value of CQI indicates use of higher modulation scheme while BS can use higher coding rate for increased energy efficiency [78]. An energy efficient (EE) link adaptation scheme, which combines traditional link adaptation with power control resulting in improved energy efficiency at BS is presented in [78]. This scheme uses BS transmitted power as a new feedback parameter and predicts optimal parameters that maximize the BS energy efficiency and satisfy the BLER constraint used for demodulation tests in multipath conditions during radio link monitoring. This scheme can be best described with the help of the LTE based downlink transmission model shown in Figure 5.4. UE estimates channel gain between BS and itself to calculate the parameters RI, PMI, CQI and transmit power. These parameters are then fed back to BS through the feedback channel as shown in Figure 5.4. The BS uses feedback received from UEs as an input parameter to adjust its transmission power; where RI helps to determine the code word, CQI helps to select MCS scheme for each transmission and PMI is used by the BS to select the precoding matrix.

5.4.3 Cross Layer Resource Allocation Schemes

A cross layer based EE resource allocation scheme for multi-cells OFDMA-based LTE networks is presented in [79]. This technique encompasses physical and medium access control (MAC) layers combining dynamic RB allocation at MAC layer with EE power allocation at Physical Layer (PHY), thus resulting in reduced overall power consumption by the BS. Dynamic RB allocation is based on feedback (energy efficiency indicator) that is used to adjust scheduling process. This method also promotes the user's fairness through allocating equal resources to all users either with good and bad quality channels.

5.4.4 MBSFN Resource Allocation Scheme

The MBFN predicts the future traffic load that needs to be served in the next frame, this predicted load is used to calculate the required RBs while turning off the unused resources. The future load prediction is made using previously served load information exchanged between BSs through X2 interface (standard interface used for BS

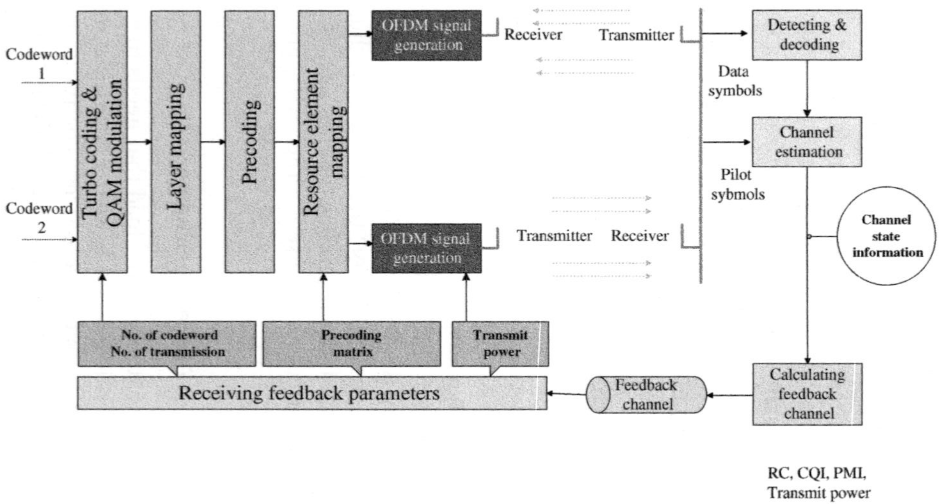

Figure 5.4 Link adaptation scheme – LTE based downlink transmission.

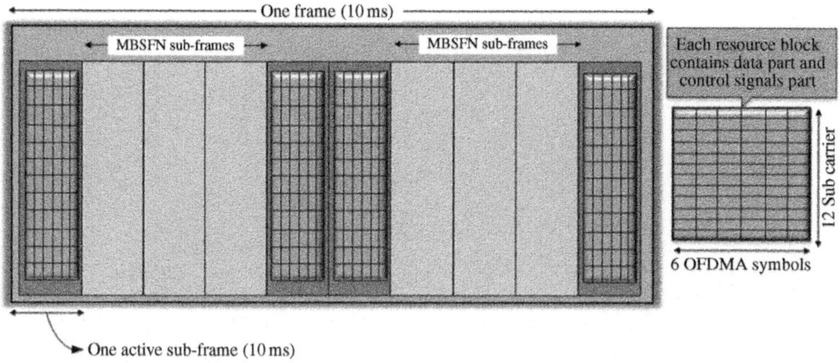

Figure 5.5 MBSFN based frame architecture.

communication in LTE). An interesting MBSFN based ES scheme in [61] configures MBSFN SFs that help to provide and setup transmitter switching off periods. Additionally, this method estimates the resources required to serve the predictive load for effective resource allocation resulting in enhanced power saving by turning off the idle resources. Based on LTE specifications, 6 out of 10 SFs can be configured as MBSFN (Figure 5.5). Importantly, MBSFN SFs carry fewer Reference Signals (RS) compared to the standard subframe. Therefore, in a case where no data is available, MBSFN SFs can be turned off resulting in reduced energy consumption.

5.5 Bandwidth Expansion Schemes

The EE LTE networks can also be realized through bandwidth expansion. Several proposed techniques employing bandwidth expansion for improved energy efficiency are presented here.

5.5.1 CoMP Based Coverage Expansion

Work in [92] uses Coordinated Multiple Point (CoMP) for improved ES. CoMP expands cell coverage thus resulting in better expansion compared to antenna adjustments and transmission power measurements (see Section 5.5.6). The proposed work employs link budget and SINR as input parameters and then divides networks in clusters on the basis of the equivalent cell principle with a distributed method

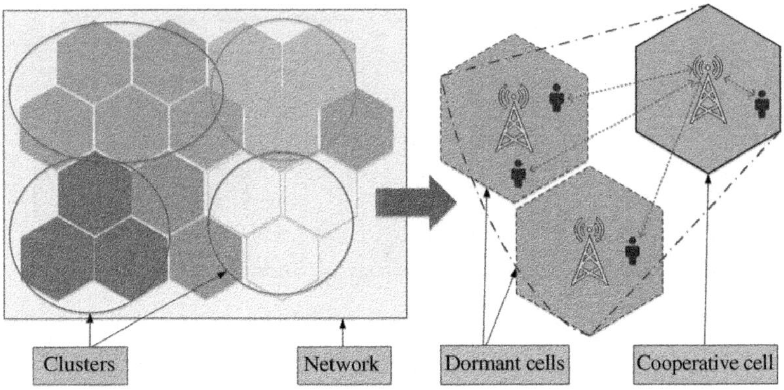

Figure 5.6 CoMP based coverage expansion.

(Figure 5.6). Cells in this scheme are divided into two main categories, that is cooperative and dormant cells, which is decided by a Joint Processing (JP) cooperative cell selection model. During off peak traffic time periods, cooperative cells expand their coverage to serve dormant cells that are turned off for ES purposes.

5.5.2 Time Compression (TCoM) Scheme

The 10 ms frame in OFDMA consists of 10 SFs. Each SF includes two slots of 0.5 ms each and each time slot consists of 12 subcarriers and seven symbols as shown in Figure 5.7. Subcarriers of each symbol can be allocated to multiple users thereby making efficient use of radio resources. TCoM tries to reduce the power consumption caused by the usage of higher order modulation schemes in OFDMA through a decrease in control channel overhead [85]. RBs in TCoM are compressed together and ES is achieved through reduced overhead signalling by appropriately turning off unused RBs during the idle state. The time and frequency implementations of TCoM do not differ in performance because of the fact that changes in either length or bandwidth of a transmission have the same impact on the transmitter's energy. A compression factor to represent the number of RBs to be pooled together is introduced in [85]. It also uses Shannon's capacity to derive required Signal-to-Noise Ratio (SINR). TCoM is found to be around 26% more EE compared to the LTE benchmark standard.

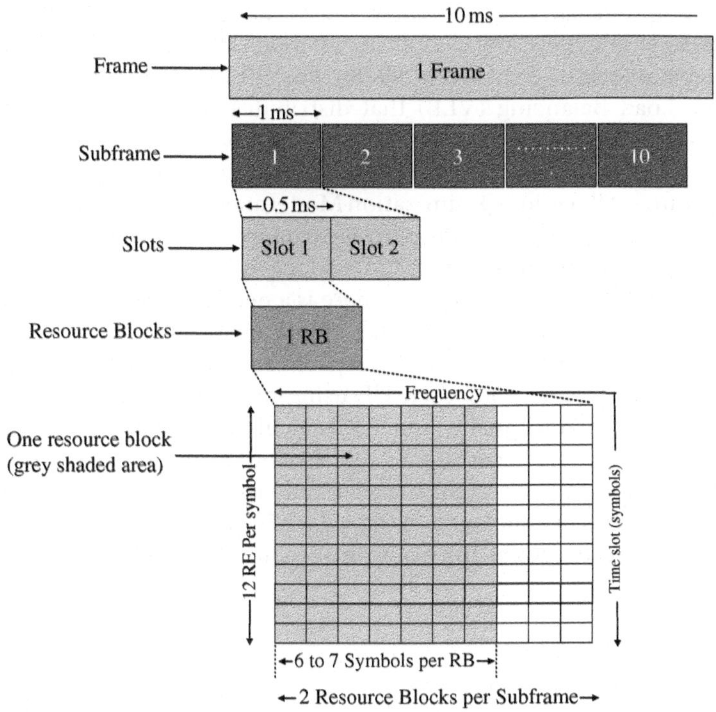

Figure 5.7 OFDMA frame architecture.

5.5.3 Bandwidth Expansion Mode (BEM) Scheme

Another Bandwidth Expansion Scheme (BEM) is described in [83]. This method is based on the concept that when the network is lightly loaded (larger number of RBs is free), in this scenario bandwidth allocation can be increased to reduce power consumption at BS. In LTE systems, minimum resource allocation is one RB for each user and allocation is done by schedulers. Expanded RB allocation (allocating more than 1 RB per user) reduces the MCS and SINR per frequency channel for each user, which in turn provides more EE transmission. Work in [83] is especially recommended for low loaded networks, because extra RBs that are idle during off peak traffic helps in bandwidth expansion. BEM addresses two important factors; EE and Mobility Load Balancing (MLB) in networks. This work proposes an effective EE resource allocation optimization model by employing a low complexity method called Energy Efficient Virtual Bandwidth Expansion Mode (EE-VBEM). The concept of Virtual Load Balancing (VLB) that distributes some of the traffic (users) from highly loaded cells to the lightly loaded cells is used as shown in Figure 5.8. The EE-VBEM consists of two major parts: (1) EE Resource Allocation Optimization Model and (2) Low Complexity Method to achieve (1). Firstly, all BS exchange load information of neighbouring cells through the X2 interface. Based on this information, each BS determines whether there is a need for load balancing. In case load balancing is required, VLB automatically start shifting users from overlapping area to lightly loaded BS. BEM then calculates the required RBs for each UE using minimum required data rate and user channel quality. Once RB calculation is done, the BEM prioritizes the users according to SINR value. A higher SINR indicates higher BEM priority for the user and vice versa. After priority assignment, RBs are allocated to the UEs. BEM saves energy by allocating extra resources at the expense of reduced overall capacity of the BSs [83].

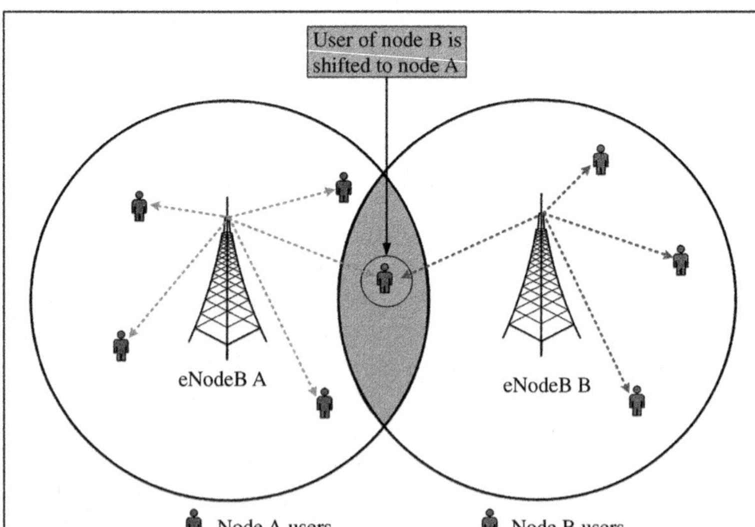

Figure 5.8 Resource allocation through load balancing.

5.5.4 Component Carrier Based Schemes

Carrier aggregation is a well-known technology used in LTE networks for the effective use of bandwidth. Each aggregated carrier is known as a Component Carrier (CC) that can have bandwidths ranging from 1.4, 3, 5, 10, 15 or 20 MHz, while a maximum of five carriers can be aggregated at a time. Carrier aggregation can be achieved through three methods as shown in Figure 5.9. The simplest method is known as Intra-Band Contiguous, which uses contiguous carrier aggregation at the same frequency band. The second method is known as Non-Contiguous Intra-Band Carrier Aggregation in which the CC operates at the same frequency band but has gaps as shown in Figure 5.9. The third method is Non-Contiguous Inter-Band Carrier Aggregation in which carriers operate at different frequency bands. To achieve EE communication in LTE networks, more CCs can be jointly utilized in a BS for enhanced ES opportunities. In [65], authors recommended OFDMA-based multiple CC technique for EE transmission that uses two CCs for data transmission. The main idea is to transmit only

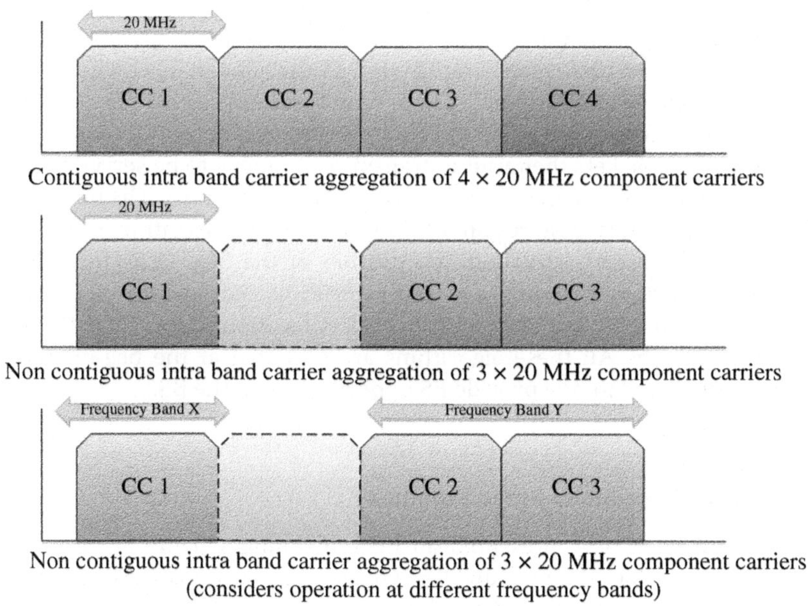

Figure 5.9 Carrier aggregation.

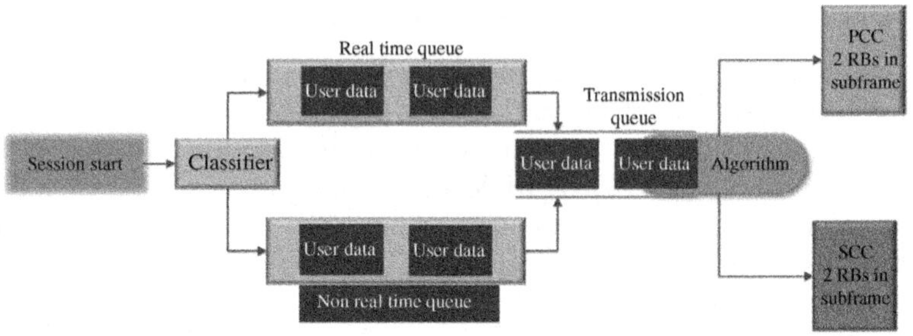

Figure 5.10 OFDMA-based CC ES scheme.

necessary CCs thus providing opportunities for appropriate deactivation of idle CCs to reduce the power consumption. The ES scheme in [65] works in the downlink in BS and supports both real and non-real-time traffic simultaneously as shown in Figure 5.10. The ES scheme consists of two CCs operating at same frequency band and can be jointly utilized in BS for data transmission. The two CCs are called the Primary Component Carrier (PCC) and Secondary Component Carrier (SCC), respectively. Normal data transmission uses PCC while SCC is only used during high traffic conditions. During transmission, a user's data packets are transmitted to the session level where they are classified as Real Time (RT) or Non-Real-Time (NRT) by the classifier and forwarded to RT and NRT Queues, respectively (Figure 5.10). The data packets then wait in transmission queue to be served by the proposed ES scheme, which consists of two algorithms. The first algorithm allocates radio resources, while the second algorithm is used for the appropriate activation/deactivation of the SCC. The first algorithm further contains two sub-algorithms; Bandwidth Allocation Algorithm (BAA) and Resource Block Allocation Algorithm (RBAA), respectively. All these algorithms are executed at the beginning of every SF and jointly provide ES opportunities at the BS.

5.5.5 Scheduling Based Schemes

Videv et al. have presented an EE scheduling scheme in [96]. The method is based on bandwidth expansion through allocation of extra resources to the UEs and uses lower order modulation schemes for ES. This scheme reduces power consumption by 44% while maintaining throughput and QoS constraints. It uses an energy-aware scoring scheduler, which considers best channel conditions and allocates additional resources to the UEs. The scheduler allocates resources by following the integer factor defined for bandwidth expansion. This method is effective only for networks where traffic is low and more free resources are available to be allocated to the UEs. This scheme provides ES at the cost of overall system capacity and therefore is not effective in a practical RT environment.

5.6 Load Balancing Schemes

Research has shown that traffic load varies significantly at the BSs and a lot of energy is wasted during low load operation. Load balancing is a part of Radio Resource Management (RRM). The term *load balancing* presents any method that could be used to transfer load from highly loaded cells to lightly loaded neighbour cells for the efficient use of radio resources. The user's distribution and traffic flow are irregular in cells, which can cause an unbalanced load condition in the network. In wireless cellular networks with unequal traffic load distribution, some of the users at the edges of cells can be transferred from highly loaded cells to the lightly loaded cells thereby providing opportunities for efficient resources utilization. When UEs detect that neighbour cells can provide better signal quality than its current serving BS, they are handed over to that neighbour cell. During load balancing, if the cell is desirable or already in ES mode and it is selected as a candidate for load balancing from a nearby heavily loaded cell, then two options exist. Firstly, to prioritize the load balance without considering the ES and, secondly, focus is made only to prioritize the ES. In the second case, UEs are not allowed to be handed over to the cells, which are desirable, or already in ES mode and the heavy loaded cell has to find another neighbour cell for load balancing. In this case, edge users may not be served efficiently but power saving could be improved.

5.6.1 Distance Aware Schemes

Work in [68] has introduced distance aware schemes that involve switching off a BS that is a greater distance from UEs. This work reduces energy consumption by appropriate activation/deactivation of BS through information of varying distance and load. Each BS in a seven-cell-based cluster calculates its average distance from associated UEs and adjacent cells UEs as shown in Figure 5.11. Since the larger average distance between BS and UEs leads to the higher power consumption, the appropriate BS (with greater average distance) is selected for switching off. If the bandwidth requirements to serve associated UEs are less than the total available capacity supported by adjacent cells, then the selected BS is switched off and traffic is allocated to the neighbour cells resulting in reduced power consumption. Moreover, the BSs in sleep mode can be activated if the network becomes busy due to high volumes of traffic. The ES scheme aims to turn off as much BSs as possible without any degradation of QoS. This scheme divides the day into two zones, a night zone (7 pm to 7 am) and a day zone (8 am to 6 pm). Turning off the BS is performed in the night

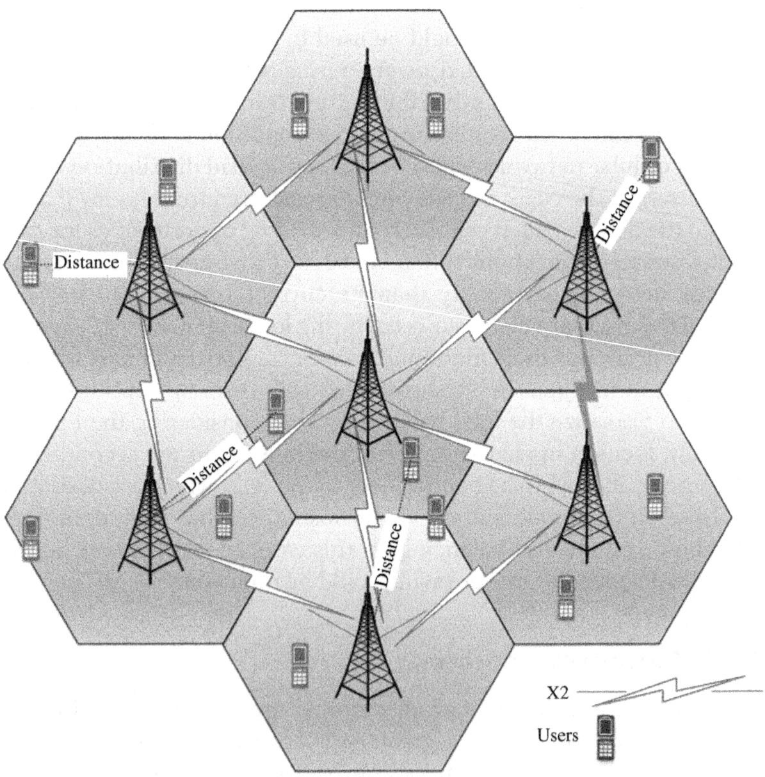

Figure 5.11 Distance aware based BS communication.

zone to achieve ES during 12 hours of low traffic load conditions. The BS is switched on in the day zone when traffic load increases and network becomes busy. In high traffic load conditions, a number of BSs should remain switched on in order to serve the UEs appropriately without affecting the QoS. The ES scheme proposed in [68] significantly reduces power consumption by deactivating BSs while neighbour cells can send activation instruction back to the BS in sleep mode through an X2 interface.

5.6.2 Coverage Expansion Based Schemes

A centralized ES algorithm is proposed in [72] that provides ES by turning off the lightly loaded BSs. This scheme is based on the idea of shifting the traffic towards the highest loaded BS using load and coverage information of the network and switches off a lightly loaded BS. The main idea lies in the fact that all UEs of a lightly loaded BS are served by the neighbouring busiest BS, thus it permits a lightly loaded BS to switch off for ES. Initially, a neighbour BS sectorizes its coverage, then extends coverage of the appropriate sector through transmission power adjustment and reconfiguration of the antenna as shown in Figure 5.12. The extended sector coverage helps BSs to serve UEs where a lightly loaded BS is turned off. The proposed algorithm, while deployed at every BS, sectorizes and extends its coverage for ES purposes. It uses two algorithms; first, one monitors network for load information while a second operates on individual BS and manages its sectorization and transmission expansion process. Initially, on the basis of load information, a centralized algorithm selects the busiest BS and analyses its resources availability. If the selected BS has enough resources to serve the neighbour BS's users, then one of its sector transmission coverage is expanded to serve UEs of the neighbour cell that is being switched off for ES as shown in Figure 5.12 [72].

5.6.3 Distributed Schemes

In [71], a distributed self-organized sectorization of BSs is presented for EE communication. Based on the varying load information, each BS reconfigures itself in RT, thus utilizing the minimum number of sectors for ES while promising adequate QoS. Since each BS dynamically reconfigures itself and no correspondence is required with neighbour BSs, this scheme is inherently distributed and self-organized.

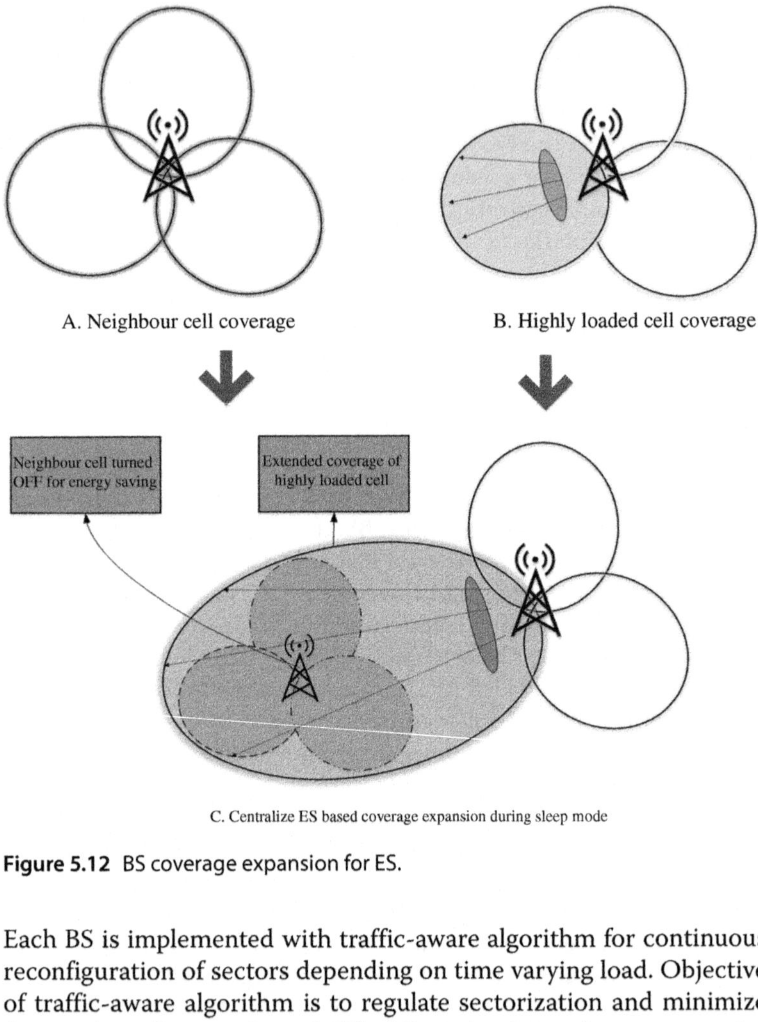

Figure 5.12 BS coverage expansion for ES.

Each BS is implemented with traffic-aware algorithm for continuous reconfiguration of sectors depending on time varying load. Objective of traffic-aware algorithm is to regulate sectorization and minimize the number of sectors in each BS while maintaining the necessary signal power required for each UE and QoS constraints. It is worth noting that during low traffic durations, a lower number of sectors is sufficient to serve BS users, thus unused sectors are turned off to achieve ES as shown in Figure 5.13. This scheme estimates the required number of RBs for each UE and uses both time-inhomogeneous and time-homogeneous traffic models for performance analysis. It also employs interference-managing arrangements to handle inter-cell

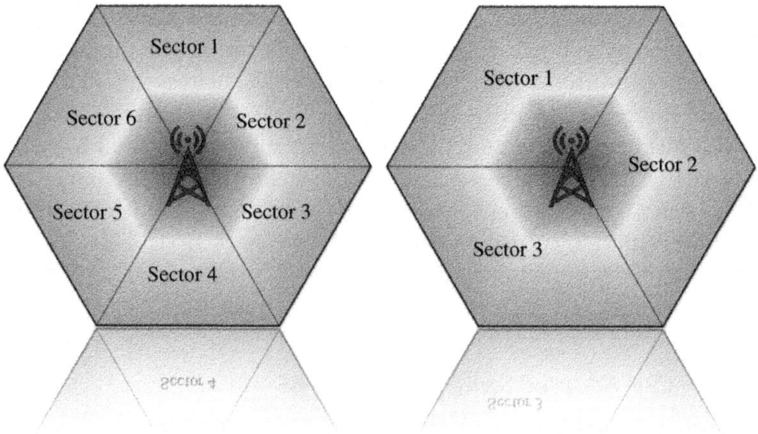

Figure 5.13 Distributed schemes: sectorization in BS.

interference and significantly reduces overall system dynamic power consumption by turning off the unused sectors in each BS (Figure 5.13). However, one of the major disadvantages of this technique is that a sector can be turned off only if it does not serve even a single UE in low traffic durations.

5.6.4 Shared Relay Based Schemes

Researchers have proposed a shared relay based load balancing ES scheme for the LTE networks in [77]. The operators or service providers share their network resources to accommodate additional users and support their demand of increased voice and data services through load balancing. This scheme, however, needs reasonable investment in the network infrastructure and is based on two assumptions. First assumption states that two different network operators jointly provide coverage to the service area through service level agreement, which allows UEs to communicate with operators through a load balancing algorithm. A second assumption says that a centralized SON algorithm is used for optimization of communication between UEs for ES. It lays the foundation for a relay based shared network based on two LTE networks belonging to two different operators with their own BSs. UEs from both operators can freely communicate with any BS regardless of their operator. BSs of both operators are connected through backhaul link, which is monitored and controlled by a remote entity called a RAN. Using information about the load and channel conditions, the SON algorithm calculates the Reference Signal Received Power (RSRP) of both BSs for each user. Once the RSRP has been calculated, the UE is then allocated to the BS having better RSRP for it. However, if the RSRP of both BSs is the same then UEs prefer to communicate with their own operator BS because both operators prefer to utilize their own resources first. This scheme reduces power consumption by 15–20% with the help of SON based load balancing.

5.6.5 CRN Adopted Switching Off of a BS

The work presented in [91] incorporates CRNs with LTE and turns off BSs for ES purposes. The proposed algorithm employs three modes of operation, namely sleep, awake and listening modes, respectively (Figure 5.14). During the awake mode, PU data is transmitted using pre-emptive priority while SU data is sent using unused remaining spectrum.

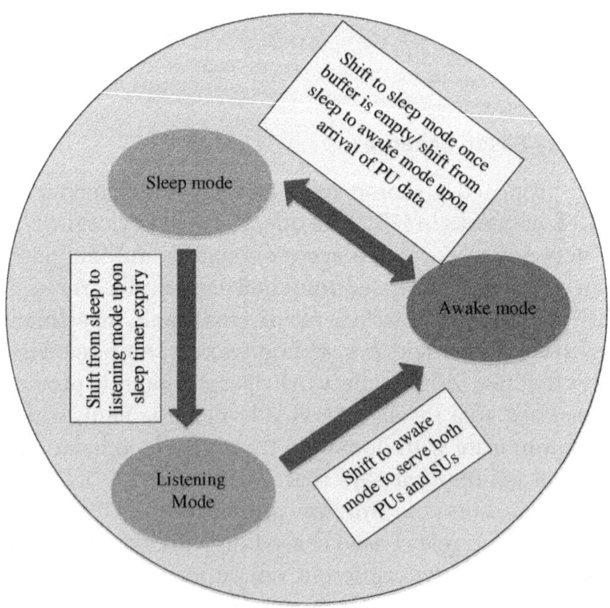

Figure 5.14 State diagram for CRN based ES.

Once all packets have been transmitted and the buffer becomes empty, the BS is turned in to sleep mode for ES purposes. Importantly, arrival of PU data can shift a BS from sleep mode straight back to awake mode. Otherwise, the BS remains in sleep mode to conserve energy and shifts to the listening mode upon expiry of sleep mode timer. The BS listens to data traffic in listening mode before it repeats the whole cycle.

5.6.6 Reduced Early Handover (REHO) Scheme

Taking into account challenges and open research issues, we have proposed a reduced early handover (REHO) energy scheme in [97]. REHO merges bandwidth expansion with RBs switching off for enhanced ES purposes. The REHO scheme, while deployed at every BS, relocates users from overlapping areas of seven neighbour cells to the one centre cell through load balancing thus enabling neighbour cells to turn off freed RBs for ES. REHO employs the concept of time compression, thus combining two RBs to form one incremental RB and allocates to a single user resulting in reduced control channel overhead transmission; further, it initiates early handover using a reduced value of hysteresis. REHO achieves ES through fine-tuning of hysteresis, offset and is explained with help of Figure 5.15. A BS transmits cell specific Reference Signals to all users within its coverage area, which are used by users to measure RSRP. When the RSRP of a target cell becomes better than the serving cell, then the user triggers an A3 event and sends a measurement report (best candidate BS information) to the

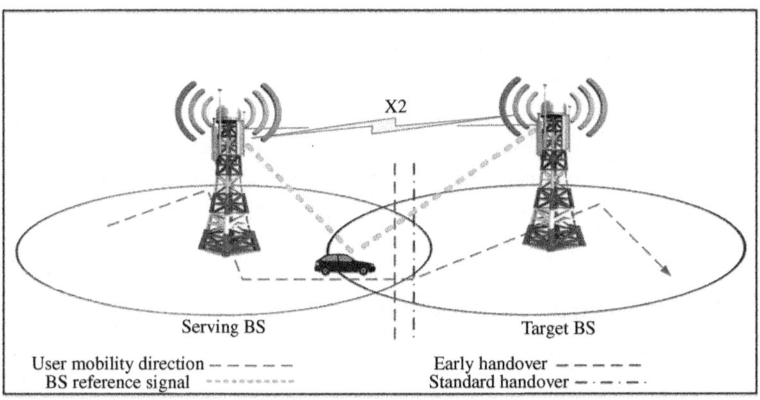

Figure 5.15 REHO ES scheme.

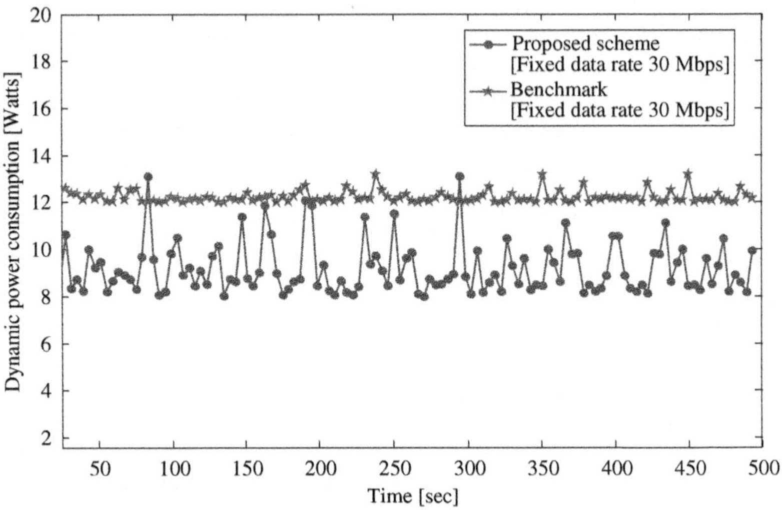

Figure 5.16 REHO dynamic power consumption.

serving cell to initiate handover [98]. The hysteresis and offset are used to push a user closer to the target cell thus ensuring minimal radio link failure. REHO uses the minimum reduced value of hysteresis thus resulting in an early handover initiation compared to standard handover for ES while maintaining radio link failure at acceptable levels. Systems level simulations are performed to demonstrate the behaviour of REHO. The chosen network scenario consists of seven overlapping cells with 50 users randomly distributed in each cell. Figure 5.16 compares REHO with standard LTE handover for dynamic power consumption. Clearly, REHO outperforms standard handover in terms of dynamic power consumption where the reduction in power consumption in REHO is achieved by early turning off of RBs.

5.7 Comparative Analysis

Table 5.2 critically compares existing ES schemes in terms of their pros and cons followed by detailed discussion and analysis. Table 5.2 shows that distance aware scheme [68] operates during 12 hours

Table 5.2 Critical analysis of existing ES schemes.

ES Schemes	Advantages	Disadvantages
Distance Aware [68]	• Power saving up to 30% compared to an always ON network.	• Runs during a limited time period (8:00 pm – 8:00 am). • No power saving during the peak traffic time period.
Dynamic Distance aware [69]	• Power saving up to 70% compared to an always ON network. • Low blocking probability. • Runs every hour.	• Exchange of information overhead between cells every hour. • Low power saving 7:00 pm to 11:00 pm.
Micro DTX scheme [60]	• Power saving up to 61% compared to a cell without any DTX. • No need to power off the whole BS. • Uses MBSFN subframes for power saving. • Creates empty transmission intervals during which PA can be deactivated.	• Longer sleep mode increases delays; 10–20 s in going back to active mode. • Increased number of MBSFN subframes decreases the capacity and bandwidth. • In LTE rel-8, information could change at the system broadcast channel only once in every 6 min.
Enhanced DTX scheme [60]	• Power saving up to 89% compared to a cell without any DTX. • Only synchronization and other secondary signals transmitted.	• Increased number of MBSFN subframes decreases the capacity and bandwidth • In LTE rel-8, information could change at the system broadcast channel only once every 6 min.
Energy efficient bandwidth expansion scheme [96]	• Saves power up to 44% • Effective for a lightly loaded network.	• Reduces the actual capacity of bandwidth. • As traffic load increases, bandwidth decreases.
Centralized Algorithm [73]	• Uses the load information scope from the entire network. • More effective with a lower number of users.	• Lower transition cost with low bandwidth requirements • Higher worst-case complexity due to binary heaps.

(Continued)

Table 5.2 (Continued)

ES Schemes	Advantages	Disadvantages
TCoM [85]	• Provide ES up to 26% compared to an always ON System. • Deactivation of RBs is a very effective ES technique.	• Ineffective ES at cell edges. • Suffers from capacity limitation.
EE Link Adaptation Scheme [78]	• More effective for the UEs closer to the BS.	• Limited reduction. • Increased feedback overhead.
BEM [83]	• Significantly reduced power consumption at BS in low load cells. • Distributes users only from the overlapping area between two cells, thus reducing overheads. • Allows the use of lower order modulation schemes, which consume less power.	• Not suitable for highly loaded cells. • ES in trade-off with more bandwidth used. • Distributes users to those cells that are already desirable for ES mode, thus reduce power saving opportunities in the overall network.
CC Based EE Scheme [65]	• Supports both real and non-real time traffic simultaneously. • Reduces power consumption by 50% compared to an always ON CC Network.	• Only considers two CCs. • Session blocking increases the delay in a high traffic period.
Energy Efficient BS deployment [87]	• Provides static hardware based ES.	• Scheme does not provide further ES opportunities once BSs have deployed.
Power Aware allocation of MBSFN subframes [61]	• Uses two power saving concepts. • Deactivates unused subframes. • Allocates RBs as much as required depending on load.	• Increased delays. • Only a few subframes can be switched off because control signals require capacity in a few subframes.

Table 5.2 (Continued)

ES Schemes	Advantages	Disadvantages
Coverage Based Scheme [72]	• Significantly reduces power consumption. • Recommended for a lightly loaded network. • Only one partition of BS could be expended rather than full BS coverage area.	• Execution of multiple algorithms increases processing computation. • Challenging for a BS to use one sector to provide coverage to the full area of neighbour BSs. • Load information overheads.
Sector Based Scheme [71]	• SON based ES scheme. • Distributed in nature; each BS provides ES without communicating with other BSs. • Divides coverage in different numbers of sector depending on load.	• Additional processing computation for sectorization of BS coverage. • Even existence of users in each sector reduces ES opportunities for BS. • Challenging for BSs to manage varying sectors in their coverage.
Relay Based Scheme [77]	• Significantly reduces power consumption from around 15–20%. • UEs can freely access recourses of two different operators. • Does not require load information exchange through an X2 interface.	• Difficult for two operators to work together • Allocation of resources to another operator's UEs may cause capacity limitation for their own associated UEs.
CRN Based Schemes [91]	• Turns off only unused resources during idle time frames.	• Secondary users wait until sleep mode has completed thus, resulting in delay.
REHO Scheme [97]	• Early handover helps resources to become free earlier and are turned off for ES.	• Increased radio link failure due to early handover.

and save energy up to 30% as compared to always on network during night zone. Since traffic load is high in the daytime, distance aware schemes fail to turn off BSs during day time and are only effective in the night zone when traffic load is low. The dynamic distance aware approach achieves ES up to 70% in comparison to an always-on network and operates every hour in contrast to the distance aware scheme [69].

Since each BS is required to exchange load information every hour with other neighbour cells, this results in an increased overhead in the network. DTX is one of the most interesting ES schemes. The main advantage of DTX is that it targets operational ES where there is no need of turning off whole BS and only unused RBs are switched off [60, 61]. On the other hand, the main disadvantage of DTX is the long sleep mode of unused RBs that increases delay time required by RBs to go back to the active mode. Distributed schemes also contribute in ES in LTE networks by effectively migrating UEs to the neighbour cells [72, 73]. In these types of schemes, BSs keep exchanging load information with each other resulting in an increased traffic information overhead. Bandwidth expansion is also used to achieve 44% of ES in lightly loaded networks. However, allocation of extra RBs results in reduced available capacity of the BS and thus is not very effective during the peak hour time period [96]. A combination of load balancing with bandwidth expansion is also used to reduce power consumption in the network [83]. However, this scheme could migrate UEs to those cells that are already desirable for ES mode, thus reducing ES opportunities in the overall network. The centralized schemes also provide ES but suffer from high traffic load similar to the distributed scheme. The TCoM scheme provides 26% ES by cutting down control channel signalling [85]. The main idea is similar to bandwidth expansion; however, it reduces the control channel overhead by transmitting two RBs jointly to a single user. TCoM suffers from the drawback of being not effective at cell edges and also requires a reduced overall system capacity. The EE link adaptation scheme is only effective for UEs located closer to the BS and saves 9.4% of energy while increasing the feedback overhead in the network [78]. The carrier aggregation approach is also used for ES reducing power consumption by 50% compared to an always-on network [65]. One of the disadvantages of this scheme is session blocking, which may increase the delay during high traffic time period. Coverage expansion is also used as a means to realize ES in the LTE networks [72]. It is, however, very complicated for the BS to make partitions to expand their transmission power and provide coverage to full neighbour BSs using one partition. On the other hand, execution of two algorithms to implement this scheme also increases overall computation overhead.

Division of the BS in different sectors and turning off unused or free sectors is also employed to attain ES [71]. This scheme, however, only works for a completely free sector and existence of even a single user would not allow the BS to turn off that sector. A shared relay ES scheme based on the idea of sharing resources of two different operators is proposed in [77]. However, it is very difficult for two different operators to work and integrate their operations under a shared environment.

Table 5.3 summarizes the performance of the discussed ES schemes in relation to other QoS issues. Figure 5.17 presents the ES percentage achieved by different ES schemes discussed before. It can be observed that dynamic distance aware scheme is the most EE technique of other ES schemes.

5.8 Open Research Issues

Comparative study of various existing ES schemes has shown that most of these schemes are only effective for lightly loaded networks and energy is not saved during highly loaded network conditions. DTX based schemes affected from delay that occur for RBs to come back in active mode [60]. Further research work is needed to reduce these delays. Reduced delay could have significant effect on overall performance of the system. The distance aware and bandwidth expansion based schemes fail to reduce power consumption during peak traffic hours. Therefore, these schemes could be further explored to provide enhanced ES during highly loaded traffic [68, 69]. Bandwidth expansion schemes could work more effectively in a balanced network. Therefore, load balancing could be further exploited with bandwidth expansion [83, 96]. On the other hand, centralized and distributed schemes exchange load information between the entire BS, which increases load information overheads in a network and reduces system efficiency [72, 73]. Means should be devised to reduce the load information overhead. Similarly, a link adaptation based ES scheme also suffers from overheads produced by energy consumption feedback sent to the BS [78]. Feedback overhead reduction could be exploited for an improved ES in the LTE networks. EE BS deployment provision could be integrated with any other dynamic ES based schemes for enhanced EE systems [87]. An MBSFN based ES scheme suffers from control signals that basically reduce the opportunities for

Table 5.3 QoS factors involved in ES schemes.

	QoS Issues				
ES Scheme	**Lightly loaded**	**Heavily loaded**	**Reduced capacity**	**Increased delay**	**Increased overhead**
Distance Aware [68]	✓	X	X	X	✓
Dynamic Distance aware [69]	✓	X	X	X	✓
Micro DTX scheme [60]	✓	✓	X	✓	X
Bandwidth expansion [83]	✓	X	✓	X	X
Centralized Algorithm [73]	✓	X	X	X	✓
TCoM [85]	✓	✓	X	X	X
EE Link Adaptation [78]	✓	✓	✓	X	✓
Component Carrier [65]	✓	✓	✓	X	X
EE BS deployment [87]	✓	✓	X	X	X
Power Aware MBSFN [61]	✓	✓	✓	✓	X
Coverage Based Scheme [72]	✓	X	X	✓	✓
Sector Based Scheme [71]	✓	X	X	✓	X
Relay Based Scheme [77]	✓	✓	X	✓	✓
CRN Based Scheme [91]	✓	X	X	✓	X
REHO Scheme [97]	✓	✓	✓	X	X

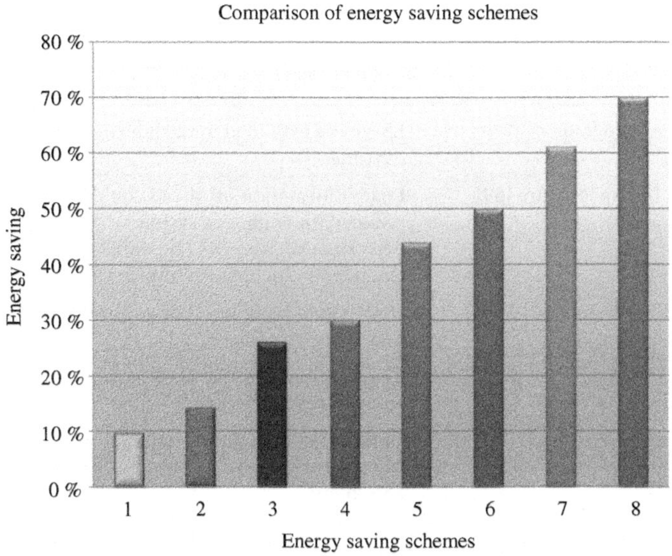

1: EE Link adaptation ES scheme [78]
2: Relay based ES scheme [77]
3: TCoM ES scheme [85]
4: Distance aware based ES scheme [68]
5: Bandwidth expansion based ES scheme [83]
6: Component carrier ES scheme [65]
7: Micro DTX ES scheme [60]
8: Dynamic distance aware based ES Scheme [69]

Figure 5.17 Percentage of energy saved in each ES scheme.

turning off the unused RBs [61]. ES through control signals could be further explored for enhanced MBSFN based ES. In other words, a few aspects of both the TCoM and MBSFN schemes can be taken into account to develop a hybrid ES scheme that may provide a better EE system compared to systems using TCoM and MBSFN schemes on an individual basis [61, 87]. Table 5.4 presents open research areas for ES in LTE networks.

Though various schemes have been discussed that help to achieve ES, thereby improving energy efficiency, significant lessons, however, have been learnt. DTX provides significant ES; however, it does not consider any delay related issues that may occur due to the longer sleep mode from a 10–20 ms cycle. Similarly, the distance aware switching off of BSs is effective for lightly loaded networks; however, these schemes are not at all suitable for heavily loaded networks. Since

Table 5.4 Open research issues.

ES Scheme	Open Research Areas
Distance Aware [68]	ES can be extended for 24 h including daytime.
Dynamic Distance aware [69]	Load information overhead could be reduced for enhanced system performance, whereas ES could be extended to include daytime period.
Micro DTX scheme [60]	Delay could be reduced or controlled for better performance.
Enhanced DTX scheme [60]	Capacity limitation could be explored as an open research issue.
Centralized Algorithm [73]	Complex system due to binary heaps could be explored for better performing ES scheme.
TCoM [85]	ES could be further improved at cell edges for enhanced system performance.
EE Link Adaptation Scheme [78]	Feedback overhead could be considered as an open research issue.
BEM [83]	Further enhancement can be done offering ES during peak load hours.
Component Carrier Based EE Scheme [65]	More CCs can be considered in future research work.
Energy Efficient BS deployment [87]	ES could be extended towards dynamic part of an enhanced system.
Power Aware allocation of MBSFN subframes [61]	Capacity limitations can be studied as an open research issue.

it is not possible to turn off a BS during peak load hours, therefore these schemes fail to save energy. Bandwidth expansion schemes allocate extra RBs to UEs that reduce overall available capacity of the system. Accordingly, BEM does not provide ES opportunities during peak hour traffic. Further exploring the literature review, TCoM considered constant SINR and provides ES through reduced control channels overheads, yet RBs still consume power during idle mode and TCoM loses ES opportunities. Importantly, centralized schemes have also proven to be effective for lightly loaded networks only; these schemes fail to provide ES in highly loaded networks. Link adaptation schemes reduce power consumption by only 9.4%, while it considerably increases overhead of feedback sent back to BS for ES purpose. The carrier aggregation based ES scheme combines two carrier components whereas it suffers from session blocking resulting in increased delay during high traffic periods. Most existing ES proposals mostly work in lightly loaded networks while they do not provide adequate and considerable ES during peak hours. Indeed, plenty of room and a need for additional research work exists that could be carried out to provide ES during peak traffic time periods too. The mentioned drawbacks and learnt lessons can indeed be carefully employed and guide future researchers to conduct and develop novel robust EE schemes in the light of discussed open research issues (Table 5.4).

5.9 Summary

Due to increased global warming and worldwide climate change, energy consumption has become a major hurdle. The ICT contributes approximately 2% to global warming, while a major part is attributed to telecommunications in ICT. In cellular networks, energy consumption is affected by growing mobile users and their data requirements. Moreover, further deployment of additional and enhanced BSs to fulfil ever growing UEs also adds in ICT contribution. Therefore, the increasing trend for energy consumption has become a major challenge for vendors thus affecting both economic and environmental aspects. The rapid increase in energy consumption not only increases OPEX but also effects climate change. Research work has proven that BSs in LTE networks consume a lot of dynamic power during the idle state, which could be saved by appropriate ES schemes. The reduced power consumption enhances the LTE system performance through cutting down OPEX and carbon emissions, thus also helping vendors to have a high profile in green communication. This chapter has provided a detailed discussion of existing ES schemes developed for LTE networks. Critical analysis of the schemes has been presented before open research issues were discussed. Finally, the authors' novel REHO ES scheme was also briefly described that takes into account challenges and builds itself on research issues. The chapter is a comprehensive account of the existing ES schemes for LTE networks and can help researchers to understand the current state-of-the-art, open research issues to come up with innovative solutions resulting in optimized system performance.

References

1 M. A. Garrett, "Radio astronomy transformed: Aperture arrays – Past, present and future," AFRICON, 2013, Pointe-Aux-Piments, 2013, pp. 1–5.
2 Q. Fan, H. Lu, P. Hong and Z. Zhu, "Throughput-power tradeoff association for user equipment in WLAN/cellular integrated networks," in IEEE Transactions on Vehicular Technology, vol. 66, no. 4, pp. 3462–3474, 2017.
3 Y. W. Blankenship, "Achieving high capacity with small cells in LTE-A," Communication, Control, and Computing (Allerton), 2012 50th Annual Allerton Conference on, Monticello, IL, 2012, pp. 1680–1687.
4 M. Agiwal, A. Roy and N. Saxena, "Next generation 5G wireless networks: A comprehensive survey," in IEEE Communications Surveys & Tutorials, vol. 18, no. 3, pp. 1617–1655, third quarter 2016.
5 A. Damnjanovic, J. Montojo, and Y. Wei et al., "A survey on 3GPP heterogeneous networks," in IEEE Wireless Communications, vol. 18, no. 3, pp. 10–21, June 2011.
6 F. Adachi, "Keynote talk #1: Wireless evolution and challenges for 5G wireless networks," Information and Computer Science (NICS), 2015 2nd National Foundation for Science and Technology Development Conference on, Ho Chi Minh City, 2015, pp. xxi–xxii.
7 N. Chandran and M. C. Valenti, "Three generations of cellular wireless systems," in IEEE Potentials, vol. 20, no. 1, pp. 32–35, Feb/Mar 2001.
8 B. Esmailpour, S. Salehi and N. Safavi, "Quality of service differentiation measurements in 4G networks," Wireless Telecommunications Symposium (WTS), 2013, Phoenix, AZ, 2013, pp. 1–5.
9 Y. Tian, A. Nix and M. Beach, "4G femtocell LTE base station with diversity and adaptive antenna techniques," Wireless Communications, Networking and Mobile Computing (WiCOM 2014), 10th International Conference on, Beijing, 2014, pp. 216–221.
10 3GPP, "Evolved Universal Terrestrial Radio Access (E-UTRA): Base Station (BS) radio transmission and reception," TS 36.104, V11.9.0.
11 M. Nasimi, F. Hashim and C. K. Ng, "Characterizing energy efficiency for heterogeneous cellular networks," IEEE Student Conference on Research and Development (SCOReD), 2012, Pulau Pinang, 2012, pp. 198–202.

12 Shayea, M. Ismail and R. Nordin, "Advanced handover techniques in LTE-advanced system," International Conference on Computer and Communication Engineering (ICCCE), 2012, Kuala Lumpur, 2012, pp. 74–79.
13 U. Dampage and C. B. Wavegedara, "A low-latency and energy efficient forward handover scheme for LTE-femtocell networks," 2013 IEEE 8th International Conference on Industrial and Information Systems, Peradeniya, 2013, pp. 53–58.
14 P. Skocir, D. Katusic, I. Novotni, I. Bojic and G. Jezic, "Data rate fluctuations from user perspective in 4G mobile networks," 22nd International Conference on Software, Telecommunications and Computer Networks (SoftCOM), 2014, Split, 2014, pp. 180–185.
15 S. Abeta, "Toward LTE commercial launch and future plan for LTE enhancements (LTE-Advanced)," IEEE International Conference on Communication Systems (ICCS), 2010, Singapore, 2010, pp. 146–150.
16 M. F. L. Abdullah and A. Z. Yonis, "Performance of LTE Release 8 and Release 10 in wireless communications," Cyber Security, Cyber Warfare and Digital Forensic (CyberSec), 2012 International Conference on, Kuala Lumpur, 2012, pp. 236–241.
17 N. Mehra, and A. Noliya, "Performance analysis of OFDMA, MIMO and SC-FDMA technology in 4G LTE networks," 2016 6th International Conference – Cloud System and Big Data Engineering (Confluence), Noida, 2016, pp. 554–558.
18 N. Takaharu, "LTE and LTE-advanced: Radio technology aspects for mobile communications," General Assembly and Scientific Symposium, 2011 30th URSI, Istanbul, 2011, pp. 1–4.
19 H. Lee, S. Vahid and K. Moessner, "A Survey of Radio Resource Management for Spectrum Aggregation in LTE-Advanced," in IEEE Communications Surveys & Tutorials, vol. 16, no. 2, pp. 745–760, Second Quarter 2014.
20 N. Becker, A. Rizk and M. Fidler, "A measurement study on the application-level performance of LTE," Networking Conference, 2014 IFIP, Trondheim, 2014, pp. 1–9.
21 S. B. Manir, M. M. Rahman and T. Ahmed, "Comparison between FDD and TDD frame structure in SC-FDMA," International Conference on Informatics, Electronics & Vision (ICIEV), 2012, Dhaka, 2012, pp. 795–799.
22 L. Wan, M. Zhou and R. Wen, "Evolving LTE with flexible duplex," 2013 IEEE Globecom Workshops (GC Wkshps), Atlanta, GA, 2013, pp. 49–54.

23 R. Ratasuk, A. Ghosh, Weimin Xiao, R. Love, R. Nory and B. Classon, "TDD design for UMTS long-term evolution," 2008 IEEE 19th International Symposium on Personal, Indoor and Mobile Radio Communications, Cannes, 2008, pp. 1–5.
24 R. Zheng, X. Zhang, X. Li, Q. Pan, Y. Fang and D. Yang, "Performance evaluation on the coexistence scenario of two 3GPP LTE systems," IEEE 70th Vehicular Technology Conference Fall (VTC 2009-Fall), 2009, Anchorage, AK, 2009, pp. 1–6.
25 S. S. Prasad, C. K. Shukla and R. F. Chisab, "Performance analysis of OFDMA in LTE," 2012 Third International Conference on Computing Communication & Networking Technologies (ICCCNT), Coimbatore, 2012, pp. 1–7.
26 L. A. M. R. de Temino, G. Berardinelli, S. Frattasi, K. Pajukoski and P. Mogensen, "Single-user MIMO for LTE-A Uplink: Performance evaluation of OFDMA vs. SC-FDMA," 2009 IEEE Radio and Wireless Symposium, San Diego, CA, 2009, pp. 304–307.
27 3GPP, "Evolved Universal Terrestrial Radio Access (E-UTRA) and Evolved Universal Terrestrial Radio Access Networks (E UTRAN): Overall description," V10.4.0., TS 36.300.
28 3GPP, "Technical Specification Group Radio Access Network; Further advancements for E-UTRA – LTE-Advanced feasibility studies in RAN WG4," V9.0.0., TR 36.815.
29 3GPP, "General Packet Radio Service (GPRS) enhancements for Evolved Universal Terrestrial Radio Access Network (E-UTRAN) access," 3rd Generation Partnership Project (3GPP), TS 23.401, Jun. 2011.
30 S. B. H. Said, M. R. Sama, K. Guillouard, et al., "New control plane in 3GPP LTE/EPC architecture for on-demand connectivity service," IEEE 2nd International Conference on Cloud Networking (CloudNet), 2013, San Francisco, CA, 2013, pp. 205–209.
31 C. Cox, An Introduction to LTE: LTE, LTE-Advanced, SAE and 4G Mobile Communications, Chichester, UK: John Wiley & Sons, Ltd, 2012, pp. 21–28.
32 K. P. Makhecha and K. H. Wandra, "4G wireless networks: Opportunities and challenges," 2009 Annual IEEE India Conference, Gujarat, 2009, pp. 1–4.
33 A. Pande, V. Ahuja, R. Sivaraj, E. Baik and P. Mohapatra, "Video delivery challenges and opportunities in 4G networks," IEEE MultiMedia, vol. 20, no. 3, pp. 88–94, July–Sept. 2013.

34 M. Iwamura, K. Etemad, M. H. Fong, R. Nory and R. Love, "Carrier aggregation framework in 3GPP LTE-advanced [WiMAX/LTE Update]," in IEEE Communications Magazine, vol. 48, no. 8, pp. 60–67, August 2010.

35 Y. Yuan, S. Wu, J. Yang, F. Bi, S. Xia and G. Li, "Relay backhaul subframe allocation in LTE-Advanced for TDD," 5th International ICST Conference on Communications and Networking in China (CHINACOM), 2010, Beijing, 2010, pp. 1–5.

36 S. Ahmadi, LTE-advanced: A practical systems approach to understanding 3GPP LTE Release 10 and 11 Radio Access Technologies, Waltham, MA, USA, Elsevier Inc., 2014, pp. 61–65.

37 G. Cili, H. Yanikomeroglu and F. R. Yu, "Cell switch off technique combined with coordinated multi-point (CoMP) transmission for energy efficiency in beyond-LTE cellular networks," 2012 IEEE International Conference on Communications (ICC), Ottawa, ON, 2012, pp. 5931–5935.

38 M. Pickavet, W. Veerecken, S. Demeyer, et al., "Worldwide energy needs for ICT: The rise of power-aware networking," 2008 2nd International Symposium on Advanced Networks and Telecommunication Systems, Mumbai, 2008, pp. 1–3.

39 Y. L. Chung, "Energy-saving transmission for green macrocell–small cell systems: A system-level perspective," IEEE Systems Journal, vol. 11, no. 2, pp. 706–716, 2017.

40 E. Oh, B. Krishnamachari, X. Liu and Z. Niu, "Toward dynamic energy-efficient operation of cellular network infrastructure," in IEEE Communications Magazine, vol. 49, no. 6, pp. 56–61, June 2011.

41 R. Maihaniemi, "ICT Getting Green," 4th International Conference on Telecommunication – Energy Special Conference (TELESCON), 2009, Vienna, Austria, 2009, pp. 1–6.

42 H. O. Scheck, "ICT & wireless networks and their impact on global warming," European Wireless Conference (EW), 2010, Lucca, 2010, pp. 911–915.

43 Y. Chen, S. Zhang, S. Xu and G. Y. Li, "Fundamental trade-offs on green wireless networks," in IEEE Communications Magazine, vol. 49, no. 6, pp. 30–37, June 2011.

44 G. Fettweis and E. Zimmermann, "ICT energy consumption – Trends and challenges," in Proc. 11th Int. Symp. WPMC, vol. 2, no. 4, Sep. 2008.

45 M. Griffiths, "ICT and CO_2 Emissions," Parliamentary Office Sci. Technol., London, Postnote No. 319, Dec. 2008. [Online]. Available:

www.parliament.uk/documents/post/postpn319.pdf (accessed December 2017).

46 T. M. Knoll, "A combined CAPEX and OPEX cost model for LTE networks," Telecommunications Network Strategy and Planning Symposium (Networks), 2014 16th International, Funchal, 2014, pp. 1–6.

47 J. Moysen and L. Giupponi, "Self coordination among SON functions in LTE heterogeneous networks," 2015 IEEE 81st Vehicular Technology Conference (VTC Spring), Glasgow, 2015, pp. 1–6.

48 T. M. Knoll, "Life-cycle cost modelling for NFV/SDN based mobile networks," Telecommunication, Media and Internet Techno-Economics (CTTE), 2015 Conference of, Munich, 2015, pp. 1–8.

49 N. Zhang and H. Hämmäinen, "Cost efficiency of SDN in LTE-based mobile networks: Case Finland," International Conference and Workshops on Networked Systems (NetSys), 2015, Cottbus, 2015, pp. 1–5.

50 S. M. Azzam and T. Elshabrawy, "Re-dimensioning number of active eNodeBs for green LTE networks using genetic algorithms," Proceedings of European Wireless 2015; 21th European Wireless Conference, Budapest, Hungary, 2015, pp. 1–6.

51 O. Arnold, F. Richter, G. Fettweis and O. Blume, "Power consumption modelling of different base station types in heterogeneous cellular networks," 2010 Future Network & Mobile Summit, Florence, 2010, pp. 1–8.

52 K. Hiltunen, "Utilizing eNodeB sleep mode to improve the energy-efficiency of dense LTE networks," 2013 IEEE 24th Annual International Symposium on Personal, Indoor, and Mobile Radio Communications (PIMRC), London, 2013, pp. 3249–3253.

53 T. Chen, H. Zhang, Z. Zhao and X. Chen, "Towards green wireless access networks," 5th International ICST Conference on Communications and Networking in China (CHINACOM), 2010, Beijing, 2010, pp. 1–6.

54 S. Srikanth, P. A. Murugesa Pandian and X. Fernando, "Orthogonal frequency division multiple access in WiMAX and LTE: a comparison," in IEEE Communications Magazine, vol. 50, no. 9, pp. 153–161, September 2011.

55 S. Janaaththanan, C. Kasparis and B. G. Evans, "Comparison of SC-FDMA and HSUPA in the return-link of evolved S-UMTS architecture," Satellite and Space Communications, 2007. IWSSC '07. International Workshop on, Salzburg, 2007, pp. 56–60.

56 M. Salem, A. Adinoyi, H. Yanikomeroglu and Y. D. Kim, "Radio resource management in OFDMA-based cellular networks enhanced with fixed and nomadic relays," 2010 IEEE Wireless Communication and Networking Conference, Sydney, NSW, 2010, pp. 1–6.
57 M. M. Matalgah, B. Paudel and O. M. Hammouri, "Cross-layer resource allocation approach in OFDMA systems with multi-class QoS services and users queue status," 2013 IEEE Global Communications Conference (GLOBECOM), Atlanta, GA, 2013, pp. 1385–1390.
58 M. Deruyck, E. Tanghe, W. Joseph and L. Martens, "Modelling the energy efficiency of microcell base stations," in 1st International Conference on Smart Grids, Green Communications and IT Energy-Aware Technologies, 2012, p. 1–6.
59 W. Tomaselli, D. Sabella, V. Palestini and V. Bernasconi, "Energy efficiency performances of selective switch OFF algorithm in LTE mobile networks," 2013 IEEE 24th Annual International Symposium on Personal, Indoor, and Mobile Radio Communications (PIMRC), London, 2013, pp. 3254–3258
60 P. Frenger, P. Moberg, J. Malmodin, Y. Jading and I. Godor, "Reducing energy consumption in LTE with cell DTX," 2011 IEEE 73rd Vehicular Technology Conference (VTC Spring), Yokohama, 2011, pp. 1–5.
61 A. Migliorini, G. Stea, M. Caretti and D. Sabella, "Power-aware allocation of MBSFN subframes using discontinuous cell transmission in LTE systems," Vehicular Technology Conference (VTC Fall), 2013 IEEE 78th, Las Vegas, NV, 2013, pp. 1–5.
62 S. Herrería-Alonso, M. Rodríguez-Pérez, M. Fernández-Veiga and C. López-García, "Adaptive DRX scheme to improve energy efficiency in LTE networks with bounded delay," in IEEE Journal on Selected Areas in Communications, vol. 33, no. 12, pp. 2963–2973, Dec. 2015.
63 K. Davaslioglu, C. C. Coskun and E. Ayanoglu, "Energy-efficient resource allocation for fractional frequency reuse in heterogeneous networks," in IEEE Transactions on Wireless Communications, vol. 14, no. 10, pp. 5484–5497, Oct. 2015.
64 X. Xiao, X. Tao and J. Lu, "Energy-efficient resource allocation in LTE-Based MIMO-OFDMA systems with user rate constraints," in IEEE Transactions on Vehicular Technology, vol. 64, no. 1, pp. 185–197, Jan. 2015.
65 A. T. Tung, Y. L. Chung and Z. Tsai, "An efficient power-saving downlink transmission scheme in OFDM-based multiple component

carrier systems," Advanced Communication Technology (ICACT), 2012 14th International Conference on, PyeongChang, 2012, pp. 116–120.
66 K. Sundaresan and S. Rangarajan, "Energy efficient carrier aggregation algorithms for next generation cellular networks," 2013 21st IEEE International Conference on Network Protocols (ICNP), Goettingen, 2013, pp. 1–10
67 A. Liu, K. Zheng, W. Xiang and H. Zhao, "Design and performance analysis of an energy-efficient uplink carrier aggregation scheme," in IEEE Journal on Selected Areas in Communications, vol. 32, no. 2, pp. 197–207, February 2014.
68 A. Bousia, A. Antonopoulos, L. Alonso and C. Verikoukis, ""Green" distance-aware base station sleeping algorithm in LTE-Advanced," 2012 IEEE International Conference on Communications (ICC), Ottawa, ON, 2012, pp. 1347–1351.
69 A. Bousia, E. Kartsakli, L. Alonso and C. Verikoukis, "Dynamic energy efficient distance-aware base station switch on/off scheme for LTE-advanced," Global Communications Conference (GLOBECOM), 2012 IEEE, Anaheim, CA, 2012, pp. 1532–1537.
70 L. Li, Y. Zhang, B. Fan and H. Tian, "Mobility-aware load balancing scheme in hybrid VLC-LTE networks," in IEEE Communications Letters, vol. 20, no. 11, pp. 2276–2279, Nov. 2016.
71 M. F. Hossain, K. S. Munasinghe and A. Jamalipour, "Toward self-organizing sectorization of LTE eNBs for energy efficient network operation under QoS constraints," 2013 IEEE Wireless Communications and Networking Conference (WCNC), Shanghai, 2013, pp. 1279–1284.
72 K. Samdanis, D. Kutscher and M. Brunner, "Self-organized energy efficient cellular networks," IEEE 21st International Symposium on Personal Indoor and Mobile Radio Communications (PIMRC), 26–30 Sept. 2010, pp. 1665, 1670.
73 K. Samdanis, T. Taleb, D. Kutscher and M. Brunner, "Self organized network management functions for energy efficient cellular urban infrastructures" Mob. Netw. Appl., vol. 17, pp. 119–131, 2012.
74 S. A. Ahmad and D. Datla, "Distributed power allocations in heterogeneous networks with dual connectivity using backhaul state information," IEEE Transactions on Wireless Communications, vol. 14, no. 8, pp. 4574–4581, Aug. 2015.
75 H. H. Nguyen and W. J. Hwang, "Distributed scheduling and discrete power control for energy efficiency in multi-cell networks," IEEE Communications Letters, vol. 19, no. 12, pp. 2198–2201, Dec. 2015.

76 S. A. Ahmad and D. Datla, "Distributed power allocations in heterogeneous networks with dual connectivity using backhaul state information," IEEE Transactions on Wireless Communications, vol. 14, no. 8, pp. 4574–4581, Aug. 2015.
77 V. Monteiro, T. Ramrekha, D. Yang, J. Rodriguez, S. Mumtaz and C. Politis, "An energy efficient proposal in shared relay-based LTE network," Wireless Communication Systems (ISWCS 2013), Proceedings of the Tenth International Symposium on, Ilmenau, Germany, 2013, pp. 1–5.
78 A. Li, S. Jin, F. Zheng, X. Gao and X. Wang, "Energy efficient link adaptation for downlink transmission of LTE/LTE-A systems," Vehicular Technology Conference (VTC Fall), 2013 IEEE 78th, Las Vegas, NV, 2013, pp. 1–5.
79 D. D. Ling, Z. Lu, W. Zheng, X. Wen and Y. Ju, "Energy efficient cross-layer resource allocation scheme based on potential games in LTE-A," 15th International Symposium on Wireless Personal Multimedia Communications (WPMC), 2012, Taipei, pp. 623–627.
80 Z. Zhou, M. Dong, K. Ota, G. Wang and L. T. Yang, "Energy-efficient resource allocation for D2D communications underlaying cloud-RAN-based LTE-A networks," IEEE Internet of Things Journal, vol. 3, no. 3, pp. 428–438, June 2016.
81 S. Rostami, K. Arshad and P. Rapajic, "Energy-efficient resource allocation for LTE-A networks," IEEE Communications Letters, vol. 20, no. 7, pp. 1429–1432, July 2016.
82 A. Z. Kaddour, E. Vivier, L. Mroueh, M. Pischella and P. Martins, "Green opportunistic and efficient resource block allocation algorithm for LTE uplink networks," IEEE Transactions on Vehicular Technology, vol. 64, no. 10, pp. 4537–4550, Oct. 2015.
83 Y. Li, W. Liu, B. Cao and M. Li, "Green resource allocation in LTE system for unbalanced low load networks," 2012 IEEE 23rd International Symposium on Personal, Indoor and Mobile Radio Communications – (PIMRC), Sydney, NSW, 2012, pp. 1009–1014.
84 S. Almowuena, M. M. Rahman, C. H. Hsu, A. A. Hassan and M. Hefeeda, "Energy-aware and bandwidth-efficient hybrid video streaming over mobile networks," IEEE Transactions on Multimedia, vol. 18, no. 1, pp. 102–115, Jan. 2016.
85 S. Videv, H. Haas, J. S. Thompson and P. M. Grant, "Energy efficient resource allocation in wireless systems with control channel

overhead," Wireless Communications and Networking Conference Workshops (WCNCW), 2012 IEEE, Paris, 2012, pp. 64–68.
86 A. Han and S. Armour, "Energy efficient radio resource management strategies for green radio," IET Communications, vol. 5, no. 18, pp. 2629–2639, Dec. 16 2011.
87 C. Coskun and E. Ayanoglu, "Energy-efficient base station deployment in heterogeneous networks," IEEE Wireless Communications Letters, vol. 3, no. 6, pp. 593–596, Dec. 2014.
88 P. Van, B. P. Rimal, S. Andreev, T. Tirronen and M. Maier, "Machine-to-machine communications over WiFi enhanced LTE networks: A power-saving framework and end-to-end performance," Journal of Lightwave Technology, vol. 34, no. 4, pp. 1062–1071, Feb.15, 15 2016.
89 S. Mumtaz, K. M. Saidul Huq, J. Rodriguez and V. Frascolla, "Energy-efficient interference management in LTE-D2D communication," IET Signal Processing, vol. 10, no. 3, pp. 197–202, 5 2016.
90 J. Liu, N. Kato, J. Ma and N. Kadowaki, "Device-to-device communication in LTE-advanced networks: a survey," IEEE Communications Surveys & Tutorials, vol. 17, no. 4, pp. 1923–1940, Fourthquarter 2015.
91 S. Jin, X. Ma and W. Yue, "Energy-saving strategy for green cognitive radio networks with an LTE-advanced structure," Journal of Communications and Networks, vol. 18, no. 4, pp. 610–618, Aug. 2016.
92 F. Zheng, W. Li, L. Meng, P. Yu and L. Peng, "Distributed energy saving mechanism based on CoMP in LTE-A system," China Communications, vol. 13, no. 7, pp. 39–47, July 2016.
93 V. J. Kotagi, R. Thakur, S. Mishra and C. S. R. Murthy, "Breathe to Save Energy: Assigning downlink transmit power and resource blocks to LTE enabled IoT networks," IEEE Communications Letters, vol. 20, no. 8, pp. 1607–1610, Aug. 2016.
94 L. You, L. Lei and D. Yuan, "Optimizing power and user association for energy saving in load-coupled cooperative LTE," 2016 IEEE International Conference on Communications (ICC), Kuala Lumpur, 2016, pp. 1–6.
95 A. Kalogridis and O. Georgiou, "CreD2D: A credit-driven self-evolving D2D towards LTE HetNet energy saving," 2016 IEEE International Conference on Communications Workshops (ICC), Kuala Lumpur, 2016, pp. 742–748.
96 S. Videv and H. Haas, "Energy-efficient scheduling and bandwidth-energy efficiency trade-off with low load," 2011 IEEE International Conference on Communications (ICC), Kyoto, 2011, pp. 1–5.

97 K. Kanwal and G. A. Safdar, "Reduced early handover for energy saving in LTE Networks," IEEE Communications Letters, vol. 20, no. 1, pp. 153–156, Jan. 2016.

98 G. A. Safdar, M. Ur-Rehman, M. Muhammad, M. A. Imran and R. Tafazolli, "Interference mitigation in D2D Communication underlaying LTE-A Network," IEEE Access, vol. 4, pp. 7967–7987, 2016.

6

Spectrum Sensing Mechanisms in Cognitive Radio Based LTE Femtocells

Tazeen S. Syed and Ghazanfar Ali Safdar

School of Computer Science and Technology, University of Bedfordshire, Luton, UK

The ever-increasing demand for wireless technology has rendered the spectrum a restricted resource. Most research work has shown that efficient usage of spectrum can be accomplished by employing the emerging cognitive radio (CR) technology. The secondary users (SUs) utilize the spectrum when it is unoccupied and vacate it when the primary users (PUs) return. Sensing the spectrum holes is the most vital process of the CRs. When it comes to cellular communication indoors, the coverage is substantially poor. To overcome this issue, Long-Term Evolution (LTE) femtocells are employed, the main drawback being, however, cross- and co-tier interference. This issue could be resolved by integrating LTE femtocells with CR technology innovations. The LTE cognitive femtocells have CR capabilities of sensing the spectrum and preventing interference by proper allocation of the spectrum. This chapter provides an insight into fundamental signal processing used for spectrum sensing and explores the conventional spectrum sensing techniques used in CR networks. We then investigate different models and statistics used for performance analysis of these schemes and present the state-of-the-art research results, which outline the widely used models and statistics for specific sensing techniques. Further, a comparison of the conventional spectrum sensing techniques used by CRs is presented.

LTE Communications and Networks: Femtocells and Antenna Design Challenges, First Edition. Edited by Masood Ur Rehman and Ghazanfar Ali Safdar.

6.1 Fundamentals of Signal Processing

A functional cognitive femtocell network encounters several challenges that need to be resolved by employing various signal processing techniques before being fully implemented [1]. A comprehensive knowledge of signal processing is essential for exploiting its capabilities for use in CR technology. In this section, fundamental concepts of signal processing that are pertinent to CR are described. Generally, the communication systems are designed such that they achieve the desired limit of data transmission by considering varying operating conditions such as available bandwidth (BW), channel capacity and signal to noise ratio (SNR). When designing communication systems, it is convenient to implement mathematical models to statistically characterize the transmission medium. When a signal is transmitted through a channel the output is a modified signal. The process that causes this modification of the signal can be evaluated by modelling the channel. In communication systems, the channel model facilitates the connection of the transmitter and the receiver mathematically. There are both analogue and digital channel models. Some of the most commonly used channel models are discussed in the next section.

6.1.1 Channel Model

6.1.1.1 Additive Gaussian Noise Channel

Noise is undesirable as it distorts the information carried by the original signal. Noise can originate from electronic components in the system, interference during transmission and so on. [2]. The additive noise channel is one of the simplest mathematical models that can be implemented for analysis of a communication channel. The stimuli of both noise and signal have the same distribution (bell-shaped curve) and if the signal is present, then statistical parameters such as mean and standard deviation of the signal distribution are changed and the curve is shifted to the right relative to the noise distribution [3]. The additive Gaussian noise channel model in shown in Figure 6.1. When the signal $s(t)$ is transmitted through the channel and additive random noise $n(t)$ is added to the signal and the output signal $r(t)$ is given as

$$r(t)=s(t)+n(t) \tag{6.1}$$

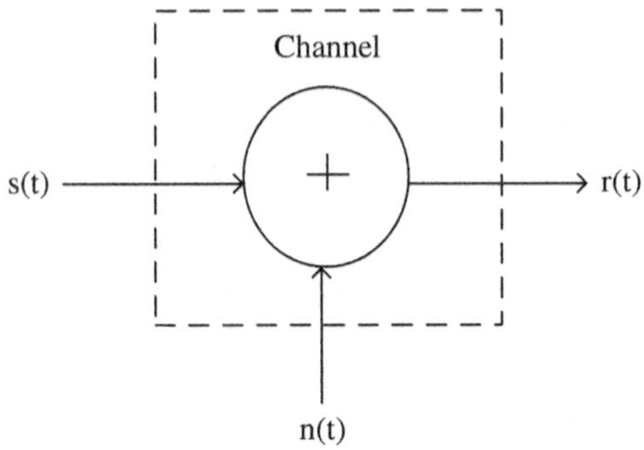

Figure 6.1 Additive Gaussian noise channel.

This channel's assumption is that all noise that is received at any frequency range is always the same and hence is termed the additive white Gaussian noise (AWGN) channel. The complexity of this mathematical model is fairly reduced since the Gaussian statistics of noise (Gaussian probability distribution) are assumed.

6.1.1.2 Linear Filter Channel

In order to avoid interference between signals, the signals are band limited using filters. Figure 6.2 shows a channel model with impulse response *h(t)* of linear filter, input signal *s(t)* and additive noise *n(t)*. The output signal *r(t)* is obtained as a result of the convolution of *s(t)* and *h(t)* added with noise. It is given as

$$r(t) = s(t) * h(t) + n(t) \tag{6.2}$$

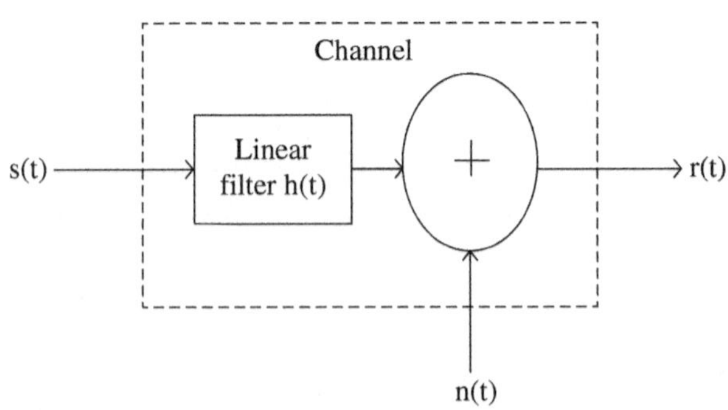

Figure 6.2 Linear filter channel with additive noise.

The linear filter channel can be further categorized depending on whether the filter is time variant or time invariant. If the characteristics of the linear filter do not change with time, then it is a linear time invariant filter. On the other hand, if the characteristics of the linear filter change with time then such filters are known as linear time variant filter.

- Linear time variant channel
- Linear time invariant channel

6.1.1.3 Band Limited Channel

Due to BW restrictions in the RF spectrum, signals cannot be allowed to exceed a band limit. Narrowband filters are employed at the transmitter and the receiver side of the communication system to allow only the modulated signal to pass through [2].

6.1.2 Modulation Technique

After detecting and selecting an available spectrum segment, the CR system should essentially use modulation schemes that provide the best spectrum utilization and capacity while avoiding interference to any primary user. The desired transmission system should be flexible and scalable with the number of users and bands. The digital modulation scheme based on orthogonal frequency division multiplexing (OFDM) has the natural approach to avoid interference due to inherent frequency subbanding. OFDM is one of the best modulation techniques that can be used in CR networks since it has the underlying capabilities of spectrum sensing and spectrum shaping [4–6]. For opportunistic access, multicarrier methods are employed in the cognitive femtocell network. OFDM is a multicarrier modulation technique; due to its considerable advantage of allocating radio resources flexibly, OFDM is regarded as a potential air interface for LTE femtocells [7]. With OFDM modulation, the interference can be eliminated by exploiting orthogonal radio resources among femtocells. The OFDM modulation technique addresses the problems that result from data transmission of high bit rate by splitting it in to lower bit rate streams and transmitting using the orthogonally overlapped sub-carriers in parallel. This modulation technique uses Fast Fourier transform (FFT) and Inverse Fast Fourier transform (IFFT) mathematical tools for spectral analysis, and another advantage of using these operations means that the sub-channels avoid interference with other. Figure 6.3 shows the operation of OFDM transmitter and receiver.

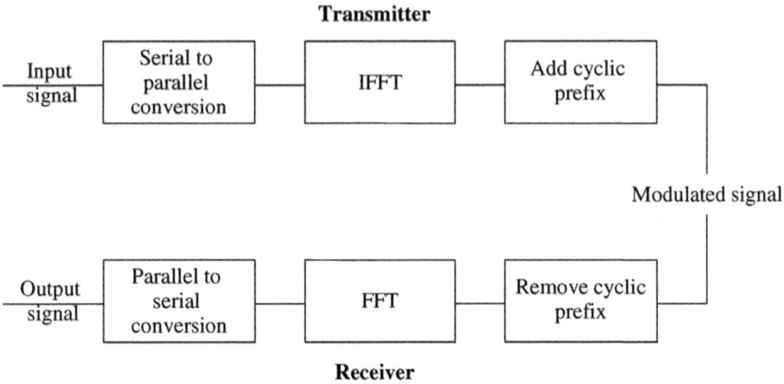

Figure 6.3 OFDM block diagram.

The main advantage of OFDM is that it combats the problem of multipath fading and avoids inter-symbol interference (ISI) by using the cyclic prefix, which maintains the sub-carriers to be orthogonal with each other [5]. The drawback of OFDM is the overhead that is introduced due to the cyclic prefix. Moreover, if the cyclic prefix length is not optimal then it will not capture the effects of channel impulse response and is thereby unable to prevent symbol distortions [2].

6.1.3 Error Probability

Receivers in communication systems have the ability to distinguish between noise and the actual signal. The probability of error in distinguishing the signals from noise is the key performance characteristics of receivers. In binary systems, the binary digits (0, 1), denoted by pulse levels 0 and A, correspond to the absence and presence of signal, respectively. The decision about the presence or absence of the signal can be deduced using a threshold value μ. Assuming Gaussian distribution, with zero mean and variance, σ^2, for the actual waveform $y(t)$, the probability density in the absence of signal is given as

$$P_0\left(y\right)=\frac{1}{\sigma\sqrt{2\pi}}e^{-y^2/\left(2\sigma^2\right)} \tag{6.3}$$

In the presence of signal, the probability density is

$$P_1(y) = \frac{1}{\sigma\sqrt{2\pi}} e^{-(y-A)^2/(2\sigma^2)} \tag{6.4}$$

The probability of detecting a signal when it is actually absent is the probability of false alarm, given as

$$P_{FA} = \int_{\mu}^{\infty} \frac{1}{\sigma\sqrt{2\pi}} e^{-y^2/(2\sigma^2)} dy \tag{6.5}$$

Similarly, probability of not detecting the signal when it is actually present is the probability of missed detection given as

$$P_{MD} = \int_{-\infty}^{\mu} \frac{1}{\sigma\sqrt{2\pi}} e^{-(y-A)^2/(2\sigma^2)} dy \tag{6.6}$$

The probability of error P_E can be defined as

$$P_E = P_0 P_{FA} + P_1 P_{MD} \tag{6.7}$$

In this section, the fundamental concepts of digital communication systems relevant to CR networks were illustrated. A comprehensive description of the basic channel models was provided in this section. Digital modulation such as OFDM has also been introduced. Furthermore, the probability of error has been discussed.

6.2 Spectrum Sensing Techniques

Spectrum sensing is the initial process in CR technology. The major function of CRs is to detect the empty spaces for the SUs to use the spectrum and vacate the band upon arrival of primary users without any interference. Typically, spectrum sensing can be categorized as follows (Figure 6.4).

6.2.1 Primary Transmitter Detection

The primary transmitter detection technique relies on the received signal for PU detection. Different sensing techniques will require knowing different degrees of a PU signal's characteristics. Conventional

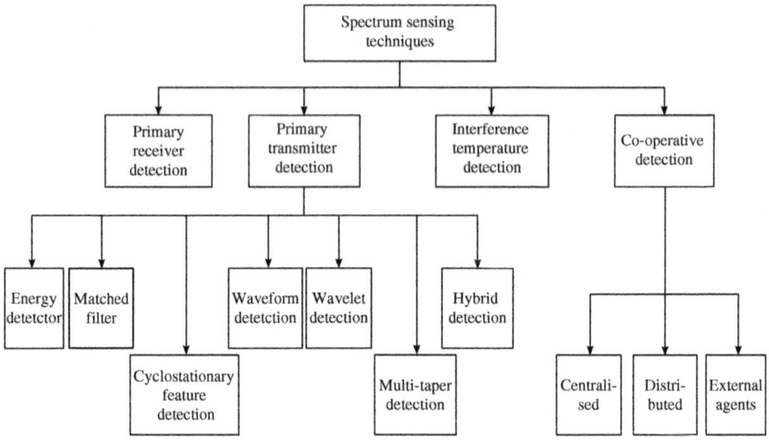

Figure 6.4 Spectrum sensing mechanisms.

sensing techniques such as the energy detector technique do not require any knowledge of PU, while matched filter detection and cyclostationary feature detection techniques require adequate knowledge of the characteristics of a PU signal.

6.2.1.1 Energy Detector

The energy detection technique is the one of the simplest spectrum sensing techniques and is commonly used because of low computation and easy implementation [8–10]. This method does not require any prior information about the primary user's signal such as preamble signalling for synchronization, pilot patterns for channel estimation and even modulation orders [11]. A fixed threshold based on noise variance, which is the noise power, is set and this is compared with the strength of the perceived signals evaluated by the energy detector. If the perceived signal strength is above the threshold, then the signal is considered as the primary user signal and suggests that the channel is occupied by the primary user. If the detector senses a signal whose signal strength is below the threshold, then it implies that the perceived signal is noise and the channel is free for the secondary user. Consider the signal $y(t)$ given as

$$y(t)=r(t)+n(t) \tag{6.8}$$

where $r(t)$ is the signal that needs to be detected and $n(t)$ is the white Gaussian noise, which is a sequence of uncorrelated random values and

t is the sample index. When channel is not occupied by the primary user, then $r(t) = 0$ and $y(t) = n(t)$. The probability of detecting a signal in the frequency band when the primary user signal is present is the probability of detection, P_D and the probability of detecting a signal in the channel, when it is actually free is known as the probability of missed detection P_F. For good performance of the energy detector, it is essential to have the P_F low to prevent underutilization of the channel when free.

The energy detector implementation shown in Figure 6.5a consists of a low pass filter, to filter out of band noise signal, the analogue to digital (ADC) converter, square-law device, to measure the energy of the received signal and an integrator to evaluate the energy in the observation time interval, t [12]. A square-law device is often found in the receiver front end of communication devices and it gives an output voltage, which is proportional to the square of the range of input voltages. According to [13], the implementation shown in Figure 6.5b is more suitable for narrow band signals and sine waves in which the magnitude of FFT is squared and then averaged. Then the test statistics will be compared to a threshold and make decisions about the presence or absence of the PU signal. In [14] and [15], improved versions of energy detector have been proposed where, instead of the traditional squaring operation on the received signal, the power operation is implemented and the reduction of instantaneous signal energy drop, respectively. The main drawback of this method is that it becomes difficult to detect weak primary user's signal as this would be assumed to be noise. When the secondary user tries to occupy the band unaware of primary user's presence, this results in interference. Moreover, noise and interference levels have an impact on the selection of the threshold level.

Energy detector method cannot be used as the standalone method. Instead, it can be used in conjunction with other spectrum sensing methods. However, in [16–18], better performance of energy detector has been suggested at moderate to low SNR conditions. In [19, 20], a sliding window function is applied to the energy detector, which suggests that the actual power of the PUs can be estimated. The performance of energy detector usually degrades due to the dynamic nature of the sensing environment. According to [21] and [22], an adaptive sensing approach has been suggested to overcome the difficulties faced by conventional energy detectors in a dynamic environment. A reliable energy detector in case of low SNR was suggested in [23], where the first order cyclostationary features such as periodic training sequences of PU signal are used instead of the conventional second order cyclostationary features to reduce the noise uncertainty. In [24],

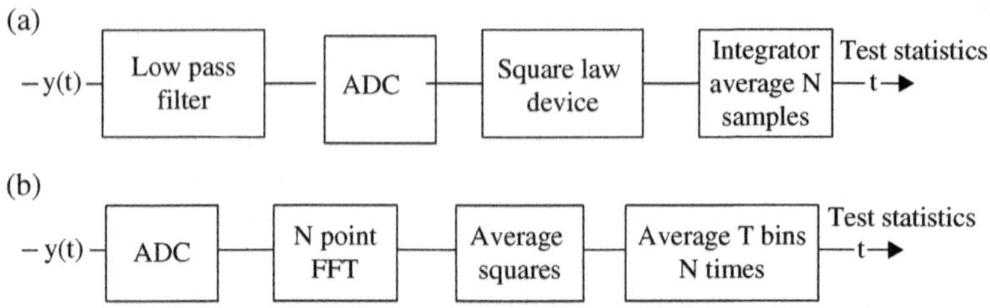

Figure 6.5 (a) Energy detector implementation using analogue pre-filter and square-law device. (b) Energy detector implementation using periodogram FFT and averaging.

there is an enhancement in reliable spectrum sensing by utilizing multiple verifications by the energy detector using time delays to come to a proper decision.

6.2.1.2 Matched Filter Detection

The matched filter detection technique requires the receiver to have comprehensive information about the characteristics of the primary user's signal in order to provide an optimum detection of primary signal in the licensed band [25]. The matched filter is a linear filter that maximizes the received SNR. Basically, a linear filter is an operation L, which transforms a time series X = {X (t)} into another time series Y = {Y (t)}. The implementation of the matched filter technique in Figure 6.6 consists of a matching filter that has prior information about the transmitted signal parameters such as bandwidth, type and order of modulation, frame format and pulse shaping [2].

This technique gives results that are far better than the other spectrum sensing techniques since the probability of false detection P_F can be achieved in short period [26], as the speed and accuracy of spectrum sensing is high. The limitation of matched filter detection is the designing of the filter, which requires high computation power and increased complexities. If the primary user signal parameters are not accurate, then the matched filter does not operate optimally.

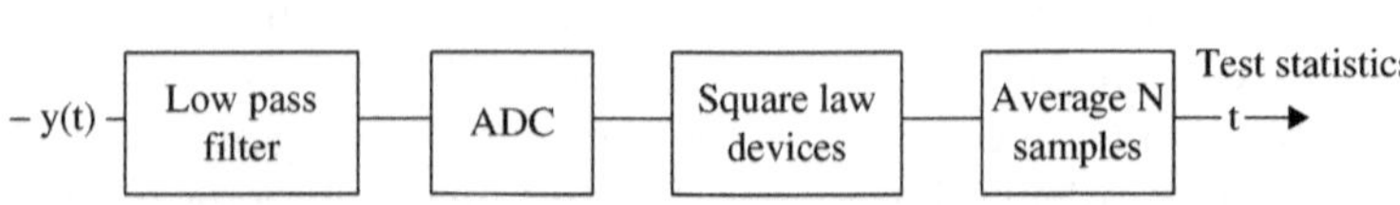

Figure 6.6 Main blocks of matched filter spectrum sensing technique.

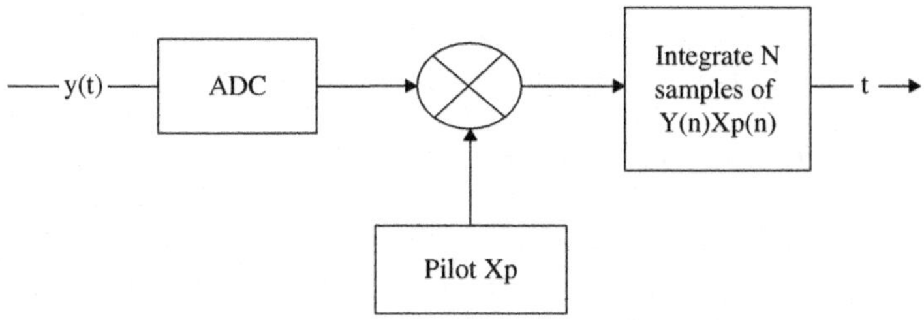

Figure 6.7 Implementing pilot detection using matched filter technique.

However, the performance of matched filter spectrum sensing can be enhanced by coherent detection, where pilot patterns for channel estimation, preamble signalling for synchronization, modulation orders and even spreading codes of the PU signal can be used [27]. The preamble signal is used to synchronize two or more different systems. Pilot symbols are the training symbols, which are distributed in the time and frequency domain to form a pattern known as the pilot pattern. These patterns are used to estimate the time and frequency related characteristics of the wireless channel, which has an influence on the transmitted signal. Figure 6.7 shows the implementation of pilot detection using a matched filter technique.

6.2.1.3 Cyclostationary Feature Detection

Communication signals, when subjected to operations such as modulation, sampling, coding and multiplexing, exhibit periodicity of the spectral correlation features, which are basically second order statistical parameters. These recurrent properties can be used effectively for sensing the presence of PUs in the spectrum. This built-in periodicity is not present in noise and this is used to differentiate between the PU signal and noise [28, 29]. In order to avoid interference, the spectrum, apart from being continuously monitored, should also be detected and identified.

The inherent presence of periodicity in modulated signal is due to the carrier wave, cyclic prefixes, repeating spreading codes, hopping sequence and pulse trains. Due to this, there is also periodicity in the second order statistical values of mean and autocorrelation. Autocorrelation is the measure of how rapidly a signal changes over time and correlation is the measure of the relationship between two time signals. The cyclostationary feature detection technique needs to know at least some information about the PU signal. The implementation of cyclostationary feature detection

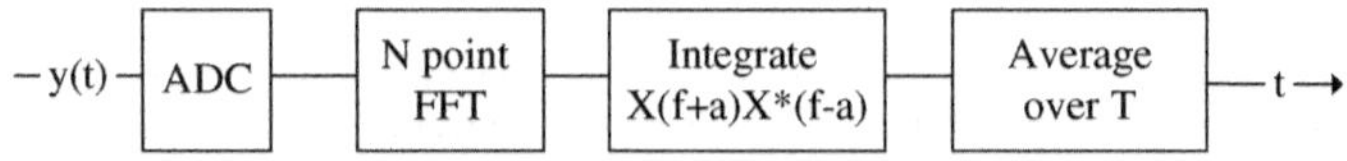

Figure 6.8 Implementation of cyclostationary feature detection.

shown in Figure 6.8 includes an evaluating cyclic autocorrelation function (CAF), computing spectral correlation function (SCF) by using discrete Fourier transformation (DFT) of CAF and, finally, detection [27].

The signal $y(t)$ is cyclostationary if its second order statistical properties are a periodic function of time. $R_y(\tau)$ is the CAF of the observed signal and is represented as in [30],

$$R_y(\tau) = E\left[y(t+\tau)y^*(t+\tau)e^{-j2\pi\alpha t}\right] \tag{6.9}$$

where $E[\ldots]$, denotes the expectation operation and α is the cyclic frequency. By using FFT on CAF, the SCF, also known as the cyclic spectrum, is computed. It is a two-dimensional function. Different modulated signals with the same power spectral density will have dissimilar SCFs. It is easier to distinguish noise and interference from the primary user's signal since noise and interference do not show any spectral correlation. If the received signal is stationary, then it represents noise and if the signal is cyclostationary then it suggests the presence of PU signal. The method used in [31] suggests an improvement in detection performance in low SNR and signal-interference-noise ratio (SINR), where the pilot information of the PU signal is injected into the CR signal. When both PU and CR are active, the intentional feature of this results in easy detection of PU signals by active nodes using the SCF. The advantage of this spectrum sensing technique is that it is robust in the presence of noise signals and interference from other signals. The performance of the detector is better even in cases of low SNR values [32, 33]. The limitations of this technique are that, firstly, some information of the primary user's signal has to be known and secondly, the cost incurred due to computation is very high.

6.2.1.4 Waveform Detection

In this spectrum sensing technique, the known patterns of the signal such as preambles, spreading sequences, midambles and redundant pilot patterns are taken and utilized to correlate with the received signal patterns to detect the presence of the signals [10]. Consider the received signal $y(n)$ as

$$y(n) = x(n) + w(n) \tag{6.10}$$

where $x(n)$ is the signal to be observed, $w(n)$ is the white Gaussian noise and n being the sample index. Assuming the known time domain patterns to contain NB signal samples, the waveform sensing metric can be given as in [10] and [34]:

$$S = \text{Re}\left[\sum_{n=1}^{NB} y(n)\, x^*(n)\right] \tag{6.11}$$

$$= \sum_{n=1}^{NB} \left|s(n)\right| {}^\wedge 2 + \text{Re}\left[\sum_{n=1}^{NB} w(n)\, x^*(n)\right] \tag{6.12}$$

The sensing metric consists of two terms. The first term is the signal component and the second term is the noise component. In the absence of the signal, when $s = 0$, then the sensing metric becomes [34]:

$$S = \text{Re}\left[\sum_{n=1}^{NB} y(n)\, x^*(n)\right] \tag{6.13}$$

According to [35], if the value of NB is large, then the known signal pattern in the primary user's signal will allow the waveform based sensing technique to perform well, even at low SNRs.

6.2.1.5 Wavelet Detection

The wavelet detection approach is used mostly for wideband signals where the wide frequency band is subdivided into sub-bands. The incoming signal is split into different sub-bands, which can be exploited for opportunistic usage of the spectrum by SUs [36]. The signal is divided into many parts using a modulated window that is fully scalable and then each part is analysed separately. The window can be positioned in the area of interest. The edges of the band are determined by taking advantage of the first and second order derivative of the convolution, which is derived by convolving the wavelet with the power spectral density of the received signal, the location of

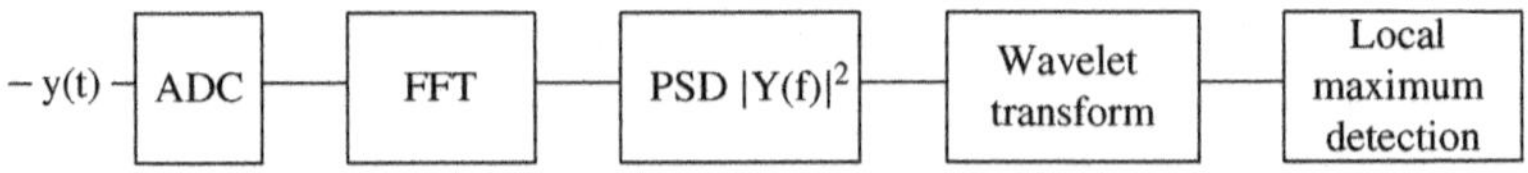

Figure 6.9 Implementation of a wavelet detector [34].

the frequency boundaries of each band within the wide band of interest is found [2]. The transition between the occupied and empty bands can be determined by the irregularities at the edge of the PSD. Estimating the power within each band and determining the edges will allow the characterization of the band as being occupied or empty [37]. Figure 6.9 shows the implementation of the wavelet detection.

Another approach of using wavelet detection is suggested in [38], where the received signal is correlated with the modulated wavelet in order to determine the spectral characteristics of the received signal. In [39], the entropy of the received signal wavelet is compared to a threshold to determine the presence of the PU signal.

6.2.1.6 Hybrid Sensing

Conventional spectrum sensing techniques, such as the energy detector, have some drawbacks and in order to overcome them hybrid sensing techniques have been proposed in [11, 40, 41]. The energy detector and cyclostationary sensing techniques are integrated to come up with an efficient hybrid sensing technique to sense the presence of PUs in a noise uncertain environment. The received signal energy is compared with the two threshold levels x1 and x2; if the value is above the threshold x1, it suggests the presence of a primary signal and if the value is lower than the threshold value x2 then the signal is noise [40]. This is known as coarse sensing. When the value lies in the range of the two thresholds, then a cyclostationary sensing technique known as fine sensing is employed. In [41], hybrid sensing employs an energy detector and covariance absolute value (CAV) method, which uses the covariance of signal and noise to detect the presence of a PU. The hybrid sensing technique uses the energy detection technique in cases of lower correlation and, on the other hand, uses the CAV method in cases of high correlation [41]. In [42], the hybrid method uses only the energy detection method except that it does the sensing in two steps. It has suggested an additional hypothesis along with the conventional binary hypothesis in an energy detector.

6.2.1.7 Multi-Taper Spectrum Sensing

The multi-taper spectrum sensing technique is one of the best for a CR network since it is very efficient and accurate in sensing and performs faster computations. This sensing technique uses multiple sets of tapers and orthogonal filters to reduce variance [43].

6.2.2 Collaborative/Cooperative Detection

Enhanced and reliable detection of PUs can be achieved with CR users interacting with each other in the collaborative detection technique. Problems such as fading, shadowing and hidden node problems can be addressed to a greater extent using this technique.

The accuracy of sensing the presence of primary users by a single CR is usually hindered by factors such as multipath fading, shadowing and receiver uncertainty. This severely degrades the performance of the CR [2]. The cooperative sensing method overcomes this issue to a greater extent [44–47]. An effective decision regarding the presence of PUs is obtained by analysing the observations of different CRs. Cooperative sensing can be modelled as centralized, distributed and external sensing approaches based on the type of data processing by the local sensors [48].

1) *Centralized*: A fusion/central node identifies the PU signal and the availability of the spectrum based on the sensing data transferred from the participating CRs to the fusion node. It sends the information to the CRs for opportunistic utilization of the spectrum (Figure 6.10).
2) *Distributed*: A decision about the spectrum availability is made by the individual nodes based on the information shared among different CRs (Figure 6.11).
3) *External Agents*: Spectrum sensing is performed by the external agents and the channel occupancy information is broadcasted to the CRs. This scheme overcomes the problem of hidden PUs. One of the main issues is the power consumption of the internal detectors for spectrum sensing and this addressed by external agents. This scheme has been proposed for implementation to identify the PUs in the IEEE 802.22 standard. The sensing techniques used by the CRs can be different or the same. Mostly, the common method used in cooperative sensing is energy detection due to its simplicity. It can be seen that, in [49–54], the CRs participating in cooperative sensing only use the energy detection technique.

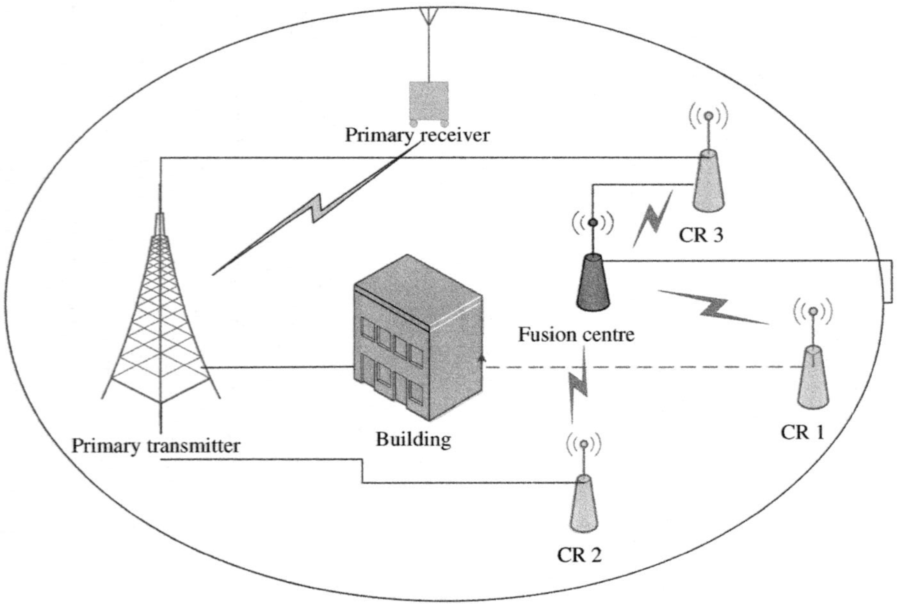

Figure 6.10 Centralized cooperative network.

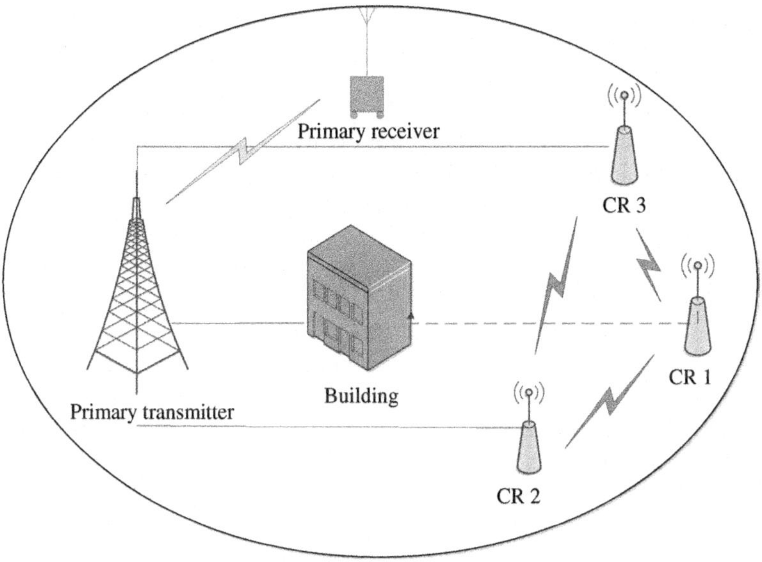

Figure 6.11 Distributed cooperative network.

However, in [55–57], CRs employ the cyclostationary feature detection approach. According to [48], if the sensing methods used in cooperative sensing are the same, then there are chances of increased misdetection of the PU signals in the spectrum due to large shadowing in that area and, therefore, [48] suggested the use of different spectrum sensing methods for different CR nodes: hard or coarse sensing, such as the energy detection method, and soft sensing, which involves more parts such as cyclostationary feature detection. CRs that have a higher value of SNR are usually chosen for collaboration. Apart from this, the constant false alarm rate and constant detection rate are used for filtering and optimal selection of cognitive users for collaborative sensing. This would reduce the overall sensing time. The advantages of using cooperative sensing are [58]:

- An effective solution to spectrum sensing problems such as shadowing, multipath fading, hidden node terminal and noise uncertainty.
- Probability of false alarm and probability of misdetection can be significantly reduced.
- The sensing time can be lowered.

- Reduces the sensitivity requirements of the receiver.
- The coverage area is increased and detector design is much simpler.

In [59], however, there are challenges that face cooperative sensing such as the difficulty of developing highly efficient sensing information sharing algorithms and the immense complexity involved. In [60], suggestions have made for addressing the sensing delay overheads.

6.2.3 Interference Temperature Detection

In the interference temperature detection technique, the CR users coexist with PUs. However, the CR users are restricted to transmitting using low power in order to avoid interference to PUs. An interference limit is set that will be used to determine the power intensity to be used by the transmitters. Above this limit, interference becomes inevitable and is unacceptable.

6.2.4 Primary Receiver Detection

The primary receiver local oscillator's leakage power is used to detect PUs in the primary receiver detection method.

6.3 History Assisted Spectrum Sensing

A technique is under research where the sensing of PUs is performed using the accumulated history of the spectrum usage by the PUs. This previous activity knowledge about PUs in the spectrum band makes it easier for CRs to make quick decisions about their presence. In [57], spectrum sensing techniques proposed using the past observations of the PU's state for improved performance of their detection. On the arrival of a PU, the SUs have to vacate the band and start the restoration mechanism by searching and sensing other frequency bands. In [61], the restoration mechanism performance is improved by utilizing the elementary features of learning and history awareness of CRs by providing a shorter restoration time or a restored channel with a higher quality.

History assisted energy efficient spectrum sensing is well suited to infrastructure CR networks and employs a processing database [62]. The advantage of using a database instead of agents is to reduce the duplication of learning and processing algorithms in resource constrained SUs; rather, the intelligence and processing in the history assisted scheme is delegated to the central database. Figure 6.12 shows

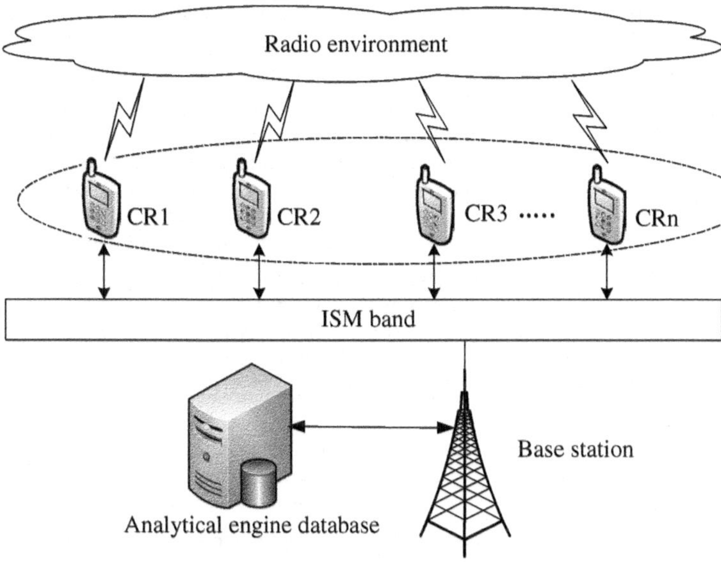

Figure 6.12 History assisted CR model using an analytical database [62].

the usage of an analytical database in a history assisted spectrum sensing scheme. The advantage of using this history for spectrum sensing techniques is that it reduces the sensing time, minimal activity in the control channel, fine tunes the threshold level to be scanned and reduces spectrum sensing.

The conventional spectrum sensing techniques and signal processing methods employed in these techniques have been presented. These sensing techniques have both advantages and disadvantages in terms of cost, complexity involved, ease of implementation, hardware realization, accuracy and sensing time, which are presented in Table 6.1.

6.4 Model- and Statistics-Based Spectrum Sensing Classification

Based on the models implemented and the statistics used in different research papers, we have come up with a state of the art method for classifying the spectrum sensing techniques. Many research papers have been explored and the mathematical models that were implemented to statistically characterize the sensing techniques are taken.

Table 6.1 Pros and cons of standard spectrum sensing techniques.

Sensing Method	*Advantages*	*Limitations*
Energy Detection	Simple to implement when noise power is known at the receiver.	Uncertainty of noise power and this results in false detection. Noise power estimation error can lead to an SNR wall. Unable to detect signals in conditions of low SNR.
Matched Filter	Optimal detection of signals in the channel. Computational cost is low.	Requires detailed information about the parameters of the primary user's signals. High implementation complexity.
Cyclostationary feature detection	Robust even in low SNR, able to differentiate noise and signal.	Computational cost is high. Some information about the primary user signal is required. Requires excessive signal processing expertise.
Waveform Detection	Performs well even in low SNRs if the known pattern is large.	Susceptible to synchronization errors.
Wavelet Detection	Performs well for the wideband signal.	High computational cost. Not very effective in case of spread spectrum signals.
Cooperative sensing	Reduced sensing time, Effective solution for problems such as shadowing, multipath fading and hidden node terminal.	Developing efficient sensing algorithms. Complex sensing technique.

The statistics used by these models are grouped together, thus showing the statistics that are commonly used for performance analysis of spectrum sensing. Figure 6.13 shows the classification of spectrum sensing techniques based on models and statistics implemented. Table 6.2 shows the conventional spectrum sensing techniques and the models useful in evaluating the performance of the various sensing technologies under different conditions. From Table 6.2, it is clearly shown that energy detection and cooperative sensing techniques have been used extensively for spectrum sensing in CR networks by most researchers. It also indicates that the model that was widely used in

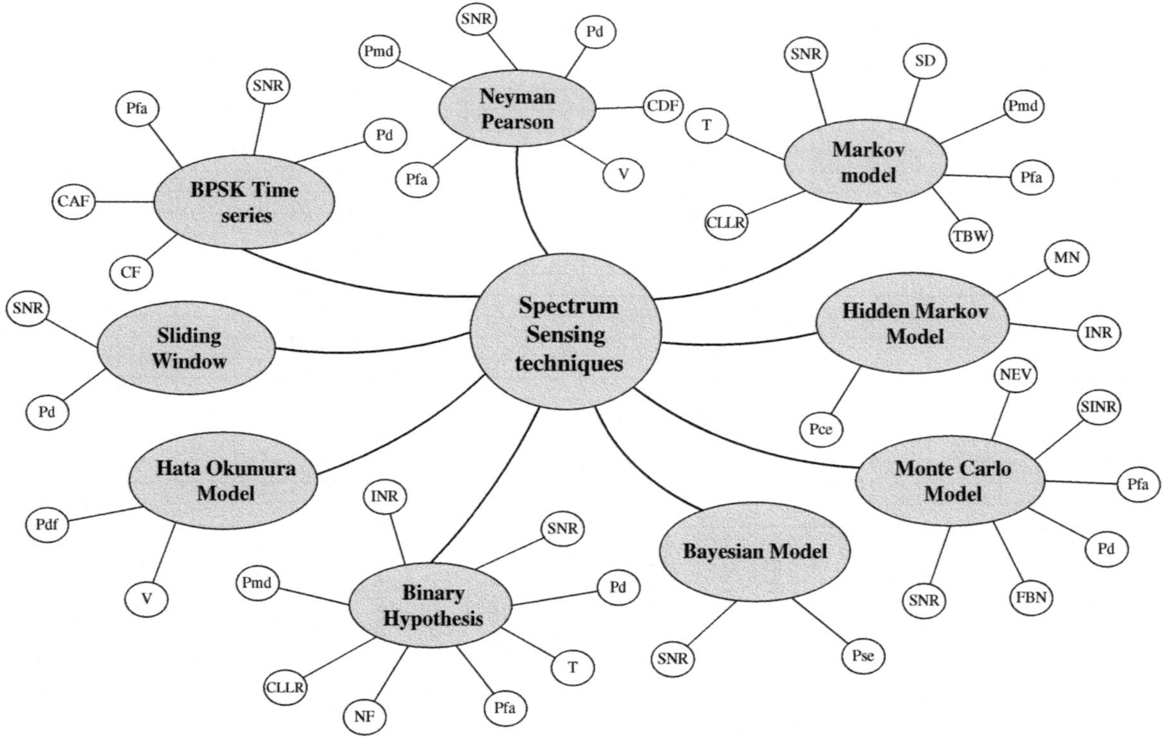

Figure 6.13 Classification of spectrum sensing techniques based on models and statistics implemented.

Table 6.2 Spectrum sensing methods and the models implemented.

	Spectrum Sensing Methods							*Model Implemented*													
Ref	**Energy detection**	**Matched filter**	**Cyclostationary feature**	**Waveform Detection**	**Hybrid Detection**	**Wavelet detection**	**Cooperative detection**	**Neyman–Pearson**	**Multi-resolution Spectrum Sensing Feature**	**Baye-sian**	**Hata–Okumura model**	**Central Limit Theorem**	**Reconfigurable Filter Implementation**	**Monte Carlo**	**Markov Model**	**Binary hypothesis**	**Rayleigh Fading channel**	**BPSK time series**	**Likelihood ratio**	**Sliding window**	**Hidden Markov Model**
6		✓											✓								
7			✓									✓									
8							✓				✓										
11				✓								✓									
13						✓			✓												
14					✓			✓													
15	✓															✓					
16	✓																				✓
18	✓														✓						
19					✓			✓													
20					✓			✓													
21															✓						
21	✓																✓				
22							✓					✓									
23			✓				✓				✓										
24			✓									✓									
25														✓							
26							✓	✓													
27							✓	✓													
28							✓									✓					
29							✓									✓					
31							✓							✓							
32	✓															✓					
33							✓	✓													
35							✓	✓													

37	✓													✓							
38						✓										✓					
39						✓										✓					
40	✓						✓														
42						✓	✓														
43						✓										✓					
44						✓							✓								
45			✓														✓				
46			✓														✓				
47			✓												✓						
48						✓			✓												
49			✓												✓						
50	✓															✓					
51					✓																
52	✓																✓				
53						✓								✓							
54					✓								✓								
55						✓															
56	✓														✓						
57						✓								✓							
58						✓					✓										
59			✓														✓				
60	✓																		✓		
61	✓														✓						
62	✓														✓						
63	✓															✓					
64	✓														✓						
65	✓															✓					
66				✓									✓								
67						✓															
68	✓															✓					
69	✓																	✓			
70	✓																				

these techniques for the statistical analysis is the Neyman–Pearson model and the binary hypothesis. The statistics used while implementing different models has been presented in Table 6.3. The commonly used statistics are the probability of false alarm (PFA), SNR and the PD. For an optimal sensing technique, the PFA should be low and the PD should be high. The information obtained from these tables has been used to develop a novel way of classifying the spectrums sensing techniques based on the models implemented and the statistics used as shown in Figure 6.12. Tables 6.2 and 6.3 provide information after examining a number of research papers about the models implemented for different sensing techniques and the relevant statistics utilized. It is found that the Neyman–Pearson model has been widely used for the energy detection technique.

6.5 Challenges and Issues

The major issues with LTE femtocells has been with interference mitigation. Interference results in poor reception in a network. Attaining an efficient spectrum sensing technique is a significant factor in detecting unoccupied channels that can be utilized by cognitive femtocells to reduce interference. However, there are certain issues relating to implementation and complexity involved in the spectrum sensing procedure. Further, the other main concerns in spectrum sensing are the sensing time and power consumption utilized due to complex sensing algorithms. The regulatory requirements are also an issue in the implementation of CR networks. Most research work has suggested that standalone spectrum sensing schemes cannot achieve reliable detection of white spaces or spectrum holes. A cooperative mechanism can be employed that would also reduce interference.

The traditional spectrum sensing techniques and the signal processing methods employed in these techniques have been presented. An optimal sensing technique, one that has low sensing time, high accuracy, reduced complexity and minimal hardware requirements that could be applied for different types of technologies, has not yet been realized. In primary transmitter detection methods, the energy detector is the simplest of all. It requires minimal hardware implementation and the level of complexity is very low. ED works well when the noise power is known at the receiver. The drawback of ED is that the sensing results are not accurate and reliable. Moreover, at low

Table 6.3 Statistics used in different spectrum sensing methods.

	Statistics Used																															
Ref	P_{FA}	SNR	SD	PDF	PD	V	P_{CE}	CLLR	P_{MD}	CF	SCF	CC	P_I	No. of Users	Mean Noise	No. of Sensing slots	INR	P_{SE}	ND BW	ACL	PSD	T	RF	SINR	CDF	AEP	TBW	NEV	CAF	FBN	NF	P_E
6																							✓									
7										✓	✓																					
8													✓	✓																		
11																			✓	✓												
13																					✓											
14	✓								✓																							
15		✓			✓											✓															✓	
16							✓								✓		✓															
18								✓														✓										
20		✓			✓							✓																				
21		✓	✓																													
23				✓		✓																										
24																																
25										✓														✓								
26		✓			✓																				✓							
27	✓								✓													✓				✓						
28	✓	✓							✓																							
29	✓	✓							✓																							
31	✓				✓																											
32	✓	✓			✓																											
33	✓				✓																											
35	✓					✓			✓																							
37	✓	✓							✓																							
38		✓		✓	✓																											
39	✓	✓		✓	✓																											

(Continued)

Table 6.3 (Continued)

Ref	P_{FA}	SNR	SD	PDF	PD	V	P_{CE}	CLLR	P_{MD}	CF	SCF	CC	P_I	No. of Users	Mean Noise	No. of Sensing slots	INR	P_{SE}	ND BW	ACL	PSD	T	RF	SINR	CDF	AEP	TBW	NEV	CAF	FBN	NF	P_E
45										✓																			✓			
46	✓	✓			✓																											
47	✓	✓			✓																											
48		✓																✓														
49	✓	✓			✓				✓															✓								
50	✓	✓			✓				✓																							
51		✓																	✓													
52	✓				✓																											
53	✓				✓																											
54		✓																										✓		✓		
55	✓				✓									✓													✓					
56		✓							✓													✓										
57	✓	✓			✓																						✓					
58	✓	✓					✓		✓																							
59		✓			✓					✓	✓																					
60		✓			✓																											
61					✓											✓																
62					✓																	✓										
63	✓				✓																	✓										✓
64	✓				✓												✓															
65		✓		✓												✓																
66		✓			✓																											
67	✓	✓			✓																											
68	✓	✓			✓				✓																							
69	✓	✓			✓				✓							✓																
70		✓			✓																											

Table 6.4 Table of acronyms.

P_{FA}	**Probability of False Alarm**	P_{MD}	**Probability of Missed Detection**
SNR	Signal to Noise Ratio	CF	Cyclic frequency
SD	Standard deviation	SCF	Spectral Correlation function
PDF	Probability density Function	CC	Correlation coefficient
P_D	Probability of detection	P_I	Probability of Interference
V	Variance	INR	Interference to Noise Ratio
P_{CE}	probability of classification error	PSD	power spectral density
CLLR	cumulative log likelihood ratio	RF	reconfiguration factor
RgOSP	relative gain of optimal sensing policy	T	threshold
SINR	signal to interference to noise ratio	ACL	adjacent channel leakage
NDBW	normalized data bandwidth	CDF	cumulative distribution function
AEP	average error probability	TBW	time bandwidth product
CAF	cyclic autocorrelation function	NEV	normalized energy vector
NF	normalized frequency	P_E	probability of error
FBN	frequency band of nodes	MN	mean noise

SNR it is not possible to sense the signal using ED. Matched filter detection gives the most accurate results but requires comprehensive information about the PUs and hence the complexity involved in this technique is very high. To detect signals in low SNR the cyclostationary feature detection technique proves to be more effective. The wavelet detection technique is widely used for edge detection in image processing [69]. In spectrum sensing, the edges will behave as dividers of empty holes and the occupied bands. Standalone techniques are unable to provide accurate results in a given time. Therefore, employing hybrid detection techniques and cooperative/collaborative techniques could provide accurate results. Though the cooperative sensing technique gives effective results and reduces sensing time, the complexity involved in sharing information between the CRs is high.

6.6 Summary

The fundamental spectrum sensing techniques have been classified and analysed. Sensing the spectrum to find idle channels for gaining immediate access enhances the opportunities for SUs to access the spectrum. However, the performance of the techniques can be analysed using the performance indicators such as speed and accuracy. Of all the primary transmitter signal detection methods, ED is the most widely used sensing technique because of its simplicity and low computational cost; it does not require any information about the primary signal's characteristics. On the other hand, the cyclostationary feature detection method and matched filter detection needs comprehensive information about the primary signal of interest. This makes these techniques complex compared to ED. In the cooperative spectrum sensing method, the CRs work in collaboration using any of these sensing techniques. This technique has been widely used and the ED is at least one of the sensing techniques of the CRs.

In this chapter, we have provided an outline of the signal processing fundamentals that are relevant to CR networks. A general description of the channel models, OFDM modulation technique and the error probability are given. The need for effective utilization of the spectrum and minimal interference to primary users makes spectrum sensing highly crucial. A detailed description of various spectrum sensing techniques has been presented. We have also presented the advantages and disadvantages of the standard primary transmitter sensing methods, made a comparison and found that there is no optimal spectrum sensing technique that could be applied to all technologies in all scenarios. Some techniques are suitable for a technology while others are more suitable for some other technologies. Detailed tabulation of spectrum sensing techniques and the models implemented in those techniques has been presented. It shows that most researchers have worked on cooperative sensing techniques predominantly using energy detection as one of the sensing techniques and employing the Neyman–Pearson model. Most research papers have also shown that P_{D} and P_{FA} are the widely used statistics to access the performance of CR networks. It is desirable to have a higher value of P_{D} and a lower value of P_{FA} for obtaining accurate signal detection results. Extensive research should be done in CR systems to overcome the challenges such as implementation complexity, accuracy, sensing time and hardware implementation.

References

1 H. Zhang, C. Jiang, X. Mao and A. Nallanathan, "Resource management in cognitive opportunistic access femtocells with imperfect spectrum sensing," 2014 *IEEE Global Communications Conference*, Austin, TX, pp. 3098–3102, 2014.

2 A. M. Wyglinski, M. Nekovee and T. Hou. *Cognitive Radio Communications and Networks – Principles and Practice.* Academic Press: 2010

3 J. Ma, G. Y. Li and B. H. Juang, "Signal processing in cognitive radio," *Proceedings of the IEEE*, vol. 97, no. 5, pp. 805–823, May 2009.

4 H. Mahmoud, T. Yucek and H. Arslan, "OFDM for cognitive radio: merits and challenges," *Wireless Communications, IEEE*, vol. 16, no. 2, pp. 6–15, April 2009.

5 J. Singh, R. Garg and I. K. Aulakh, "Effect of OFDM in cognitive radio: Advantages & issues," *2016 Second International Conference on Computational Intelligence & Communication Technology (CICT)*, Ghaziabad, pp. 554–558, 2016.

6 S. Bokharaiee, H. H. Nguyen and E. Shwedyk, "Blind spectrum sensing for OFDM-based cognitive radio systems," *IEEE Transactions on Vehicular Technology*, vol. 60, no. 3, pp. 858–871, March 2011.

7 Y. Zhang and S. Wang, "Resource allocation for cognitive radio-enabled femtocell networks with imperfect spectrum sensing and channel uncertainty," in *IEEE Transactions on Vehicular Technology*, vol. 65, no. 9, pp. 7719–7728, Sept. 2016.

8 A. Ben Jemaa, M. Turki and W. Guibene, "Enhanced energy detector via algebraic approach for spectrum sensing in cognitive radio networks," *Cognitive Radio Oriented Wireless Networks and Communications (CROWNCOM), 2012 7th International ICST Conference*, pp. 113–117, June 2012.

9 G. Yang, J. Wang, J. Luo, *et al.*, "Cooperative spectrum sensing in heterogeneous cognitive radio networks based on normalized energy detection," in *IEEE Transactions on Vehicular Technology*, vol. 65, no. 3, pp. 1452–1463, March 2016.

10 G. Feng, W. Chen and Z. Cao; "A joint PHY-MAC spectrum sensing algorithm exploiting sequential detection," *IEEE Signal Processing Letters*, vol. 17, no. 8, pp. 703–706, Aug. 2010.

11 F. Moghimi, R. Schober and R. K. Mallik, "Hybrid coherent/energy detection for cognitive radio networks," *IEEE Transactions on Wireless Communications*, vol. 10, no. 5, pp. 1594–1605, May 2011.

12 D. D. Ariananda, M. K. Lakshmanan and H. Nikookar "A survey on spectrum sensing techniques for cognitive radio," *Cognitive Radio and Advanced Spectrum Management, 2009. CogART 2009*, pp. 74–79. May 2009.
13 D. Cabric, A. Tkachenko and R. W. Brodersen, "Spectrum sensing measurements of pilot, energy, and collaborative detection," *MILCOM 2006 - 2006 IEEE Military Communications Conference*, Washington, DC, 2006, pp. 1–7.
14 J. Song, Z. Feng, P. Zhang and Z. Liu, "Spectrum sensing in cognitive radios based on enhanced energy detector," *IET Communications*, vol. 6, no. 8, pp. 805–809, 22 May, 2012.
15 M. López-Benítez and F. Casadevall, "Improved energy detection spectrum sensing for cognitive radio," *IET Communications*, vol. 6, no. 8, pp. 785–796, 22 May, 2012.
16 H. Rasheed, N. Rajatheva and F. Haroon, "Spectrum sensing with Energy detection under shadow-fading condition," *Wireless Pervasive Computing (ISWPC), 2010 5th IEEE International Symposium on*, pp. 104–109, 5–7 May, 2010.
17 S. Chunyi, Y. D. Alemseged, N. R. Ha, G. Villardi, C. Sun, S. Filin, *et al.* "Adaptive two thresholds based energy detection for cooperative spectrum sensing," *7th IEEE Consumer Communications and Networking Conference (CCNC), 2010*, pp. 1–6, 9–12 Jan. 2010.
18 S. Atapattu, C. Tellambura, J. Hai, "Spectrum sensing via energy detector in low SNR," *IEEE International Conference: Communications (ICC), 2011*, pp. 1–5, June 2011.
19 Y. M. Kim, G. Zheng, S. H. Sohn and J. M. Kim, "An alternative energy detection using sliding window for cognitive radio system," *Advanced Communication Technology, 2008. 10th International Conference on*, pp. 481–485, 17–20 Feb. 2008.
20 T.-H. Yu, S. Rodriguez-Parera, M. Marković and D. Čabrić, "Cognitive radio wideband spectrum sensing using multitap windowing and power detection with threshold adaptation," *IEEE International Conference Communications (ICC), 2010*, pp. 1–6, May 2010.
21 A. Vakili and B. Champagne, "An adaptive energy detection technique applied to cognitive radio networks," *Personal Indoor and Mobile Radio Communications (PIMRC), 2011 IEEE 22nd International Symposium on*, pp. 509–514, 11–14 Sept. 2011.

22 T. S. Shehata and M. El-Tanany, "A novel adaptive structure of the energy detector applied to cognitive radio networks," *Information Theory, 2009, 11th Canadian Workshop on*, pp. 95–98, 13–15 May, 2009.

23 L. Zhengwei, M. Yi and R. Tafazolli, "A first-order cyclostationarity based energy detection approach for non-cooperative spectrum sensing," *Personal Indoor and Mobile Radio Communications (PIMRC), 2010 IEEE 21st International Symposium*, pp. 554–559, 26–30 Sept., 2010.

24 J.-H. Baek, H.-J. Oh and S.-H. Hwang, "Improved reliability of spectrum sensing using energy detector in cognitive radio system," *10th International Conference on Advanced Communication Technology, 2008. ICACT 2008*, vol. 1, pp. 575–578, 17–20 Feb., 2008.

25 T. Ikuma and M. Naraghi-Pour, "A comparison of three classes of spectrum sensing techniques," *Global Telecommunications Conference, 2008*, pp. 1–5, Dec. 2008.

26 D. Bhargavi and C. R. Murthy, "Performance comparison of energy, matched-filter and cyclostationarity-based spectrum sensing," *Signal Processing Advances in Wireless Communications (SPAWC), 2010 IEEE Eleventh International Workshop*, pp. 1–5, June 2010.

27 D. Bera, S. S. Pathak and I. Chakrabarti, "A normal factor graph approach for co-operative spectrum sensing in cognitive radio," *2012 National Conference on Communications (NCC), Kharagpur*, 2012, pp. 1–5.

28 Y. Lin, C. He, Chen, L. Jiang and D. He, "A cyclostationary-based spectrum sensing method using stochastic resonance in cognitive radio," *IEEE International Conference on Communications Workshops (ICC), 2010*, pp. 1–5, 23–27 May, 2010.

29 Y. Lin, C. He, L. Jiang and D. He, "A spectrum sensing method in cognitive radio based on the third order cyclic cumulant," *International Conference on Wireless Communications & Signal Processing, 2009*, pp. 1–5, 13–15 Nov., 2009.

30 X. Gan, M. Xu and H. Li. "Energy efficient sequential sensing in multi-user cognitive ad hoc networks: A consideration of an ADC device," *Journal of Communications and Networks*, vol. 14, no. 2, pp. 188–194, April 2012.

31 H.-W. Chen and C.-L. Wang, "Design of a new signal structure for active sensing in cognitive radio systems," *IEEE 75th Vehicular Technology Conference (VTC Spring), 2012*, pp. 1–5, 6–9 May, 2012.

32 O. Ali, F. Nasir and A. A. Tahir, "Analysis of OFDM parameters using cyclostationary spectrum sensing in cognitive radio," *IEEE 14th International Multitopic Conference (INMIC), 2011*, pp. 301–305, 22–24 Dec., 2011.

33 K.-L. Du and M. Wai Ho, "Affordable cyclostationary-based spectrum sensing for cognitive radio with smart antennas," *IEEE Transactions on Vehicular Technology*, vol. 59, no. 4, pp. 1877–1886, May 2010.

34 D. B. Rawat and G. Yan, "Spectrum sensing methods and dynamic spectrum sharing in cognitive radio networks: a survey". *International Journal of Research and Reviews in Wireless Sensor Networks*, vol. 1, Mar 2011.

35 K. Hareesh and P. Singh, "An energy efficient hybrid co-operative spectrum sensing technique for CRSN," *2013 International Multi-Conference on Automation, Computing, Communication, Control and Compressed Sensing (iMac4s), Kottayam*, pp. 438–442, 2013.

36 A. Chandran, R. A. Karthik, A. Kumar, R. C. Naidu, M. Subramania Siva, U. S. Iyer, et al. "A novel spectrum sensing technique for low Signal to Noise Ratio conditions," *International Conference on Emerging Trends in Electrical and Computer Technology (ICETECT), 2011*, pp. 681–685, 23–24 March, 2011.

37 X. Ling, B. Wu, H. Wen, P.-H. Ho, Z. Bao and L. Pan, "Adaptive threshold control for energy detection based spectrum sensing in cognitive radios," *IEEE Wireless Communications Letters*, vol. 1, no. 5, pp. 448–451, October 2012.

38 Y. Verma and N. Dewangan, "Co-operative spectrum sensing in cognitive radio under Rayleigh fading channel," *2015 International Conference on Computer, Communication and Control (IC4), Indore*, 2015, pp. 1–5.

39 Z.-L. Jiang, Q.-Y. Zhang, Y. Wang and X.-Q. Shang, "Wavelet packet entropy based spectrum sensing in cognitive radio," *Communication Software and Networks (ICCSN), 2011 IEEE 3rd International Conference on*, pp. 293–298, 27–29 May, 2011.

40 Z. Khalaf, A. Nafkha, J. Palicot and M. Ghozzi, "Hybrid spectrum sensing architecture for cognitive radio equipment," *Telecommunications (AICT), 2010 Sixth Advanced International Conference*, pp. 46–51, 9–15 May, 2010.

41 D. Simunic and T. S. Dhope, "Hybrid detection method for spectrum sensing in cognitive radio," *MIPRO, 2012 Proceedings of the 35th International Convention*, pp. 765–770, 21–25 May, 2012.

42 S. Badrinath and V. U. Reddy, "A hybrid energy detection approach to spectrum sensing," *Cognitive Wireless Systems (UKIWCWS), 2009 First UK-India International Workshop*, pp. 1–6, Dec. 2009.

43 A. Al-Mamun and M. R. Ullah, "Multitaper spectrum: A promising method in spectrum sensing cognitive radio," *13th International Conference on Computer and Information Technology (ICCIT), 2010*, pp. 556–560, 23–25 Dec., 2010.

44 M. Derakhtian, F. Izedi, A. Sheikhi and M. Neinavaie, "Cooperative wideband spectrum sensing for cognitive radio networks in fading channels," *IET Signal Processing*, vol. 6, no. 3, pp. 227–238, May 2012.

45 S. J. Zahabi, A. A. Tadaion and S. Aissa, "Neyman–Pearson cooperative spectrum sensing for cognitive radio networks with fine quantization at local sensors," *IEEE Transactions on Communications*, vol. 60, no. 6, pp. 1511–1522, June 2012.

46 J. Shen, S. Liu, L. Zeng, G. Xie, J. Gao and Y. Liu, "Optimisation of cooperative spectrum sensing in cognitive radio network," *IET Communications*, vol. 3, no. 7, pp. 1170–1178, July 2009.

47 S. Mosleh, A. A. Tadaion and M. Derakhtian, "Bayesian decision for the fusion center of a distributed network in Cognitive Radio," *IEEE International Conference on Signal and Image Processing Applications (ICSIPA), 2009*, pp. 508–513, 18–19 Nov., 2009.

48 Y. Zhang, F. Song and H. Zhao, "A mixed spectrum sensing method for Cognitive Radio," *International Conference on Computer Science and Network Technology (ICCSNT), 2011*, vol. 1, pp. 138–141, 24–26 Dec., 2011.

49 W. Ejaz, N. ul Hasan, K. Hyung Seok and M. A. Azam, "Fully distributed cooperative spectrum sensing for cognitive radio ad hoc networks," *Frontiers of Information Technology (FIT), 2011*, pp. 9–13, 19–21 Dec., 2011.

50 Q. Zhi, C. Shuguang and A. H. Sayed, "Optimal linear cooperation for spectrum sensing in cognitive radio networks," *IEEE Journal of Selected Topics in Signal Processing*, vol. 2, no. 1, pp. 28–40, Feb. 2008.

51 H. Sadeghi, P. Azmi and H. Arezumand, "Cyclostationarity-based soft cooperative spectrum sensing for cognitive radio networks," *IET Communications*, vol. 6, no. 1, pp. 29–38, January 4 2012.

52 H. Vu-Van and I. Koo, "Cooperative spectrum sensing with collaborative users using individual sensing credibility for cognitive radio network," *IEEE Transactions on Consumer Electronics*, vol. 57, no. 2, pp. 320–326, May 2011.

53 X. Zhang, Z. Qiu and D. Mu, "Asynchronous cooperative spectrum sensing in cognitive radio," *9th International Conference on Signal Processing, 2008*, pp. 2020–2023, 26–29 Oct., 2008.

54 Y. Tani and T. Saba, "Quantization scheme for energy detector of soft decision cooperative spectrum sensing in cognitive radio," *GLOBECOM Workshops (GC Wkshps), 2010 IEEE*, pp. 69–73, 6–10 Dec., 2010.

55 S. Atapattu, C. Tellambura and J. Hai "Energy detection based cooperative spectrum sensing in cognitive radio networks," *IEEE Transactions on Wireless Communications*, vol. 10, no. 4, pp. 1232–1241, April 2011.

56 H. Sadeghi, P. Azmi and H. Arezumand, "Cyclostationarity-based soft cooperative spectrum sensing for cognitive radio networks," *Communications, IET*, vol. 6, no. 1, pp. 29–38, 4 January, 2012.

57 M. Derakhshani, L.-N. Tho and M. Nasiri-Kenari, "Efficient cooperative cyclostationary spectrum sensing in cognitive radios at low SNR regimes," *IEEE Transactions on Wireless Communications*, vol. 10, no. 11, pp. 3754–3764, November 2011.

58 S. Hussain and X. Fernando, "Spectrum sensing in cognitive radio networks: Up-to-date techniques and future challenges," *IEEE Toronto International Conference on Science and Technology for Humanity (TIC-STH)*, pp. 736–741, 26–27 Sept., 2009.

59 W. Prawatmuang and D. K. C. So, "Adaptive sequential cooperative spectrum sensing technique in time varying channel," *2012 IEEE 23rd International Symposium on Personal Indoor and Mobile Radio Communications (PIMRC)*, pp. 1546–1551, 9–12 Sept., 2012.

60 B. F. Lo and IF. Akyildiz, "Reinforcement learning-based cooperative sensing in cognitive radio ad hoc networks," *Personal Indoor and Mobile Radio Communications (PIMRC), 2010 IEEE 21st International Symposium on*, pp. 2244–2249, 26–30 Sept., 2010.

61 A. Azarfar, J. F. Frigon and B. Sansò, "History-aware channel search schemes in cognitive radio networks," *2011 IEEE 22nd International Symposium on Personal, Indoor and Mobile Radio Communications*, Toronto, ON, pp. 379–383, 2011.

62 T. S. Syed and G. A. Safdar, "On the Usage of History for Energy Efficient Spectrum Sensing," in *IEEE Communications Letters*, vol. 19, no. 3, pp. 407–410, March 2015.

63 Y. Gao, H. Zhang, Y. Shen and D. Wang, "Threshold selection for energy detection in Cognitive Radio with a non-line-of-sight environment," *11th International Conference on Wireless Communications, Networking and Mobile Computing (WiCOM 2015)*, Shanghai, 2015, pp. 1–6.

64 G. Ding, Q. Wu, F. Song and J. Wang, "Spectrum sensing in opportunity-heterogeneous cognitive radio networks: Non-cooperative case," *2011 International Conference on Wireless Communications and Signal Processing (WCSP)*, Nanjing, 2011, pp. 1–5.

65 P. C. Sofotasios, L. Mohjazi, S. Muhaidat, M. Al-Qutayri and G. K. Karagiannidis, "Energy detection of unknown signals over cascaded fading channels," in *IEEE Antennas and Wireless Propagation Letters*, vol. 15, pp. 135–138, 2016.

66 M. Gaafar, Y. Hassan and T. Elshabrawy, "A hybrid cooperative spectrum sensing technique for cognitive radio networks using linear classifiers," *2014 International Conference on Engineering and Technology (ICET)*, Cairo, 2014, pp. 1–6.

67 K. Yadav, S. D. Roy and S. Kundu, "Hybrid cooperative spectrum sensing with cyclostationary detection for cognitive radio networks," *2016 IEEE Annual India Conference (INDICON)*, Bangalore, 2016, pp. 1–6.

68 A. Ranjan, Anurag and B. Singh, "Design and analysis of spectrum sensing in cognitive radio based on energy detection," *2016 International Conference on Signal and Information Processing (IConSIP)*, Vishnupuri, 2016, pp. 1–5.

69 T. Trump and I. Müürsepp, "An energy detector for spectrum sensing in impulsive noise environment," *2011 IEEE 22nd International Symposium on Personal, Indoor and Mobile Radio Communications*, Toronto, ON, 2011, pp. 467–471.

70 M. A. Mohamoud, E. M. A. Elsheikh and M. H. Habaebi, "A comparative study of energy detector performance under AWGN and fading channels," *2016 International Conference on Advances in Electrical, Electronic and Systems Engineering (ICAEES)*, Putrajaya, 2016, pp. 89–94.

Part II

Antennas for LTE Femtocells

7

Antenna Consideration for LTE Femtocells

Masood Ur Rehman

School of Computer Science and Technology, University of Bedfordshire, Luton, UK

Antennas have been one of the key building blocks in the development of wireless communications systems right from the first demonstration of wireless technology in 1886 by Heinrich Hertz and its first practical radio application by Guglielmo Marconi in 1901. It is the only component that enables the transition of the radio energy between the wired transmission lines and free space. Hence, its efficiency essentially defines the operation and robustness of the overall communication system at both the transmission and reception sides. The last decade has seen a tremendous growth in the wireless technologies, especially with the advent of 4G and LTE standards. This in turn has dictated great advancements in antenna designs from a dimple wire dipole used by Hertz to complex array designs and upcoming massive MIMO (multiple input multiple output) systems. This chapter provides an overview of the fundamental antenna properties and required specific considerations for antenna design in the context of LTE femtocells.

7.1 Antenna Fundamentals

The antenna can be considered to be a transducer between the electromagntic waves travelling in a guided radio frequency (RF) circuit and the electromagnetic waves travelling in an unguided free space.

LTE Communications and Networks: Femtocells and Antenna Design Challenges,
First Edition. Edited by Masood Ur Rehman and Ghazanfar Ali Safdar.

Careful selection of the antenna element can result in an improved overall system performance and reduced number of access points in a femtocell environment. This in turn can bring down the cost of the system installation. On the contrary, poor or wrong antenna selection can lead to a degraded system performance and increased cost.

Antenna performance is defined using a number of parameters leading to various categories and design classes. Important parameters for antenna characterization include impedance matching, radiation pattern, directivity and gain, efficiency, polarization and bandwidth. These parameters are discussed in this section.

7.1.1 Input Impedance and Matching

Input impedance of the antenna (Z_a) is defined as the impedance presented by the antenna at its terminals as shown in Figure 7.1. It is a complex number and is equal to the ratio between input voltage (V_{in}) and current (I_{in}) at its terminals. Mathematically,

$$Z_a = \frac{V_{in}}{I_{in}} = R_a + jX_a \tag{7.1}$$

Where R_a and X_a is the frequency dependent antenna resistance and reactance, respectively. The antenna resistance, R_a, consists of two components: radiation resistance (R_r) that accounts for the resistance of the antenna dissipating the power and causing radiation of equal amount and loss resistance (R_L) from the conductor and dielectric losses. Values of 50 and 75 Ω are typically used for the characteristic impedance (Z_o) of RF test equipment, connectors and cables for the

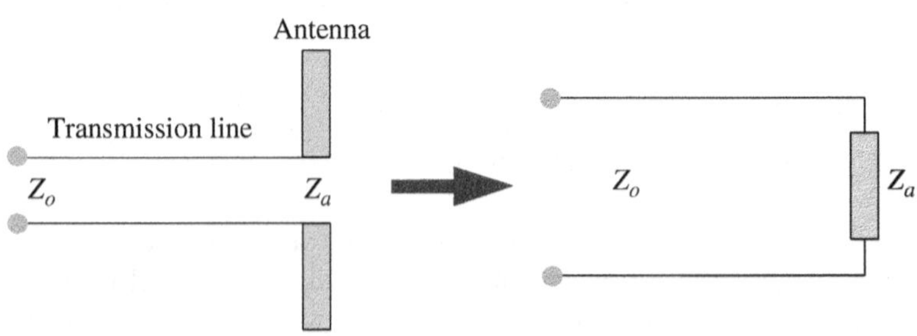

Figure 7.1 Input impedance model of an antenna.

sake of standardization. The input impedance of the antennas is, therefore, required to carry the same value in practical designs.

The antenna acts just as a load to a transmission line from the circuit point of view making impedance matching an utmost necessity in order to minimize reflections and maximize power transmission. Reflection coefficient (S_{11}), voltage standing wave ratio (VSWR) and return loss (RL), as a function of frequency, are used to assess the impedance matching of the antenna.

- Reflection coefficient:

$$S_{11} = \frac{Z_a - Z_o}{Z_a + Z_o} \tag{7.2}$$

- Return loss:

$$L_{RT} = -20log\left(\left|S_{11}\right|\right) \tag{7.3}$$

- VSWR:

$$VSWR = \frac{1+S_{11}}{1-S_{11}} \tag{7.4}$$

Commonly, an antenna is required to exhibit a reflection coefficient of < -10 dB or VSWR of < 2. These values correspond to a 11.07% of reflected power and 0.51 dB of transmission loss [1]. For mobile phone antennas, a reflection coefficient of < -6 dB or VSWR of < 3 is also usually accepted.

7.1.2 Bandwidth

Impedance bandwidth of the antenna can be defined as the frequency range over which 90% of the incident power is delivered to the antenna corresponding to the reflection coefficient value of < -10 dB and VSWR $< 2{:}1$. The absolute impedance bandwidth of the antenna is calculated as the difference between upper and lower frequencies at which these criteria are met as illustrated in Figure 7.2. Alternatively, the fractional impedance bandwidth is calculated as the percentage of the difference between the upper and lower frequency divided by the centre frequency of bandwidth, as follows:

$$BW = \frac{f_{upper} - f_{lower}}{f_{centre}} \times 100 \tag{7.5}$$

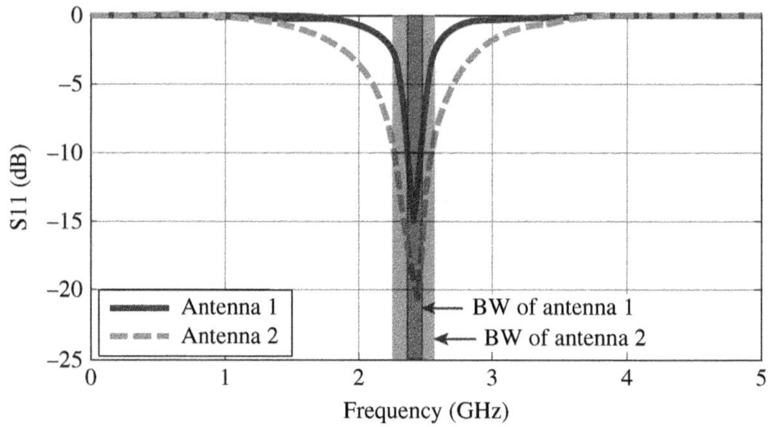

Figure 7.2 Bandwidth of two antennas using S_{11} plotted against frequency.

7.1.3 Radiation Pattern

The radiation pattern is a graphical representation of the radiation properties of an antenna as a function of spatial coordinates. It is a plot of the radiated field or power with respect to the angle at a fixed distance in the farfield region of the antenna. For an electrically small antenna with a size D, the farfield distance r can be calculated as:

$$r = \frac{2D^2}{\lambda} \tag{7.6}$$

Where λ represents the wavelength.

Spherical coordinates are typically used in antenna measurements. The 3D radiation pattern efficiently illustrates the radiated field distribution as a function of angle θ and ϕ as shown in Figure 7.3. However, measurement of 3D pattern requires sophisticated equipment. Therefore, the 2D radiation patterns in the elevation plane measuring θ with a fixed ϕ at 0° azimuth plane measuring θ with a fixed ϕ at 90° are the most useful, as illustrated in Figure 7.4.

The radiation pattern provides very useful information about the antenna operation. A scalar radiation pattern of an antenna is illustrated in Figure 7.5 [2]. The major features of an antennas's radiation pattern include:

- Major lobe: The angular region containing the direction of maximum radiation. For a symmetrical main beam, central direction is termed the boresight of an antenna.

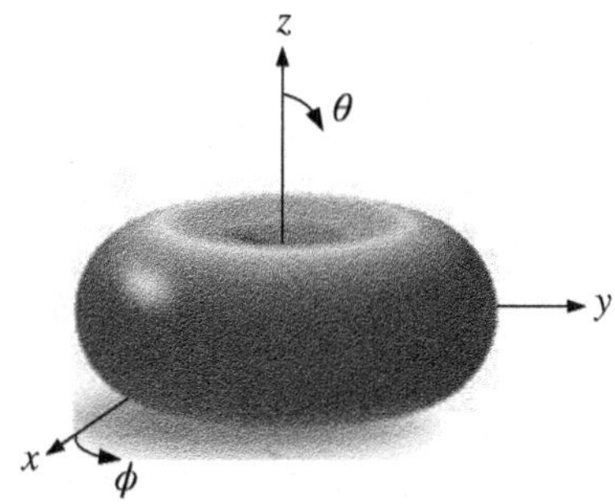

Figure 7.3 3D radiation pattern of an electrically small antenna.

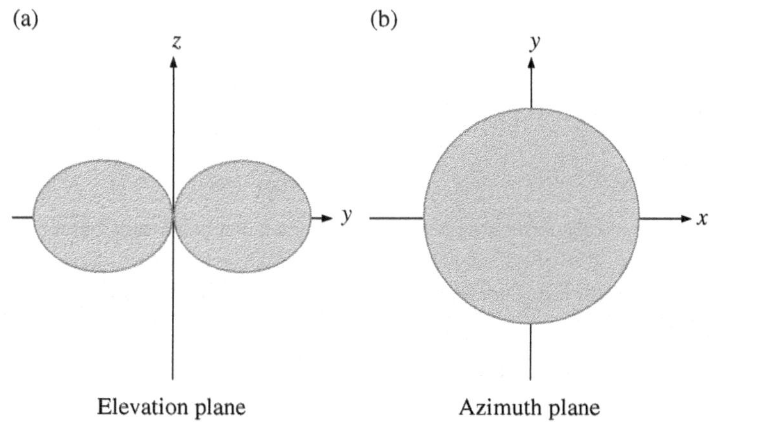

Figure 7.4 2D radiation pattern in elevation and azimuth planes of an electrically small antenna.

- Minor lobes: Any lobe apart from the main lobe.
- Back lobe: A radiation lobe that is directly opposite to the main lobe.
- Null: Angle at which the radiation pattern is zero.
- Half-power beamwidth (HPBW): The angle subtended by the half-power points of the main beam, also called 3 dB beamwidth.

7.1.4 Directivity and Gain

Directivity (*D*) is a quantitative measure of the ability of the antenna to concentrate radiated power in a particular direction. It is the ratio of the radiation intensity in a given direction from the antenna to

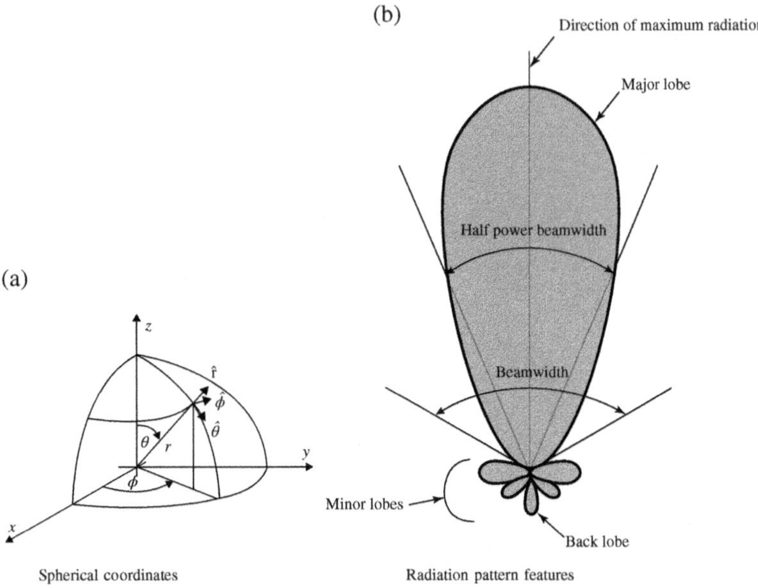

Figure 7.5 Spherical coordinate system and antenna radiation pattern features.

the radiation intensity averaged over all directions [3] and can be calculated as:

$$D = \frac{U(\theta,\phi)}{U(\theta,\phi)_{avg}} = \frac{4\pi U(\theta,\phi)}{P_{rad}} \tag{7.7}$$

where P_{rad} is the total radiated power and U is the radiation intensity per unit solid angle.

Gain (G) is defined as the ratio of the radiation intensity in a given direction from the antenna to the total input power accepted by the antenna divided by 4π. If the direction is not specified, the direction of maximum radiation is implied. It is expressed relative to a lossless isotropic and expressed in terms of dBi. Mathematically, it can be written as:

$$G = \frac{U(\theta,\phi)}{P_{in}} \tag{7.8}$$

where P_{in} is the total input power accepted by the antenna and U is the radiation intensity per unit solid angle. Gain does not include mismatch losses.

7.1.5 Efficiency

Total antenna efficiency defines the losses taking place during the process of converting the input power to the radiated power.

$$\eta_{total} = \eta_{radiation} \times \eta_{mismatch} \tag{7.9}$$

where $\eta_{radiation}$ is the *radiation efficiency* and $\eta_{mismatch}$ is the *mismatch efficiency*.

Radiation efficiency takes into account the conduction and dielectric losses of the antenna. It can be calculated as:

$$\eta_{radiation} = \frac{P_{rad}}{P_{in}} \tag{7.10}$$

Comparing Equations 7.7, 7.8 and 7.10 enables us to relate the antenna gain with the directivity as follows:

$$\eta_{radiation} = \frac{P_{rad}}{P_{in}} D = \eta_{radiation} D \tag{7.11}$$

Mismatch efficiency is also termed as *reflection efficiency*. It accounts for the power loss due to impedance mismatch and can be expressed as the ratio of the input power accepted by the antenna and the power supplied by the source (P_s):

$$\eta_{mismatch} = \frac{P_{in}}{P_s} = 1 - |S_{11}| \tag{7.12}$$

Ideally, the feed line is required to have a perfect match with the antenna so no reflections take place making S_{11} equal zero and results in a 100% matching efficiency. However, it is practically not possible due to material limitations.

7.1.6 Polarization

A radio wave propagating over a long distance takes the form of a plane wave. A typical plane electromagnetic wave has in-phase electric and magnetic fields of constant amplitude. They are orthogonal to each other as well as to the direction of propagation, z, as illustrated in Figure 7.6.

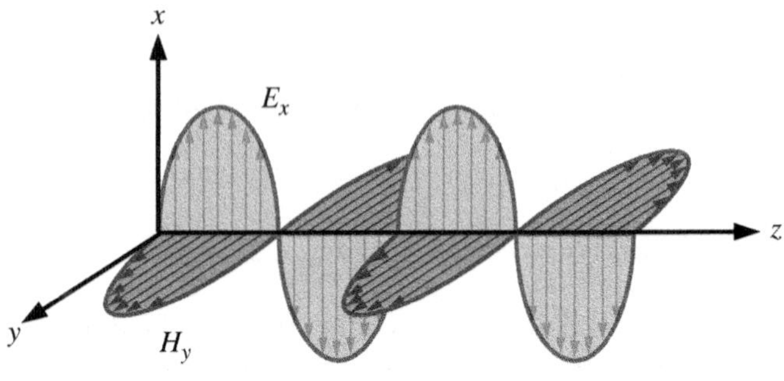

Figure 7.6 Propagation of a plane wave.

Polarization can be described as the property of the electromagnetic waves that defines the directional variation of electric field vector **E**. The electric field vector of the wave propagating in the z direction can be expressed as:

$$E = \hat{x}A\cos(\omega t - \beta z) + \hat{y}B\sin(\omega t - \beta z) \tag{7.13}$$

where A and B represent the amplitudes of the electric field components in the x and y directions, respectively.

There are three possibilities for polarization of the electromagnetic wave. A wave having $A = 0$ or $B = 0$ carries a linear polarization. If $A = B$, the wave becomes circularly polarized. If $A \neq 0$ and $B \neq 0$ and $A \neq B$, the wave tends to have elliptical polarization.

Circular and elliptical polarizations can be classified as Right Hand Polarization (RHP) and Left Hand Polarization (LHP). An RHP wave is defined as if the right hand thumb points to the direction of propagation, while the fingers curl along the direction of electric field vector rotation. An LHP wave is defined as if the left thumb points in direction of propagation while the fingers curl along the direction of electric field vector rotation. The three possible scenarios of wave polarization are shown in Figure 7.7.

The ratio of amplitudes A to B is called the *axial ratio* and is used to describe the purity of circular polarization. It is unity (0 dB) for a perfect circular polarized wave.

$$AR = \frac{Major\,axis}{Minor\,axis} = \frac{A}{B} \tag{7.14}$$

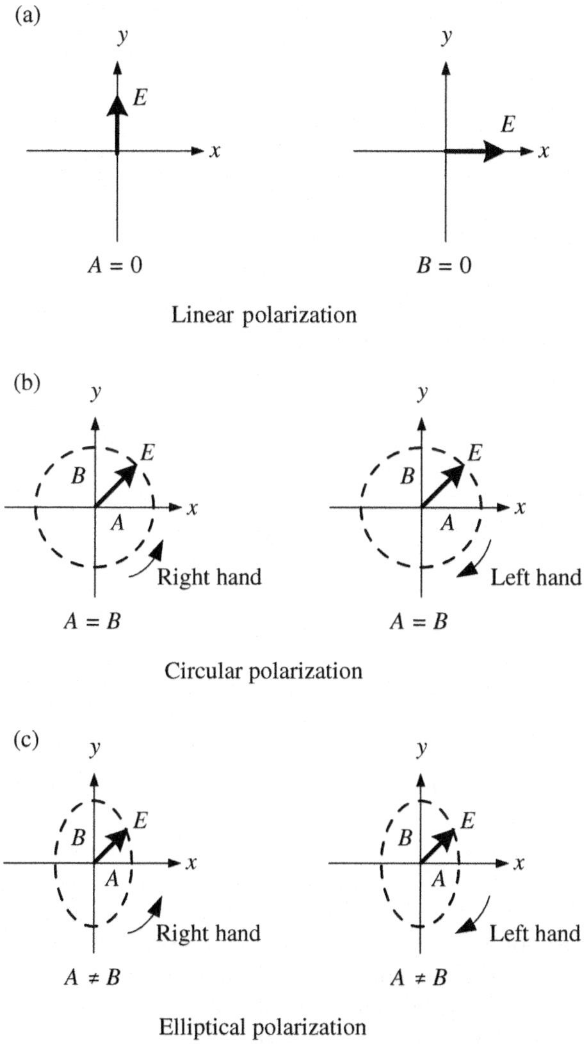

Figure 7.7 Polarization states of an electromagnetic wave.

The antenna polarization is the same as the polarization of its radiating wave. Mixed polarization may be found in many antennas since it has to meet many requirements. Moreover, some applications may also require a pure linearly polarized or circularly polarized antenna; for example, most mobile phone antennas are not purely linearly polarized as they are used to receive signals with mixed polarization. However, the polarization of the antenna needs to be matched with the incoming electromagnetic wave in order to achieve the maximum efficiency of the whole system.

7.2 Antenna Requirements for LTE Femtocells

The emergence of new mobile device categories and expansion of mobile services has brought exciting growth and innovative use of mobile broadband at an unprecedented pace. At the centre of this rapid growth is Fourth Generation Long Term Evolution (LTE). This has resulted in over 7.2 billion mobile subscription connections in the world [4]. The 4G LTE provides the foundation for 5G in the near future.

LTE is the predominant technology that emanates through the evolution of GSM and UMTS that tends towards higher data rates, reduced latency and effective connectivity. LTE is a fundamentally new technology that fully supports optimization of packet data services without any constraint of backward compatibility. The Release 8 (LTE Rel-8) provides a reasonable peak data rate of up to 300 Mbps in the downlink and 75 Mbps in the uplink. Since this did not fully capture the requirements in IMT-Advanced, Rel-10 LTE-Advanced and beyond were introduced. The increased peak data rate for downlink is 3 Gbps while the uplink is 1.5 Gbps. This is expected to increase the number of simultaneously active subscribers and improved performance at cell edges. There will be higher spectral efficiency from a maximum of 16 bps/Hz in the Rel-8 to 30 bps/Hz in Rel-10 and beyond. LTE-Advanced enhances the use of multi-antenna techniques and supports Relay Nodes and Carrier Aggregation [5–7].

LTE is expected to realize a comprehensive and secure total-IP-based mobile broadband solution to smart phones, laptop computer wireless modems and other mobile devices. Its foreseen effectiveness will help to provide facilities such as IP telephony, gaming services, ultra-broadband Internet access, stream multimedia and so on to users. The peak data rate for low and high mobility communication is 1 Gbps and 100 Mbps [8].

The requirements for LTE antennas depend on the specific application or where it will be used since there is a need to meet the increased demand for high data-rates. The antenna can be put on mobile handheld terminals, laptops, high-speed trains or cars, aboard aeroplanes and so on. Most of these applications require antennas that are low profile, low mass, can be easily fabricated, have moderate gain and efficiency, are low cost, wideband/multiband, broadband, circularly polarized (CP), ±45° dual polarized and so on.

There should always be a guide when proposing a design for suitable LTE antennas. Irrespective of the frequency bands selected, the antenna engineer must be sure of whether a multiband or broadband antenna is needed. The following requirements should be adhered to for profitable antenna design. Some of these are:

- Novel antenna technologies need to be small. Typically, the thickness of the antenna should be below 1 cm. With the need for several LTE frequency bands in handheld devices, the antenna engineer must select suitable materials that will meet the thickness requirement;
- The antenna should be simple to design and integrate with the overall system. Complex designs with difficult feeding structures should be avoided;
- Mutual coupling between different antennas must be minimized;
- Compliance with radiation hazard restrictions must be maintained;
- Channel capacity improvements using MIMO technologies are often employed and the antenna needs to facilitate that.

7.2.1 Frequency Bands

Spectrum is a big topic in LTE deployment at the moment. Over 40 frequency bands of operation have been defined in the 3GPP standard to support LTE network deployments. The International Telecommunication Union (ITU), based on the decisions reached at the World Radio Conference, allocates these frequencies. It has been up to the network operators and the regional/national telecommunications authorities to determine what frequencies will be used in practice in each nation.

Some estimates suggest a future in which LTE is deployed in many of the supported bands, possibly enabled by software-defined radio devices that can easily switch between bands. However, for the time being, most networks and devices are constrained by RF limitations to a limited number of bands.

The frequency bands allocated by the World Radio Conference for LTE are presented in Table 7.1. Some of which have been recently

Table 7.1 LTE frequency bands.

Release	Band	Common name	Uplink (MHz)	Downlink (MHz)
Rel-8	1	IMT 2.1 GHz	1920–1980	2110–2170
	2	PCS 1900 MHz	1850–1910	1930–1990
	3	DCS 1800 MHz	1710–1785	1805–1880
	4	AWS	1710–1755	2110–2155
	5	850 MHz	824–849	869–894
	6	UTRA only	830–840	875–885
	7	IMT extension 2.6 GHz	2500–2570	2620–2690
	8	GSM 900 MHz	880–915	925–960
	9	1700 MHz	1749.9–1784.9	1844.9–1879.9
	10	Extended AWS	1710–1770	2110–2170
	11	1500 MHz lower	1427.9–1447.9	1475.9–1495.9
	12	700 MHz US digital dividend (lower blocks A + B + C)	699–716	729–746
	13	700 MHz US digital dividend (upper block C)	777–787	746–756
	14	700 MHz US digital dividend (public safety)	788–798	758–768
	17	700 MHz US digital dividend (lower blocks B + C)	704–716	734–746
Rel-9	18	Japan 800 MHz lower	815–830	860–875
	19	Japan 800 MHz upper	830–845	875–890
	20	800 MHz European digital dividend	832–862	791–821
	21	1500 MHz upper	1447.9–1462.9	1495.9–1510.9
Rel-10	22	3.5 GHz	3410–3490	3510–3590
	23	2 GHz S-Band	2000–2020	2180–2200
	24	L-Band	1626.5–1660.5	1525–1559
	25	PCS 1900 + G block	1850–1915	1930–1995
Rel-11	26	Extended 850 MHz	814–849	859–894
	27	850 MHz lower	807–824	852–869

Table 7.1 (Continued)

Release	Band	Common name	Uplink (MHz)	Downlink (MHz)
	28	700 MHz APAC	703–748	758–803
	29	700 Lower SMH	N/A	717–728
Rel-12	30	2300	2305–2315	2350–2360
Rel-12	31	450	452.5–457.5	462.5–467.5
Rel-13	32	1500 L-Band	N/A	1452–1496
Rel-8	33	TDD 1900 MHz	1900–1920	1900–1920
Rel-8	34	TDD 2 GHz	2010–2025	2010–2025
Rel-8	35	TDD 1900 MHz (PCS lower)	1850–1910	1850–1910
Rel-8	36	TDD 1900 MHz (PCS upper)	1930–1990	1930–1990
Rel-8	37	TDD 1900 MHz (PCS band gap)	1910–1930	1910–1930
Rel-8	38	TDD 2.6 GHz	2570–2620	2570–2620
Rel-8	39	China TDD 1.9 GHz	1880–1920	1880–1920
Rel-8	40	China TDD 2.3	2300–2400	2300–2400
Rel-10	41	TDD 2.5 GHz	2496–2690	2496–2690
Rel-10	42	TDD 3.4 GHz	3400–3600	3400–3600
Rel-10	43	TDD 3.6 GHz	3600–3800	3600–3800
Rel-11	44	TDD 700 MHz APAC	703–803	703–803
Rel-13	45	1500 L-band(China)	1447–1467	1447–1467
Rel-13	46	5200 NII	5150–5925	5150–5925
Rel 13	65	2100 Extended IMT	1920–2010	2110–2200
Rel-13	66	1700 Extended AWS blocks A–J	1710–1780	2110–2200
Rel-13	67	700 EU 700	N/A	738–758
Rel-13	68	ME 700	698–728	753–783
Rel-13	69	2600 IMT-E (Duplex spacing)	N/A	2570–2620
Rel-13	70	1700 AWS-3	1695–1710	1995–2020

assigned, such as the 2.6 GHz IMT by the ITU to support the so-called Digital Dividend bands due to the availability from the global switch from analogue to digital television. The LTE standard defines a large number of frequency bands to allow global deployment in a variety of different spectra. This is designed to be bandwidth agnostic, permitting channel bandwidths ranging from 1.4 to 20 MHz [6].

In LTE, there is the paired spectrum that is the Frequency Division Duplex (FDD) mode and the unpaired spectrum that is the Time Division Duplex (TDD) mode. The FDD has both the downlink (DL) and uplink (UL) transmitted simultaneously on different frequencies while the TDD has the downlink and uplink transmitted at different time using the same frequency band. Mostly, the DL uses a higher frequency than UL but in Bands 13, 14, 20 and 24 the UL uses higher frequency than the DL. LTE frequency bands can be categorized into two groups; those that provide coverage and those that provide capacity. In terms of coverage, lower frequencies are needed for their propagation characteristics that accommodate larger cell sizes and better building penetration for indoor coverage. For the capacity, higher frequencies are used; they tend to include wider spectra and less spectrum fragmentation between different network operators.

To provide a comprehensive nationwide coverage, the network operator needs to use the lower frequency band, for example 700 MHz for broad coverage at a moderate data rate, and a higher frequency band, for example 2600 MHz for capacity/high data rates in a dense urban area where small cell sizes and indoor relays can be utilized. The data rate of LTE must be able to achieve peak download at 100 Mbps and peak upload of 50 Mbps. Many mobile operators are now inculcating multiple frequency bands from the LTE for effective service. In the Samsung A3 released in 2016, the LTE (FDD) Bands 20, 5, 8, 3, 1 and 7 and LTE (TDD) Band 40, LTE (Data) CAT 4 were introduced.

The antenna element needs to be designed such that it can support the device's operation in the required frequency band. If more than one frequency band is being used by the LTE device, multiple antennas resonating at different frequencies can be used. However, this method becomes trivial when strict size limitations are needed to be adhered with. Use of advanced solutions such as multiband, multimode or reconfigurable antennas turn into a more practical approach in such scenarios [9, 10]. These antennas are discussed in detail in Chapters 8, 9 and 10.

7.2.2 Form Factor and Size Limitation

The sizes of antennas range from micro-miniature to gigantic. A general proportionality between the antenna size and the wavelength at the resonant frequency exists. An antenna appreciably less than a half wavelength is termed electrically small. As the size of average handheld wireless devices continues to reduce, the space for installing the antenna on the PCB is really limited.

A major technical challenge in the LTE antenna design comes from the limitation of the size and shape of antennas and their integration with other electronic components particularly in portable devices. To tackle this challenge, the antenna designers have to exploit novel solutions that not only meet the size limitations but also offer acceptable impedance matching and radiation characteristics such as space-saving PIFAs (planar inverted-F antennas).

7.2.3 Impedance Matching, Directivity, Gain and Efficiency

The LTE antennas are typically designed to have a return loss of better than 10 dB. This corresponds to a VSWR of 2:1 and a reflected power of 11%. In the case of mobile phone applications, a 6 dB return loss or 3:1 VSWR with 25% of the reflected power is also considered acceptable. It is a relatively high level of reflection that reflects the difficulty of achieving a good match over a number of frequency bands with an electrically small antenna.

Often the device aesthetic design takes the priority over electrical considerations to appease the end user. There is no theoretical upper limit on the LTE antenna's directivity, but it again depends on the intended application. For handset and portable LTE antennas, it is assumed that the small antenna's directivity is limited by the small dipole and monopole directivities of 1.8 and 4.8 dB, respectively [11].

Practically, the antenna gain has limitations due to the issues related to minimizing the losses within the impedance matching network, achieving maximum radiation efficiency with suitable choice of conductor dimensions and dielectric materials and reducing radiation levels in undesired directions.

While matching the antenna impedance mitigates mismatch loss, there are ohmic loses due to specific implementation of a matching network. These losses reduce the radiation efficiency of the antenna, but ideally they are less than the mismatch losses. The antenna radiation efficiency is generally required to be greater than 50%. It often includes the mismatch loss while occasionally it reflects a figure averaged over each frequency band and depends on the particular requirements of the manufacturer [10].

These requirements may change depending on the usage and mechanical design, especially when a particular design of an LTE device is specified. In this case, the overall efficiency requirement may be relaxed in order to accomplish the bandwidth, which is a strong function of the size of the printed circuit board's ground plane size.

7.2.4 Directionality

Most portable LTE devices operate in a cluttered environment such as city streets and indoors where the transmitted electromagnetic waves reach the receiving terminal through multipath by going through reflections, refractions and scattering, as illustrated in Figure 7.8. This means that the signal could be arriving at the LTE device from a multitude of different directions and direction-of-arrival (DoA) cannot be defined. This scenario requires an antenna with an omni-directional pattern enabling it to receive the signal from any of these directions

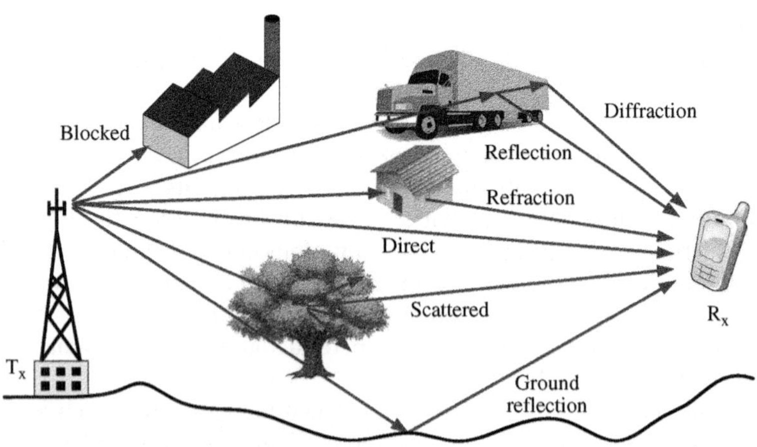

Figure 7.8 Multipath propagation.

and establish the communication link. However, such antennas usually offer limited gain values.

In scenarios where the direction of the incoming radio signal is known, for example a fixed receiving terminal or LOS links, directional antenna solutions improving the radiation in the desired direction and suppressing signals at unwanted angles become practically more viable. These antennas also offer high gain values and low interference to the neighbouring devices. An antenna designer, therefore, has to select the antenna element that suits the intended application best, given the application scenario.

7.2.5 Polarization

In many wireless communications systems where the communications link may be established through multipath, some randomness in the antenna's polarization properties may be desirable. Ideally, the LTE antennas should have a uniform radiation pattern over an entire upper hemisphere to receive the incoming communications signal efficiently. A good rejection of cross polarization is also required to avoid multipath interference [12, 13]. However, these requirements are difficult to fulfil in portable devices, especially smart phones that are required to allow maximum mobility of the user and flexibility of use with multiple functions such as Wi-Fi, Bluetooth, digital camera, mobile TV and navigation [14].

In the common scenarios of cluttered environments including indoors and on city streets, Line-of-Sight (LOS) signals arriving at the mobile device are weak while the reflected signals may have arbitrary polarizations. Moreover, mobile phones are rarely used in a fixed position and orientation of the antenna is not strictly defined. Furthermore, the antenna suffers from electromagnetic absorptions and shielding of clear sky view in hand-held positions. Establishing a quick communications link is, therefore, a difficult task in such devices. Use of wide-beam linearly polarized LTE antennas could help to address the uncertainty of antenna orientation, blockage of LOS signal and clear sky view, as well as losses due to user's body. Hence, linearly polarized antennas are a preferred choice for mobile terminals as they give better performance compared to the circular and elliptical antennas [15]. Fixed terminals, such as access points, however, can make use of omni-directional or directional radiation patterns with a choice of circular polarization or linear polarization depending on the location of the access point and intended coverage area. Therefore, antenna designers have to adhere to the required polarization requirements when designing the antenna.

7.2.6 Human Body Effects and Specific Absorption Rate (SAR)

The human body is an inherent part of portable LTE applications. Electromagnetic distortions caused by lossy human body tissues is an important issue to deal with. Performance of the antennas operating in close proximity to the human user is degraded due to losses caused by varying electric properties of human tissues resulting in distortion in radiation pattern, reduction in radiation efficiency and de-tuning of input impedance [16–19]. The antenna designers have to consider the potential electromagnetic interaction of the LTE antennas with the human user and its aftermaths in order to ensure overall system reliability and robustness.

Moreover, the fears for the safety of the human user have resulted in standardization of maximum levels of expositions of human tissues to electromagnetic radiations emitted by the portable antennas defined in terms of Specific Absorption Rate (SAR) [20–23].

The SAR is used as a figure of merit to define the level of safety of human body exposure to the electromagnetic devices including antennas. Average absorption of RF energy over a volume is termed volume-averaged SAR and can be calculated as follows:

$$SAR = \frac{\sigma}{\rho}|E|^2 \left(W/kg\right) \tag{7.15}$$

where σ is conductivity of the tissue in S/m, ρ is mass density of the tissue in kg/m^3 and E is total RMS electric field strength in V/m.

Different standardization bodies use certain SAR limits for wireless devices to ensure the safety of the user in terms of maximum absorption in a 1 or 10 g cube anywhere in a given volume, termed the spatial-peak SAR. The US and a number of other countries follow The Federal Communications Commission's (FCC) recommended SAR limit of 1.6 W/kg for a 1 g volume-averaged SAR. European countries have adopted an SAR limit of 2 W/kg for 10 g volume averaged SAR as recommended by the International Commission on Non-Ionising Radiation Protection (ICNIRP) and The Institute of Electrical and

Electronics Engineers (IEEE). The FCC limit is more stringent because it is volume-averaged over a smaller amount of tissue. The whole body averaged SAR limit is 0.08 W/kg for both the FCC and ICNIRP/IEEE [20, 21, 24].

LTE devices, especially phone and laptop manufacturers, are required to ensure that their devices comply with these objective limits for safe exposure of users. The antenna designers have to test their designs against these levels and make sure that they are not posing any health hazards.

7.2.7 Multiple Input Multiple Output (MIMO)

Multiple Input Multiple Output (MIMO) antennas are another consideration for LTE systems that require high data rates, improved system capacity and wider coverage. The MIMO approach makes use of spatial diversity and requires two or more antennas at both the transmission and reception side as depicted in Figure 7.9 [25].

By using multiple antennas along with some complex digital signal processing, MIMO enables the system to set up multiple data streams on the same channel. It results in an increased data capacity of the channel. MIMO operation requires that the transmissions from each antenna are uniquely identifiable enabling each receiver to determine what combination of transmissions it has received. This identification

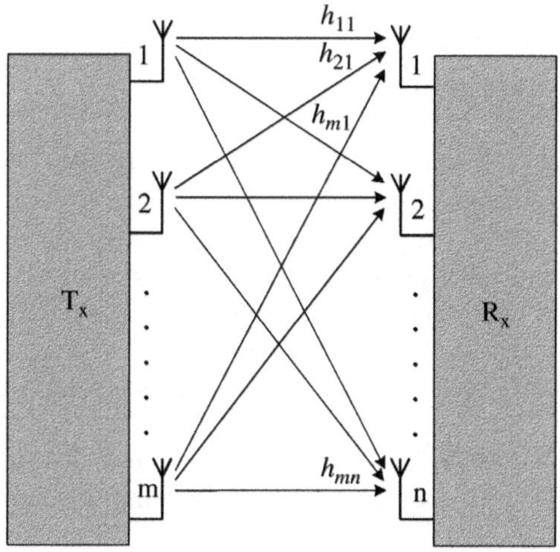

Figure 7.9 A typical MIMO arrangement in LTE systems.

can be achieved using pilot signals employing orthogonal patterns for each antenna. Although MIMO enables high data rates and improved spectral efficiency, it adds to system complexity in terms of hardware and processing.

There are a number of challenges in designing antennas for an LTE MIMO system, including antenna matching, isolation between the antenna elements, cross-correlation and interactions with other electronic components on the device. A number of techniques are proposed by the researchers to deal with these issues such as placing the antenna elements at half a wavelength apart, using branch line hybrid with passive inductors and capacitors to decouple the antenna ports and using orthogonally polarized elements for port isolation [26–28].

Along with MIMO, beamforming and diversity approaches are also recommended to improve signal robustness and the LTE system capacity [25]. Beam-forming exploits correlation such that the radiation pattern from a transmitter is directed towards the receiver. It is achieved by utilizing time delays to a calibrated phased array of antennas. The effectiveness of beamforming varies with the number of antennas. Gain improvement is small with just two antennas but considerable gain enhancements are possible with four antennas. Diversity is achieved by utilizing multiple antennas spaced or polarized appropriately in order to provide protection against fading.

These issues are further addressed in the following chapters by exploring the suitability of various antenna types to LTE solutions.

References

1 Y. Huang and K. Boyle, *Antennas: From Theory to Practice*, John Wiley & Son, Ltd, Chichester UK, 2008.
2 W. L. Stutzman and G. Thiele, *Antennas Theory and Design*, John Wiley & Son, Ltd, Chichester UK, 1981.
3 "IEEE standard definitions of terms for antennas, IEEE Std 145-1993, pp. 1–32, July 1993.
4 G. Americas, Mobile broadband evolution towards 5G rel-12 & rel-13 and beyond, 2015.
5 J. Wannstrom, "Carrier aggregation explained," 3GPP A Global Initiative: The Mobile Broadband Standard, 2013.

6 J. L. Burbank, J. Andrusenko, J. S. Everett, and W. T. M. Kasch (Eds), *Wireless Networking: Understanding Internetworking Challenges*, John Wiley & Sons, Inc., Hoboken NJ, USA, 2013.
7 S. Sesia, I. Toufik, and M. Baker, *LTE - The UMTS Long Term Evolution: From Theory to Practice* (2nd edition). John Wiley & Sons, Ltd, Chichester UK, 2011.
8 A. Anpalagan, M. Bennis, and R. Vannithamby, *Design and Deployment of Small Cell Networks*, Cambridge University Press, UK, 2015.
9 Z. N. Chen, *Antennas for Portable Devices*, John Wiley & Sons, Inc. Ltd, Chichester, UK, 2007.
10 D. A. Sanchez-Hernndez, *Multiband Integrated Antennas for 4G Terminals*, Artech House, Inc., USA, 2008.
11 A. R. Lopez, "Review of narrowband impedance-matching limitations," *IEEE Antennas and Propagation Magazine*, vol. 46, no. 4, pp. 88–90, August 2004.
12 W. Jakes, *Microwave Mobile Communications*, John Wiley and Sons Inc., New York, USA, 1974.
13 K. Kalliola, K. Sulonen, H. Laitinen, O. Kivekas, J. Krogerus, and P. Vainikainen, "Angular power distribution and mean effective gain of mobile antenna in different propagation environments," *IEEE Transactions on Vehicular Technology*, vol. 51, no. 5, pp. 823–838, September 2002.
14 G. Miller, "Adding GPS applications to an existing design," *RF Design*, pp. 50–57, March 1998.
15 T. Haddrell, N. Ricquier, and M. Phocas, "Mobile-phone GPS antennas: Can they be better?" *GPS World*, February 2010.
16 P. S. Hall and Y. Hao, *Antennas and Propagation for Body-Centric Wireless Networks* (2nd Edition), Artech House, 2012.
17 Q. H. Abbasi, M. Ur Rehman, K. Qaraqe, and A. Alomainy, *Advances in Body-Centric Wireless Communication: Applications and State-of-the-Art*, The IET, 2016.
18 X. Chen, C. G. Parini, B. Collins, Y. Yao, and M. U. Rehman, *Antennas for Global Navigation Satellite Systems*, John Wiley and Sons, Ltd, Chichester, UK, 2012.
19 S. Curto and M. J. Ammann, "Electromagnetic interaction between resonant loop antenna and simulated biological tissue," *Microwave and Optical Technology Letters*, vol. 48, no. 12, pp. 2418–2421, September 2006.

20 "IEEE standard for safety levels with respect to human exposure to radiofrequency electromagnetic fields, 3 kHz to 300 GHz," IEEE Standards, C95.1, 1999.

21 "ICNIRP guidelines for limiting exposure to time-varying electric, magnetic and electromagnetic fields (up to 300 GHz)," ICNIRP website, http://www.icnirp.org/cms/upload/publications/ICNIRPemfgdl.pdf).

22 "Considerations for the evaluation of human exposure to electromagnetic fields (EMFs) from mobile telecommunication equipment (MTE) in the frequency range from 30 MHz-6GHz," CENELEC, European Specification, Ref. No. ES-59005:1998 E, 1998.

23 A. D. Ball, N. E. Evans, S. J. Burgess, and W. G. Scanlon, "Head-SAR dependence on spectacle frame shape for operator of 450 MHz personal radio," *IET Electronics Letters*, vol. 43, no. 20, pp. 1063–1065, September 2007.

24 "Evaluating compliance with FCC guidelines for human exposure to radiofrequency electromagnetic fields," FCC, 1997.

25 J. Whitacre and M. Rumney, "Analyze antenna approaches for LTW wireless systems," *Microwaves and RF*, August 2008.

26 R. A. Bhatti, S. Yi, and S. O. Park, "Compact antenna array with port decoupling for LTE-standardized mobile phones," *IEEE Antennas and Wireless Propagation Letters*, vol. 8, pp. 1430–1433, 2009.

27 I. Dioum, A. Diallo, C. Luxey, and S. M. Farsi, "Compact dual-band monopole antenna for LTE mobile phones," in *2010 Loughborough Antennas Propagation Conference*, Nov 2010, pp. 593–596.

28 Q. Rao and D. Wang, "A compact dual-port diversity antenna for long-term evolution handheld devices," *IEEE Transactions on Vehicular Technology*, vol. 59, no. 3, pp. 1319–1329, March 2010.

8

Multiband Antennas for LTE Femtocells

Masood Ur Rehman[1] *and Xiaodong Yang*[2]

[1] *School of Computer Science and Technology, University of Bedfordshire, Luton, UK*
[2] *School of Electronic Engineering, Xidian University, Xi'an, China*

Provision of multiple services operating at different frequencies while keeping the overall size of the wireless devices within certain limitations (that are often driven by commercial aspects and user acceptability factors) is an apparent challenge in femtocell antenna design. Embedding multiple antennas makes these restrictions harder to fulfil. A single antenna supporting multiband operation is, therefore, a good alternative. This chapter explores various requirements for such antennas along with design challenges. Two antennas are studied as test cases to understand their working and benefits.

8.1 Fundamentals of Multiband Antennas

Rapid growth of wireless communications has increased the demand for mobile terminals and femtocell devices offering multiple features.

Emergence of new technologies necessitates them to integrate multiple services; for example, coverage of different cellular frequencies, availability of navigation and positioning and inclusion of Wi-Fi. It requires the device to operate at various LTE frequencies along with Wi-Fi, Bluetooth, GPS and so on as discussed in Chapter 7, but while offering a small form factor [1]. They are required to have small and

LTE Communications and Networks: Femtocells and Antenna Design Challenges, First Edition. Edited by Masood Ur Rehman and Ghazanfar Ali Safdar.

compact size, low cost and be light weight in order to guarantee the ease of mobility and flexibility of use [2].

Being an integral part of any communication device utilizing the wireless mode, antennas play a key role. Performance of the mobile terminals depends on the efficient working of the antenna. Continuous miniaturization of the RF components has enabled the realization of smaller wireless devices but has put corresponding demand for a similar decrease in the size of the antenna while maintaining high performance. The antenna design to make the device fulfil specifications of operational frequencies, size, weight and cost has added to the complexity of the antenna design process.

It is obvious that a single antenna for single service solution would not only unnecessarily add to the manufacturing cost and weight but also degrades the overall performance of the device due to mutual coupling. Therefore, provision of multiple services using a single antenna operating at different frequencies is considered a good alternative to this problem [1, 3–9].

An antenna that covers more than one of the wireless communications frequencies is termed a *multiband* antenna. For example, an antenna simultaneously operating at frequencies of 824–960 MHz and 1.71–1.99 GHz can be considered a quad-band antenna due to its coverage of the GSM800/900/1800/1900 bands.

8.1.1 Multiband Techniques

Widespread use of GSM in two frequency bands gave rise to the need for a multiband antenna that can operate at two frequencies using a single feed.

Multiband antennas can be realized by using several techniques. Two basic methods are as follows:

- Using higher order resonators;
- Employing multiple resonant structures.

8.1.1.1 Higher Order Resonances

Use of higher order resonances enables an antenna to exhibit multiband operation. This phenomenon is present in various types of resonant structures including monopoles, dipoles, patches, slots and dielectric resonators. Variations in these basic structures can be employed to achieve multiband operation.

Let us consider the example of a quarter wavelength monopole. Such a monopole resonates at a frequency f_o when the electric field is at a minimum and current is maximum at the feed point. A similar profile of the electric field and current can be achieved when the antenna length becomes $3\lambda/4$, making the antenna resonate at a second frequency that is three times that of the first, $3f_o$ [10].

Creating slots in a patch also introduces multiple resonances due to availability of multiple paths for the current. It is also a popular technique to generate higher resonance modes in multiband antennas. Figure 8.1 illustrates the higher order resonant modes for monopoles of different lengths and a slotted patch.

8.1.1.2 Multiple Resonant Structures

Using multiple resonant structures located in close proximity is the most popular method to realize a multiband antenna [10]. The structures would need to have different lengths in order to support different resonances. For example, in the case of two monopoles, the larger element will operate at a lower frequency (f_1) while the smaller element will work at the higher frequency ($3f_2$). Monopoles can be replaced by other types of structures, like meander lines, helices, spirals, rings and patches to make efficient use of space and contain the mutual coupling.

These structures can be excited in two possible ways; using a single feed to excite all of them by employing a centrally fed design or end-fire concept, or making use of proximity coupling where one of the elements is fed directly while other elements are placed in the near field of the first, acting as the parasites. Figure 8.2 shows the two feeding arrangements for the monopole example.

8.2 Types of Multiband Antennas

Patch antennas always serve as a popular choice for portable devices due to ease of design and integration and low cost [11]. Multiband antenna research can be classified into three categories; planar inverted F antennas (PIFAs), monopole antennas and slot antennas [3]. Parasitic antenna structures where the main radiator operates at the low and first high band while the shorted parasitic radiator operates at the higher bands is another popular multiband antenna design [12, 13]. A stacked patch approach to obtain distinctive multi-frequency operation is also considered an efficient technique [7].

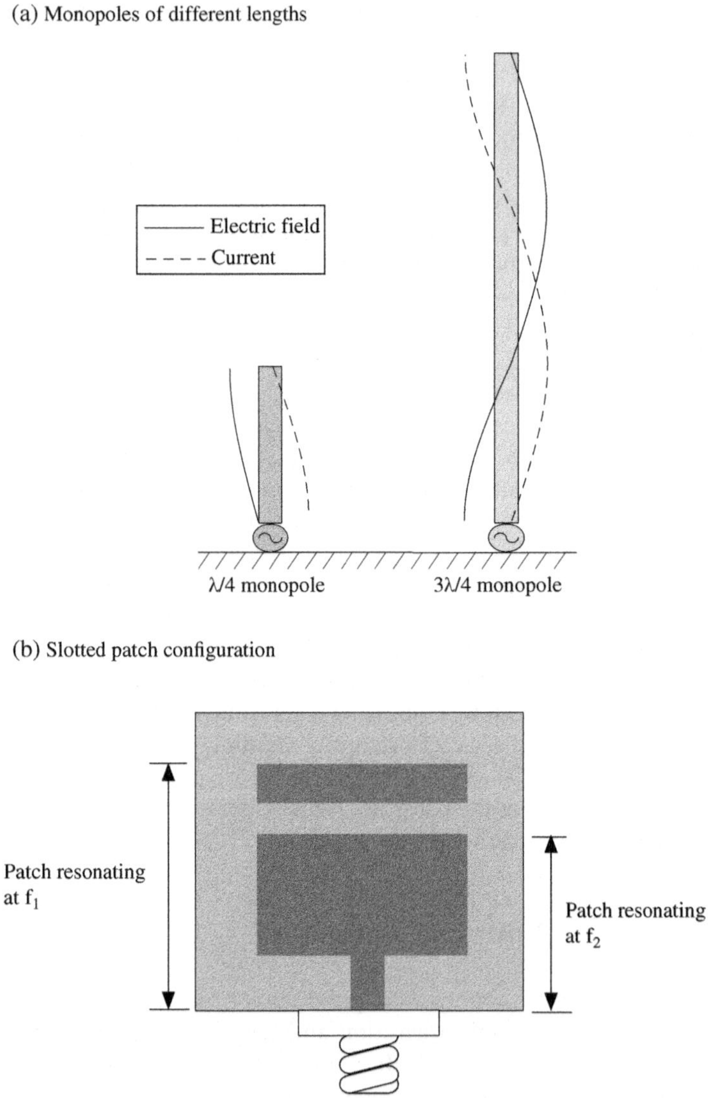

Figure 8.1 Generation of multiple resonances for multiband antenna operation via higher order resonance techniques.

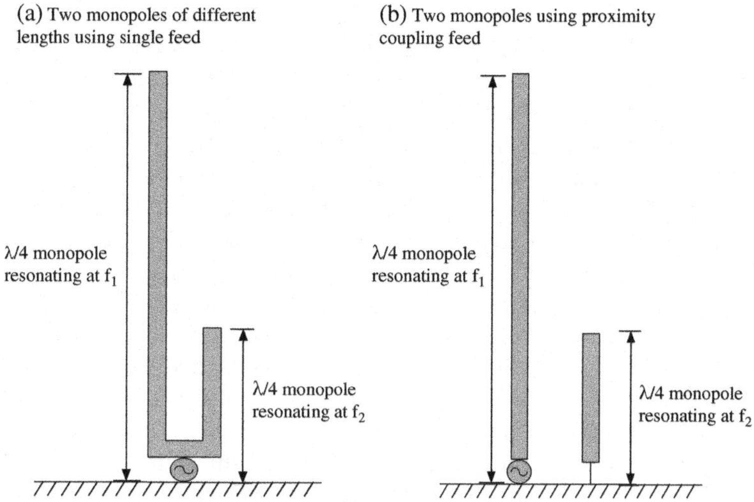

Figure 8.2 Multiple resonant structures for multiband antenna operation using single feed and proximity feed methods.

Researchers have shown deep interest in multiband antennas recently and several different multiband antennas have been proposed [5–9, 14–23]. Different designs of PIFA for multiband frequency band operation are presented in open literature [18, 19]. Wong *et al.* have proposed a thin planar monopole antenna covering four communication frequencies [20] while See *et al.* have proposed a crescent-shaped planar monopole antenna [21]. Chen *et al.* have proposed a modified T-shaped planar monopole antenna for Universal Mobile Telecommunications Systems (UMTS) and 2.4/5-GHz WLAN frequency bands [5]. This design uses two asymmetric horizontal strips as additional resonators to produce the lower and upper resonant modes for dual band operation. To make the antenna work in multiple bands, the authors proposed widening of the right horizontal strip and use an L-shaped notch in the right horizontal strip.

Hsieh *et al.* have presented a multiband antenna design combining loop and monopole structures [22]. Use of two folded L-shaped strips connected by a short stub and fed from a common excitation has been discussed by Rao *et al.* [6] for multiband operation covering the 0.9, 1.8 and 5 GHz frequency bands. By placing inverted-L- and T-shaped parasitic elements at both the radiating apertures of a microstrip patch antenna, a multiband microstrip antenna has been developed for operation at 2.0175 GHz), 2.45 and 3.5 GHz frequency bands in [8]. The resonance in the parasitic elements occurs through coupling and perturbations to the microstrip patch. A triple band antenna composed of three radiating elements: a short and a long circular radiating element to cover the 2.4, 5.2- and 5.8-GHz bands; an extra radiating element is then added to allow the antenna to cover the WiMAX band at 3.5 GHz has been proposed by Mehdipour *et al.* [9]. An additional strip is used as an interconnection between the two circular elements to control the 5.2- and 5.8-GHz performance.

Falade *et al.* have employed the concept of multi-stacked patches fed through a single coaxial probe to achieve triple operating frequency bands for GPS including L1 (1.575 GHz), L2 (1.227 GHz) and L5 (1.176 GHz). This design has three patches stacked together with a slit and symmetry I-slot [7].

Casula *et al.* have proposed an antenna fed through a proximity-coupled microstrip line. The antenna uses a flexible substrate and operates in WLAN S-band (2.45 GHz), WLAN C-band (5.2 GHz) and the Wi-Max 3.5 GHz band. Variations of general planar differential antenna topology implemented on the FR4 substrate are discussed for multiband operation in [24]. A multiband antenna made by means of compact multiresonator planar elements is discussed in [25]. The antenna employs two horizontal U-rings on an FR-4 substrate, a shorting line and a coaxial feed to generate resonances in the LTE700 (698–787 MHz), LTE2300 (2.3–2.4 GHz), LTE2500 (2.5–2.69 GHz), WiMAX3500 (3.39–3.60 GHz) bands, and a ground plane. Asghar *et al.* have proposed a multiband antenna that is comprised of a chopped circular radiator appended with a meander line and an L-strip coupled element. The antenna covers the LTE750, GSM850, GSM900, DCS, UMTS-2110, Bluetooth, WLAN and WiMAX bands [26].

These designs for multiband operation either employ complex structures or have lower values of gain and efficiency. Continuous growth of Wireless Personal Area Networks in LTE systems, however, necessitates novel solutions.

8.3 Multiband Antenna Design: Case Studies

This section explores the design challenges and performance evaluation of two multiband antennas supporting LTE operation in femtocells.

8.3.1 Multi-Slot Antenna

Design of a multiband microstrip patch antenna is studied as the first case study. The antenna make use of straight and circular slots to support the operation at 4G/LTE/CDMA (2.1 GHz) and WLAN (3.6/4.9/5 GHz) [27] frequency bands while a novel 'omega-shaped' slot is used for better impedance matching [28]. It offers compact size, good impedance matching with 2:1 VSWR and radiation pattern performance. Simple design and good gain and efficiency characteristics make this antenna a promising candidate for mobile terminal applications.

8.3.1.1 Antenna Geometry

The multi-slot antenna has a ground plane size of 120×50mm to replicate the usual size of a mobile terminal (especially the iPhone). The patch is located on the top edge of the PCB fed through a 50 omega coaxial cable. The antenna uses a FR4 board with a relative permittivity of 4.4, loss tangent of 0.0245 and thickness of 1.6 mm as the substrate. The antenna geometrical structure is shown in Figure 8.3. The dimensions of the designed antenna structure are summarized in Table 8.1.

The radiating element is 40×34mm in size. The antenna is fed using a 50 Ω coaxial feed. The position of the coaxial feed is optimized to get a good impedance match, hence it is located near the right lower corner of the radiating element. The antenna employs four slots in order to generate multiple frequency band operation. Two slots are circular in shape, one depicts the Greek letter 'omega' as illustrated in Figure 8.3. The width of each slot is 1 mm.

The multiband frequency operation is supported by the different geometrical shapes such that the rectangular patch generates the resonance at the lower frequency band, the two circular slots at middle frequencies and the straight slot at the higher frequency band. The 'Ohm Slot' is used to improve the impedance matching at the lower band.

8.3.1.2 Antenna Performance Evaluation

The CST Microwave Studio electromagnetic solver package is used to model the antenna numerically and perform the simulation. This commercial software is based on the Finite Integration Technique (FIT) for the solution of Maxwell's equations [29]. The performance of the antenna is studied and analysed using simulation results of the antenna reflection coefficient, impedance matching, bandwidth, radiation pattern, efficiency and gain.

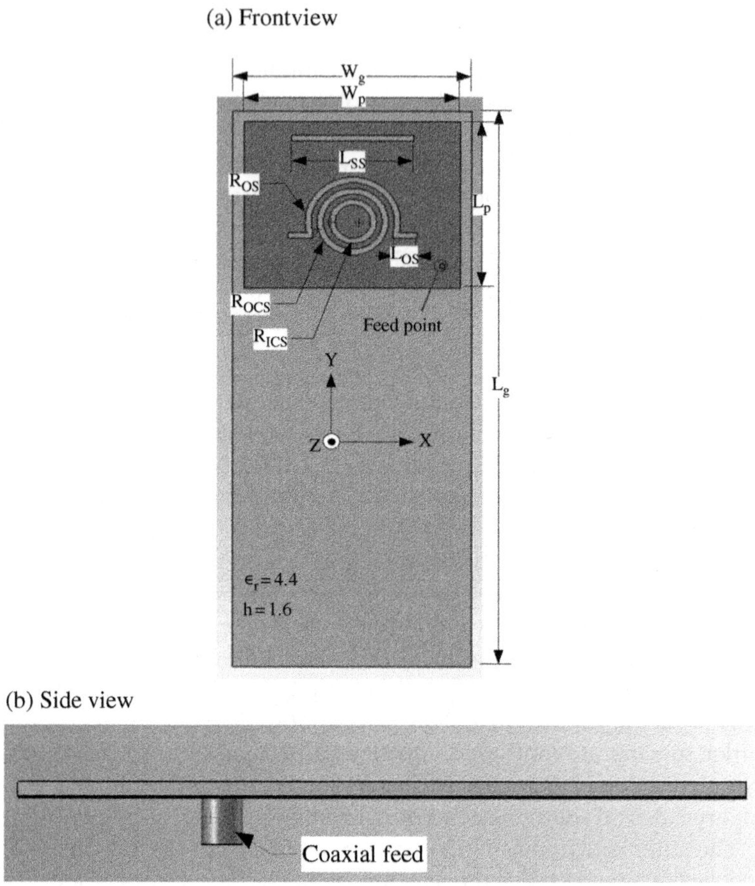

Figure 8.3 Geometry and dimensions of the multislot patch antenna for multiband operation (all units are in mm) [28].

The simulated reflection coefficient response of the antenna is depicted in Figure 8.4. It shows that the antenna performs well enough to operate at the desired frequencies for 4G/LTE/CDMA and WLAN operation. The antenna resonates at 2.1 GHz, 3.6, 4.9 and 5.7 GHz frequencies for the required services. It shows good impedance matching at −10 dB (VSWR 2:1). The observed impedance bandwidth at the different frequencies of operation is 90 MHz at 2.1 GHz (2.08–2.17 GHz), 100 MHz at 3.6 GHz (3.55–3.65 GHz), 60 MHz at 4.9 MHz (4.88–4.94 GHz) and 80 MHz at 5.7 GHz (5.68–5.76 GHz).

Table 8.1 Optimized design parameters for the multislot patch multiband antenna

Parameter	Dimension (mm)
L_g	120
W_g	50
h	1.6
L_p	40
L_p	34
L_{SS}	21
L_{OS}	4
R_{OS}	9
R_{OCS}	7
R_{ICS}	5

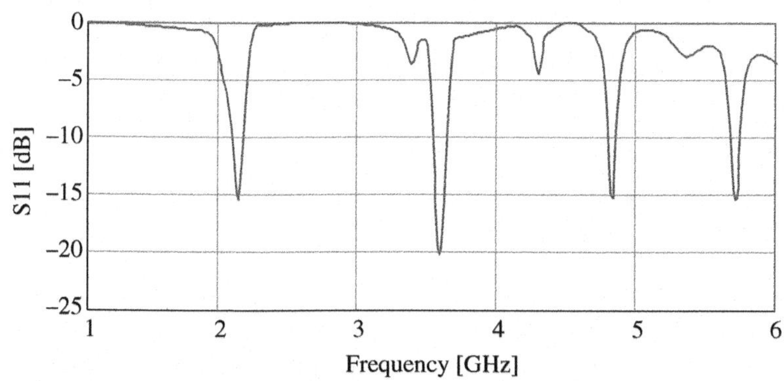

Figure 8.4 Reflection coefficient response of the multislot patch antenna [28].

The radiation characteristics of the multislot antenna are analysed by studying the 3D far-field radiation patterns at 2.1, 3.6, 4.9 and 5.7 GHz. The simulated 3D radiation patterns are illustrated in Figure 8.5. Although there is slightly directive radiation at the higher frequencies, it appears that the antenna has good broadside radiation patterns for all the desired frequency bands.

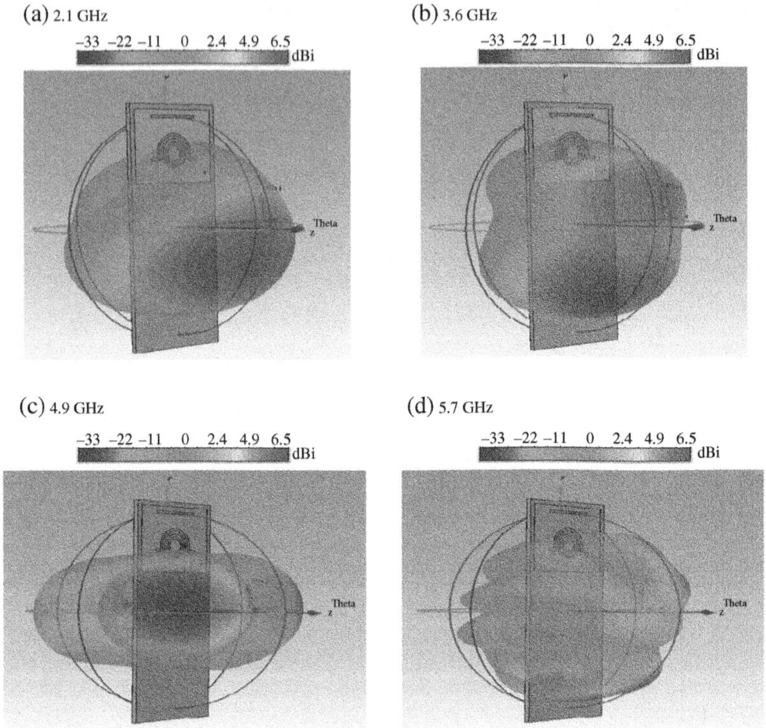

Figure 8.5 Three-dimensional radiation patterns of the multi-slot antenna for multiband operation at 2.1, 3.6, 4.9 and 5.7 GHz [28].

The simulated results for the antenna peak gain and efficiency are plotted in Figures 8.6 and 8.7 of the proposed slotted patch antenna, respectively. These results show that the antenna offers high peak gain values at all the operating frequencies. The peak gain value of 6.1 dBi at 2.1 GHz, 6.5 dBi at 3.6 GHz, 5.9 dBi at 4.9 GHz and 5.6 dBi at 5.7 GHz is observed. The antenna also exhibits good efficiency values ranging from 70 to 95% at the various resonating frequencies.

The multislot low profile slotted-patch microstrip antenna makes an efficient use of straight and circular slots arrangement to achieve multiband performance. The antenna operates at four different frequency bands with centre frequencies at 2.1, 3.6, 4.9 and 5.9 GHz for 4G/LTE/CDMA and WLAN services. The antenna excellently meets the 2:1 VSWR requirement, exhibiting –10 dB impedance bandwidth;

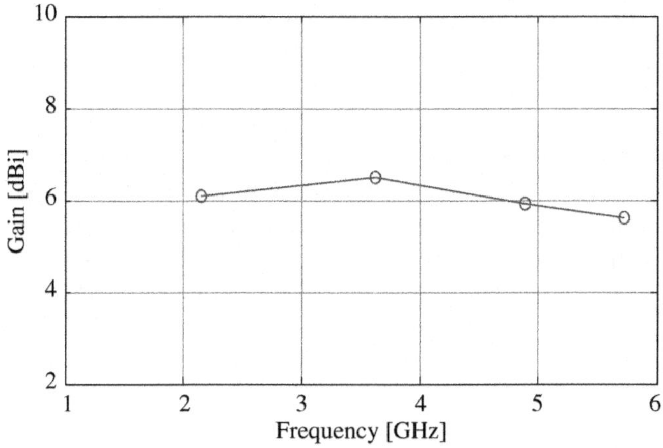

Figure 8.6 Simulated peak gain values of the proposed multiband slotted patch antenna [28].

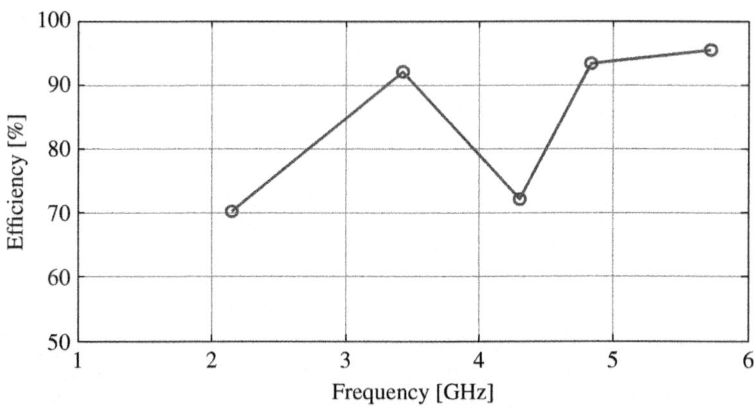

Figure 8.7 Simulated efficiency of the proposed multiband slotted patch antenna [28].

enough to support 4G/LTE/CDMA and WLAN services at these frequencies. It also offers good broadside radiation characteristics with high peak gain gain value of 6.5 dBi (maximum at 3.6 GHz) and efficiency of more than 70%. Small structure, easy design and low profile makes this antenna well-suited to support multiservice operation. It makes this antenna a promising candidate to be used in fixed and mobile wireless hand-held devices and mobile terminals.

8.3.2 Patch-Loop Combination Antenna

Design of a printed microstrip patch antenna operating at multiple frequency bands is discussed in this section. The antenna has a very simple structure combining a microstrip patch, two rings and a slot [30]. The antenna covers GPS (1.575 GHz), 4G/LTE/CDMA (2.1 GHz) and Wi-Fi (3.6/5.3 GHz) frequency bands with good impedance matching (2:1 VSWR) and radiation pattern performance. It has an omni-directional radiation pattern at most of the desired frequencies of operation. The antenna has good efficiency ranging from 60 to 98% and exhibits high peak gain values of 4.1, 5.7, 5.1 and 7.9 dBi at 1.575, 2.1, 3.68 and 5.37 GHz, respectively.

8.3.2.1 Antenna Configuration

The structure and dimensions of the patch-loop combination multiband microstrip patch antenna are depicted in Figure 8.8. The radiating element of the antenna is a combination of a microstrip patch, two rings and a slot. The antenna is printed on a commonly used FR4 substrate with a relative permittivity of 4.4 and loss tangent of 0.0245. The thickness of the substrate is 1.6 mm. The size of the ground plane is 50 × 50 mm. The antenna is fed using a 50 Ω coaxial line as shown in Figure 8.8(b).

The radiating patch has a size of 27.7 × 31 mm and it is slightly off-centre. Two rings with an inner radius (I_r) of 1.1 mm and outer radius (O_r) of 5.4 mm are placed at top and bottom edges of the diagonal. A slot is included near the top ring of the antenna. The width of the slot S_W is 2 mm. The dimensions and position of different elements of the antenna are illustrated in Figure 8.8(a).

The patch generates the resonance at the lower frequency band. The corners of the patch are truncated to improve the impedance matching and attain good axial ratio at the GPS band of operation. The rings and the slot help to generate higher resonances and improve the impedance bandwidth. The off-centre position of the patch also helps to improve the antenna performance by shifting the resonances towards required frequencies of operation.

8.3.2.2 Antenna Performance

The patch-loop antenna is modelled numerically and simulated using CST Microwave Studio [29]. The antenna performance for multiple

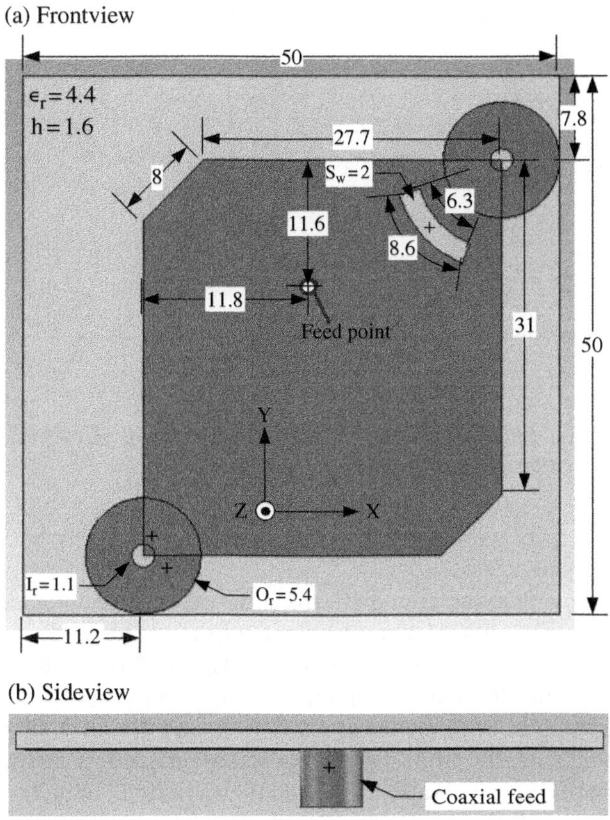

Figure 8.8 Geometry and dimensions of the patch-loop microstrip patch multiband antenna operating at GPS/4G/Wi-Fi frequencies (all units are in mm) [30].

frequency band operation is studied in terms of simulated reflection coefficient, impedance matching, bandwidth, radiation pattern, efficiency and gain.

Figure 8.9 gives the simulated reflection coefficient response. It can be observed from these results that the antenna performance is good at the desired frequencies of operation for GPS (1.575 GHz), 4G/LTE/CDMA (2.1 GHz), and Wi-Fi (3.6/5.3 GHz). It also exhibits good impedance matching of more than −10 dB (corresponding to a VSWR of 2:1) at all these frequencies.

The antenna covers a −10 dB impedance bandwidth of 30 MHz (1.56 GHz-1.59) at GPS, 40 MHz (2.07 GHz-2.11 GHz) at 4G/LTE/

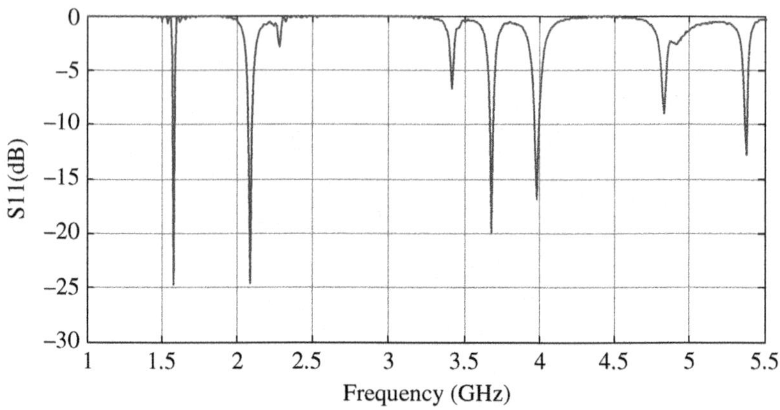

Figure 8.9 Reflection coefficient response of the patch-loop antenna for GPS/4G/Wi-Fi multiband operation [30].

CDMA, 30 MHz (3.67 GHz-3.7 GHz) and 50 MHz (5.34 GHz-5.39 GHz) at Wi-Fi frequencies.

The simulated far-field radiation patterns at the desired frequencies in the XZ (elevation) and YZ (azimuth) planes at the centre frequencies of 1.575, 2.1, 3.68 and 5.37 GHz are plotted in Figures 8.10 and 8.11. The results show that the patch-loop antenna has very good omni-directional coverage in both planes at 1.575 GHz and 2.1 GHz. The antenna exhibits a nearly omni-directional radiation pattern at the other two frequencies of operation. Overall, the radiation characteristics of the proposed antenna are very well-suited to wireless communication mobile terminals.

Figure 8.12 and Figure 8.13 present the simulated results for the peak gain and efficiency of the patch-loop antenna at GPS/4G/Wi-Fi frequencies, respectively. For GPS operation, the peak gain is observed to be 4.1 dBi. The antenna has a peak gain value of 5.7 dBi at the 4G/LTE/CDMA frequency of 2.1 GHz. Peak gain values are noted to be 5.1 dBi and 7.9 dBi for the two Wi-Fi frequencies of 3.68 and 5.37 GHz, respectively. These values show that the antenna has very good gain performance. Total efficiency values vary from 60 to 98%. At the the lower operating frequency of 1.575 GHz, antenna has a total efficiency of 60%. It is noted to be 98% at 2.1 GHz, 97% at 3.68 GHz and 88% at 5.37%.

The surface current distributions for this multiband antenna at the four considered frequencies are depicted in Figures 8.14 and 8.15. These results show that the slot helps the antenna to resonate at the

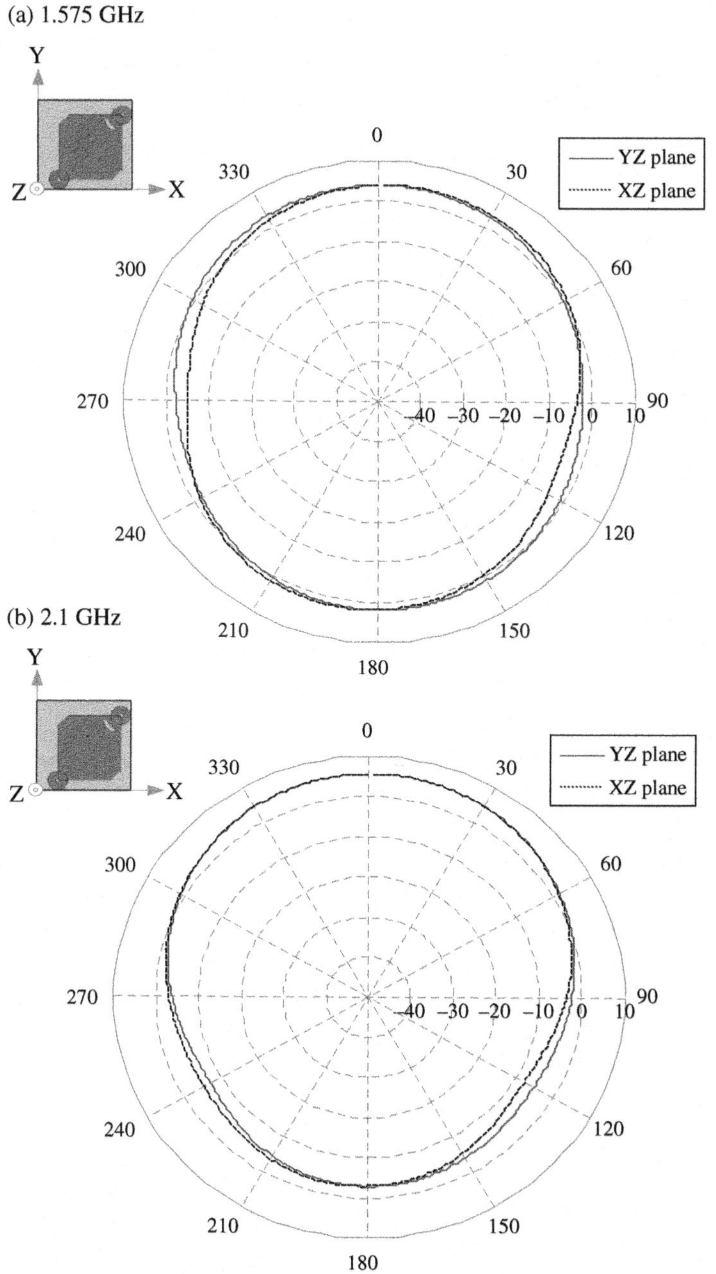

Figure 8.10 Two-dimensional radiation patterns for the patch-loop microstrip patch antenna for multiband operation at 1.575 GHz and 2.1 GHz in XZ and YZ planes [30].

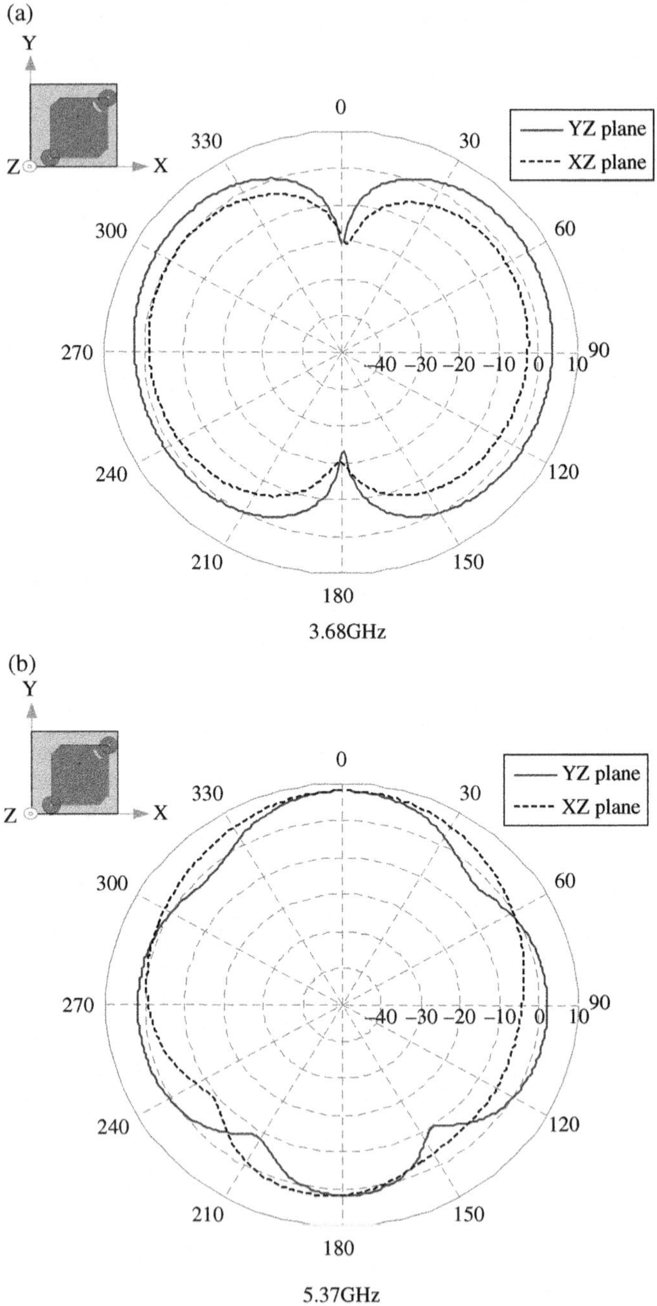

Figure 8.11 Two-dimensional radiation patterns for the proposed patch-loop antenna for multiband operation at 1.575 and 2.1 GHz in XZ and YZ planes [30].

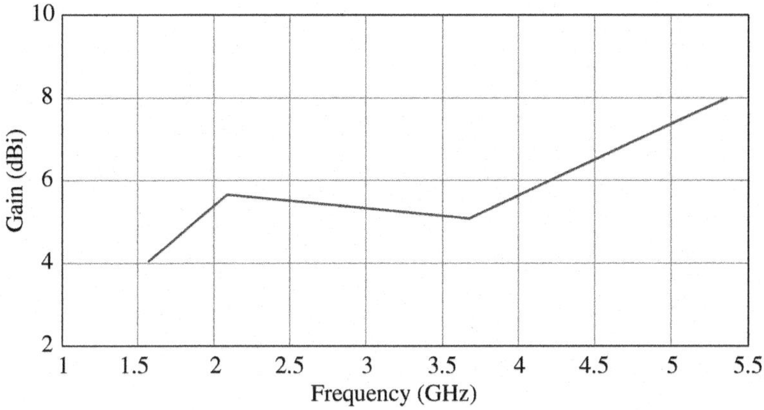

Figure 8.12 Peak gain values of the patch-loop combination antenna for GPS/4G/Wi-Fi multiband operation [30].

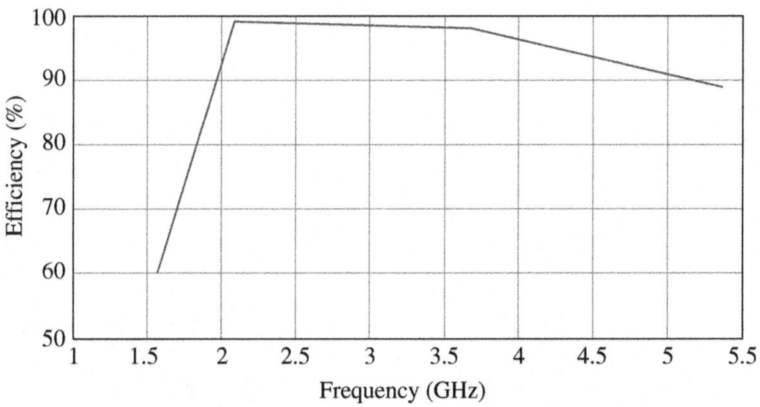

Figure 8.13 Efficiency of the patch-loop combination antenna for GPS/4G/Wi-Fi multiband operation [30].

lower frequency while the rings produce resonance at the higher frequencies, particularly at 5.37 GHz.

The analysis shows that the patch-loop antenna has a simple structure and it is easy to fabricate using widely available cost-effective FR4 PCB. The antenna has an overall compact size of $50 \times 50 \times 1.6$ mm^3. The antenna can operate at four different frequency bands for GPS, 4G/

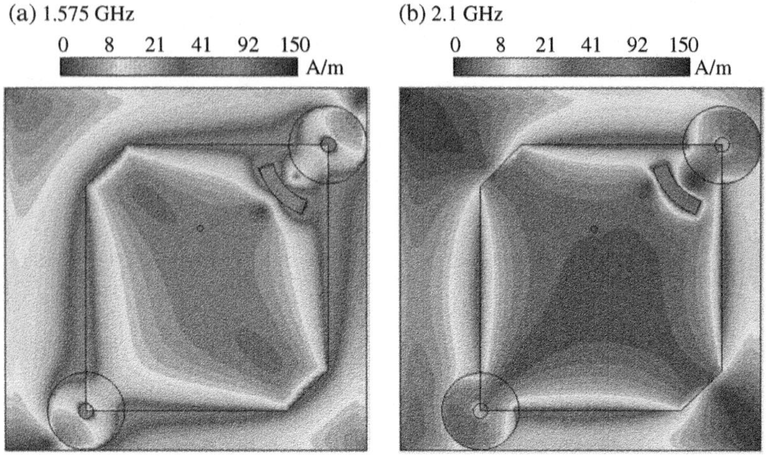

Figure 8.14 Surface current distribution at 1.575 GHz and 2.1 GHz for the patch-loop multiband antenna [30].

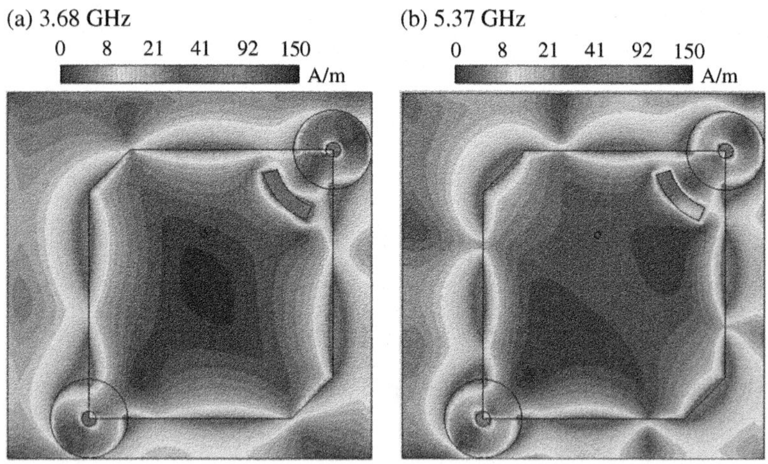

Figure 8.15 Surface current distribution at 3.68 GHz and 5.37 GHz for the patch-loop multiband antenna [30].

LTE/CDMA and Wi-Fi services with good impedance matching meeting the 2:1 VSWR threshold. Moreover, it radiates omni-directionally at the two lower frequencies while it has a nearly omni-directional pattern at the two higher frequencies. It has also achieved high peak gain and efficiency values. Good performance, simple configuration and low profile makes this antenna particularly attractive for portable as well as fixed communication devices, including femtocell access points, smart phones, personal navigation devices (PNDs), laptops and notebooks.

8.4 Open Research Issues

It has been established that multiband antennas offer a number of advantages to LTE systems. However, continuous demand for device miniaturization brings huge challenges to antenna designers. Meeting high performance thresholds with compact sizes brings great complexities. Multiband operation often brings down the antenna gain and efficiency due to mutual coupling between multiple resonators. Devising novel antenna structures offering multiband operation with high gain and efficiency is, therefore, still an open research issue.

References

1 T. Zhang, R. Li, G. Jin, G. Wei, and M. Tentzeris, "A novel multiband planar antenna for GSM/UMTS/LTE/Zigbee/RFID mobile devices," *IEEE Transactions on Antennas and Propagation*, vol. 59, no. 11, pp. 4209–4214, 2011.
2 Z. N. Chen, *Antennas for Portable Devices*, John Wiley & Sons, Ltd, Chichester, UK, 2007.
3 W. J. Liao, S. H. Chang, and L. K. Li, "A compact planar multiband antenna for integrated mobile devices," *Progress In Electromagnetics Research*, vol. 109, pp. 1–16, 2010.
4 M. Ali, G. Hayes, H.-S. Hwang, and R. Sadler, "Design of a multiband internal antenna for third generation mobile phone handsets," *IEEE Transactions on Antennas and Propagation*, vol. 51, no. 7, pp. 1452–1461, 2003.
5 S.-B. Chen, Y.-C. Jiao, W. Wang, and F.-S. Zhang, "Modified T-shaped planar monopole antennas for multiband operation," *IEEE Transactions on Microwave Theory and Techniques*, vol. 54, no. 8, pp. 3267–3270, Aug. 2006.
6 Q. Rao and W. Geyi, "Compact multiband antenna for handheld devices," *IEEE Transactions on Antennas and Propagation*, vol. 57, no. 10, pp. 3337–3339, Oct. 2009.

7 O. Falade, M. Ur Rehman, Y. Gao, X. Chen, and C. Parini, "Single feed stacked patch circular polarized antenna for triple band GPS receivers," *IEEE Transactions on Antennas and Propagation*, vol. 60, no. 10, pp. 4479–4484, 2012.

8 J.-W. Kim, T.-H. Jung, H.-K. Ryu, J.-M. Woo, C.-S. Eun, and D.-K. Lee, "Compact multiband microstrip antenna using inverted-L- and T-shaped parasitic elements," *IEEE Antennas and Wireless Propagation Letters*, vol. 12, pp. 1299–1302, 2013.

9 A. Mehdipour, A. Sebak, C. Trueman, and T. Denidni, "Compact multi-band planar antenna for 2.4/3.5/5.2/5.8-ghz wireless applications," *IEEE Antennas and Wireless Propagation Letters*, vol. 11, pp. 144–147, 2012.

10 Y. Huang and K. Boyle, *Antennas: From Theory to Practice*, John *Wiley* & Sons, Ltd, Chichester, UK, 2008.

11 J. James, P. Hall, and C. Wood, *Microstrip Antenna Theory and Design*, IET-Peregrinus, Electromagnetic Waves Series 12, 1981.

12 Y.-X. Guo, M. Y. W. Chia, and Z. N. Chen, "Miniature built-in multi-band antennas for mobile handsets," *IEEE Transactions on Antennas and Propagation*, vol. 52, no. 8, pp. 1936–1944, 2004.

13 K.-C. Lin, C.-H. Lin, and Y.-C. Lin, "Simple printed multiband antenna with novel parasitic-element design for multistandard mobile phone applications," *IEEE Transactions on Antennas and Propagation*, vol. 61, no. 1, pp. 488–491, 2013.

14 X. Q. Zhang, Y. C. Jiao, and W. H. Wang, "Compact wide tri-band slot antenna for WLAN/WiMAX applications," *Electronics Letters*, vol. 2012, no. 2, pp. 64–65, 2012.

15 G. Cappelletti, D. Caratelli, R. Cicchetti, and M. Simeoni, "Compact wide tri-band slot antenna for WLAN/WiMAX applications," *IEEE Transactions on Antennas and Propagation*, vol. 59, no. 10, pp. 3526–3535, 2011.

16 P. Liu, Y. Zou, B. Xie, X. Liu, and B. Sun, "Compact CPW-fed tri-band printed antenna with meandering split-ring slot for WLAN/WiMAX applications," *IEEE Antennas and Wireless Propagation Letters*, vol. 11, pp. 1242–1244, 2012.

17 M. J. Hua, P. Wang, Y. Zheng, and S. L. Yuan, "Compact tri-band CPW-fed antenna for WLAN/WiMAX applications," *Electronics Letters*, vol. 49, no. 18, pp. 1118–1119, 2013.

18 C. R. Rowell and R. D. Murch, "A compact PIFA suitable for dualfrequency 900/1800-MHz operation," *IEEE Transactions on Antennas and Propagation*, vol. 46, no. 10, pp. 596–598, 1998.

19 M. Martinez-Vazquez, O. Litschke, M. Geissler, D. Heberling, A. M. Martinez-Gonzalez, and D. Sanchez-Hernandez, "Integrated planar multi-band antennas for personal communication handsets," *IEEE Transactions on Antennas and Propagation*, vol. 54, no. 2, p. 384391, 2006.

20 K.-L. Wong, G.-Y. Lee, and T.-W. Chiou, "A low-profile planar monopole antenna for multiband operation of mobile handsets," *IEEE Transactions on Antennas and Propagation*, vol. 51, no. 1, pp. 121–125, 2003.

21 C. H. See, R. Abd-Alhameed, D. Zhou, T. H. Lee, and P. Excell, "A crescent-shaped multiband planar monopole antenna for mobile wireless applications," *IEEE Antennas and Wireless Propagation Letters*, vol. 9, pp. 152–155, 2010.

22 H.-W. Hsieh, Y.-C. Lee, K.-K. Tiong, and J.-S. Sun, "Design of a multi-band antenna for mobile handset operations," *IEEE Antennas and Wireless Propagation Letters*, vol. 8, pp. 200–203, 2009.

23 G. A. Casula, P. Maxia, G. Montisci, G. Valente, G. Mazzarella, and T. Pisanu, "A multiband proximity-coupled-fed flexible microstrip antenna for wireless systems," *International Journal of Antennas and Propagation*, vol. 2016, no. Article ID 8536058, 2016.

24 D. Pepe, L. Vallozzi, H. Rogier, and D. Zito, "Design variations on planar differential antenna with potential for multiple, wide, and narrow band coverage," *International Journal of Antennas and Propagation*, vol. 2015, no. Article ID 478453, 2015.

25 H. Jian-Rong, L. Jiu-Sheng, and W. Di, "A small planar antenna for 4G mobile phone application," *International Journal of Antennas and Propagation*, vol. 2016, Article ID 4791831, 2016.

26 A.M. Asghar, M. Malick, M. Karlsson, and A. Hussain, "A multi-wideband planar monopole antenna for 4G devices," *Microwave and Optical Technology Letters*, vol. 55, no. 3, pp. 589–593, 2013.

27 I.T. Union, "ITU radio regulations – Volume 1 (article 5)," *URL: http://www.itu.int.*

28 M. Ur Rehman and S. Haxha, "Design of a slotted-patch microstrip antenna for mobile terminals," in *2015 IEEE International Symposium on Antennas and Propagation USNC/URSI National Radio Science Meeting*, July 2015, pp. 1716–1717.

29 CST, "CST microwave studio user's manual," Available online. 2017.

30 M. Ur Rehman and B. Allen, "A compact multi-band slot-ring microstrip patch antenna for wireless applications," in *2013 Loughborough Antennas Propagation Conference (LAPC)*, Nov 2013, pp. 595–598.

9

Reconfigurable Antennas for LTE Femtocells

Masood Ur Rehman and Waqas Farooq

School of Computer Science and Technology, University of Bedfordshire, Luton, UK

Modern wireless devices are expected to offer a number of features and support various services that operate at distinct frequencies. They have to meet strict size limitations as well offer attractive form factor and aesthetics. Since different wireless services utilize technologies operating at different frequencies, multiple antennas need to be incorporated making the fulfilment of size limitations a challenging task. A single reconfigurable antenna supporting operation at two different frequencies using a switching mechanism is one of the available options. This chapter describes the requirements of reconfigurable antennas for LTE femtocells discussing design issues and analysis techniques.

9.1 Fundamentals of Reconfigurable Antennas

Recongurability is the ability to change fundamental operating characteristics of an antenna's radiator through electrical, mechanical or other means [1]. Reconfiguration in antennas is often achieved often by using switches, variable components or moving parts. Reconfiguration in antennas can be obtained for various antenna parameters [2–4] such as resonant frequency, polarization, impedance

LTE Communications and Networks: Femtocells and Antenna Design Challenges, First Edition. Edited by Masood Ur Rehman and Ghazanfar Ali Safdar.

bandwidth and radiation pattern. Few antennas have also been designed that can reconfigure multiple antenna parameters simultaneously [5].

Employing reconfigurability has a number of benefits. Frequency reconfigurability makes communication possible at different frequency bands using a single antenna, hence providing ease of integration, effective miniaturization, reduced interference and enhanced device performance. A frequency reconfigurable antenna can alter its resonance frequency by modifying the geometry or size of the radiating element dynamically without changing the structure of whole antenna. Such antennas are capable of operating at multiple frequency bands using an appropriate switching mechanism [6]. Moreover, the LTE portable devices are often used in cluttered environments where antenna performance is affected by the surrounding objects in an unpredicted way [7]. A reconfigurable antenna provides the ability to tune its operating frequency to the frequency band offering better coverage and filter out any interfering signals. Radiation pattern reconfiguration could be utilized to direct the antenna towards the access point resulting in low power consumption and longer battery life.

These benefits have made the reconfigurable antennas a popular choice for modern communication systems. Therefore, the development of reconfigurable antennas has seen a well deserved growth during the last two decades. They are increasingly used in applications ranging from Global Positioning System (GPS), radar and surveillance systems, satellite communication systems, cellular mobile communication systems and multi-frequency communications. However, design of these antennas is a challenging task as obtaining the desired levels of antenna operation while maintaining an overall efficient and cost-effective solution is not easy.

9.1.1 Types of Reconfigurable Antennas

Reconfigurable antennas can be characterized based on the underlying switching mechanism. There are a number of ways to make an antenna reconfigurable by changing its effective length. Some of most common techniques include:

- Use of switches;
- Structural and mechanical changes;
- Material changes.

9.1.1.1 Use of Switches

The effective length of the antenna can be changed by adding or removing part of it. It effectively shifts/tunes the operating frequency of the antenna. This can be achieved through electronic, optical and mechanical means. Commonly used switching technologies include PIN diodes, FETs, radio frequency microelectromechanical system (RF-MEMS) switches and optical switches [8–31]. Figure 9.1 shows the switching between two lengths of a simple patch antenna by means of a PIN diode providing connectivity between the two parts and, hence, changing the resonant frequency.

Capacitors can also be used as a form of variable reactive loading to change effective antenna length and realize reconfigurability [24, 32]. These allows smooth changes in the operating frequency of the antenna compared to the discrete output of optical switches, PIN diodes, FETs and RF-MEMS switches. Use of a varactor (variable capacitor) in order to tune a simple PIFA can be achieved as illustrated in Figure 9.2.

Reliability of the switching mechanism defines the efficiency of the antenna reconfiguration. Table 9.1 summarizes the pros and cons of the switching technologies that can be used in reconfigurable antennas.

9.1.1.2 Structural and Mechanical Changes

Mechanical changes in the antenna structure can also be employed to switch the operating frequency of the antenna. It can be achieved in many ways such as by using piezoelectric actuators and rotating

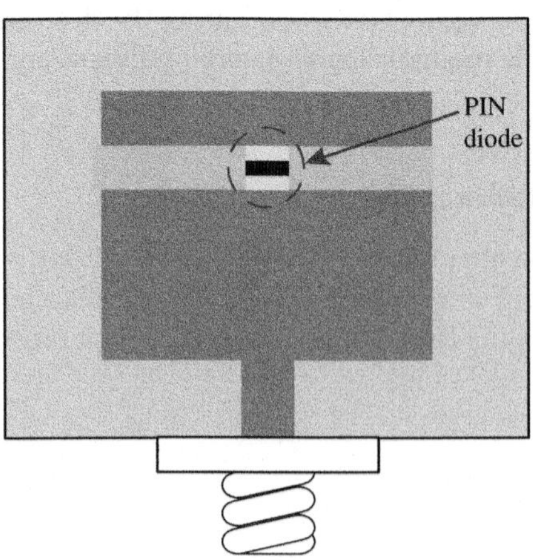

Figure 9.1 Use of PIN diode to achieve switching between two frequencies through changed effective length of a simple patch antenna.

Figure 9.2 Use of varactor to vary the effective length of a PIFA to achieve reconfigurable antenna operation.

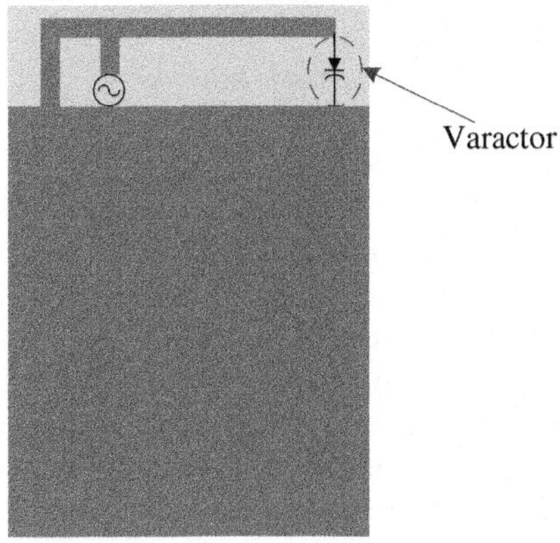

Table 9.1 Comparison of switching technologies.

Device	Advantages	Disadvantages
PIN diodes	• Low ON state loss • Low OFF state capacitance • Reasonable linearity • Mature technology	• High power consumption • Complex biasing
FETs	• Low ON state loss • Low OFF state capacitance • Good linearity • Low power consumption	• Ultra-wideband switches are not commercially available yet
MEMS	• Very low ON state loss • Very low OFF state capacitance • High linearity	• Slow switching speed • Low reliability
Varactors	• Continuous tuning • Low power consumption	• High non-linearity • Analogue bias voltage is required
Optical switches	• No biasing lines	• Difficult integration • High power optical sources are required

radiators using stepping motors [33–35]. Physical design, actuation mechanism and maintenance of the structural elements is the main challenge in this kind of antenna.

9.1.1.3 Material Changes

Changes in the material characteristics of the antenna elements, particularly relative permittivity and relative permeability, also enable variation in the effective electric length of the antenna. This results in shifting the antenna operating frequency making it reconfigurable [25, 36].

9.2 Realization of Reconfigurable Antennas

Patch antennas have copious advantages such as light weight, low cost, low profile, mechanical ruggedness and by being easy to fabricate and easy to integrate [37]. It makes them a popular choice for LTE femtocell applications. Various techniques have been proposed by researchers for achieving frequency reconfigurability including PIN diodes, RF MEMS, varactor diodes, FET switches and slotted patches [10–31].

PIN diodes are usually low cost and are available in numerous configurations and packages, therefore, the majority of frequency reconfigurable antennas are designed using PIN diodes. PIN diodes have been used as switches to transform the shape or increase the size of the radiating element an antenna [11]. Some PIN diodes are also sensitive to vibration and RF power over-stress [11]. In [12], the patch antenna was reconfigured using two different substrates and two separate feeds. The switching was performed using different PIN diodes between the two substrates. Lee *et al.* have proposed an antenna consisting of two coupled strips [13]. Reconfiguration between the lower band covering LTE700/GSM850/GSM900 and the higher band covering GSM1800/GSM1900/UMTS/LTE2300/LTE2500 has been achieved using the ON/OFF state of a PIN diode. A combination of folded loop and inverted F structure has been discussed for re-configurable operation using a PIN diode by Li *et al.* [14]. A matching bridge is used to match the impedance for both the loop and IFA modes. The antenna operates in the hepta-band, including GSM850/GSM900/GPS/DCS/PCS/UMTS/WLAN.

Combination of a U-shaped loop and monopole modes along with a diode switch is studied for the reconfigurable operation of the antenna at GSM850/GSM900/DCS/PCS/UMTS2100/LTE2300/LTE2500 in [15]. Xu *et al.* have studied a reconfigurable antenna using a T-shaped feeding strip, an encircled T-shaped parasitic shorted strip and a PIN diode [16]. In the OFF state, the antenna operates at GSM850/ GSM900/GSM1800/GSM1900/UMTS2100/LTE2300/LTE2500 bands while in the ON state, it switches to GNSS bands for positioning systems (COMPASS, GALILEO, GPS and GLONASS). Resonance frequency can also be modified by etching one or more slots of different shapes and sizes on different layers of dielectric substrates [17].

The limitations of this technique are challenging impedance matching, need for use of at least two dielectric substrates, integrating capacitors in slots for operating PIN diodes [18] and making complex shapes or slots. However, there are some demerits of PIN diodes that limit their practical use in applications such as high power loss, low linearity, low quality factor and high insertion loss. It limits their ability in terms of return loss, insertion loss, switching speed, power handling, settling time and operating life [19].

MEMS (microelectromechanical systems) switches are also being used in reconfigurable antennas instead of PIN diodes due to various advantages such as higher quality factor, lower power losses, higher linearity and low insertion loss [20–22]. MEMS switches and particle swarm optimization are used to change the slot dimensions of an E-shaped patch antenna in [23] resulting in a wide bandwidth operation in the frequency range from 2 to 3.2 GHz. The only disadvantage of MEMS switches is longer switching time.

Varactor diodes have the ability to tune operating frequency continuously and can reduce the circuit complexity in applications where multiple operating bands are required. A monopolar antenna with a patch shorted at its edges (radiating as magnetic current loop) is modified by adding four symmetrical slots on the top of the patch and two sets of varactors and two control DC voltages in order to achieve reconfigurable dual band operation [24]. The antenna operates in the range of 880–920 MHz and 1.70–1.82 GHz. However, varactor diodes have disadvantages of low dynamic range, high power loss and non-linearity.

Using more than one substrate also increases the overall thickness of the antenna. In the same way, optical switches are also used for frequency reconfiguration and this class of antennas is designated optically reconfigurable antennas (ORA) [25].

In [38], a frequency reconfigurable patch antenna was fabricated using a barium strontium titanate (BST) substrate. BST is a ceramic material belonging to the ferroelectric material family whose relative permittivity can be altered by applying different voltages. Chip capacitors have been used across the slots in patch antenna and the radiating frequency was reconfigured by using different capacitance values by Huang *et al.* [39].

A stub-loaded patch is tuned optically by changing its effective length through impedance switching by means of a photoconductive switch in [40]. The switch is illuminated with a laser to turn it ON due to increased conductivity. It results in loading the patch with the stub and decreasing the resonant frequency. A similar principle is used by Tawk *et al.* to facilitate reconfiguration [41].

In [35], a mechanically reconfigurable antenna was designed containing different movable parts. The main disadvantage of this technique is the change in antenna size with tuned frequency. In [26], a frequency reconfigurable patch antenna was designed for ultra-wideband applications by rotating different radiating elements. The rotation of the radiating elements was controlled by a stepping motor. This stepping motor had used seven open collector Darlington pairs and a power supply. The rotation of this stepping motor at an angle of 7.5° per step produced different resonant frequencies. This antenna is not suitable for small size applications due to its bulkiness and it also makes integration difficult due to its complex design. Mehdipour *at al.* have discussed probe-fed microstrip patch reconfigurable antennas based on anisotropic reinforced continuous carbon fibre composite (RCCFC) material [42]. The reconfiguration mechanism is based on the rotation of the anisotropic RCCFC ground plane that changes the current distribution to select a desirable mode of resonance and suppress other modes.

Until now, most of the frequency reconfigurable antennas have been designed using tuneable materials, diodes, switches, capacitors and mechanically moving techniques. Though effective, these technique have their limitations. Kelley *et al.* have proposed a novel switching mechanism employing liquid metal to attain the frequency reconfigurability in an edge-fed microstrip patch antenna [43]. This design used a partitioned fluidic manifold bonded to the top of the antenna and contained in a reservoir to enable the pressure-driven displacement of Eutectic Gallium Indium (EGaIn) to perform the switching. However, EGaIn is not a very commonly available material. This technique has the advantages of effective tuning with no insertion losses and power handling issues.

9.3 Rectangular Patch Reconfigurable LTE Femtocell Antenna

This section presents a study on the design of a novel frequency reconfigurable rectangular patch antenna for the LTE2400 band at 2.45 GHz and Global Positioning System (GPS) band at 1.575 GHz. In this antenna design, gaps are created between radiating elements of different lengths. To reconfigure the resonance frequency, these gaps are filled with LCP (Liquid Crystal Polymer) and mercury, a combination of low loss materials. The antenna has high directivity and efficiency because of low loss tangent ($tan\sigma$) of LCP.

9.3.1 Design Conception

The antenna has been designed using CST Microwave Studio. CST Microwave Studio is an electromagnetic simulation tool that uses finite integrating technique (FIT) to model the fields in 3D performing the calculations based on Maxwell's equations with efficient approximations [44].

The rectangular patch frequency reconfigurable antenna employs an RT-Duroid 5870 dielectric substrate. The geometrical structure of the antenna is shown in Figure 9.3. The design of conventional microstrip rectangular patch antenna was used as a basis for this antenna. First, the microstrip rectangular patch antenna was designed at the centre frequency of 2.45 GHz. The width (W) of the radiating patch was calculated using [45],

$$W = \left(\sqrt{\frac{2}{\epsilon_r + 1}}\right)\left(\frac{0.5}{f_r\sqrt{\epsilon_o \mu_o}}\right) \tag{9.1}$$

where ϵ_r is the relative permittivity of RT/Duroid 5870 with a value of 2.33, f_r is the radiating frequency (2.45 GHz), μ_o is the permeability

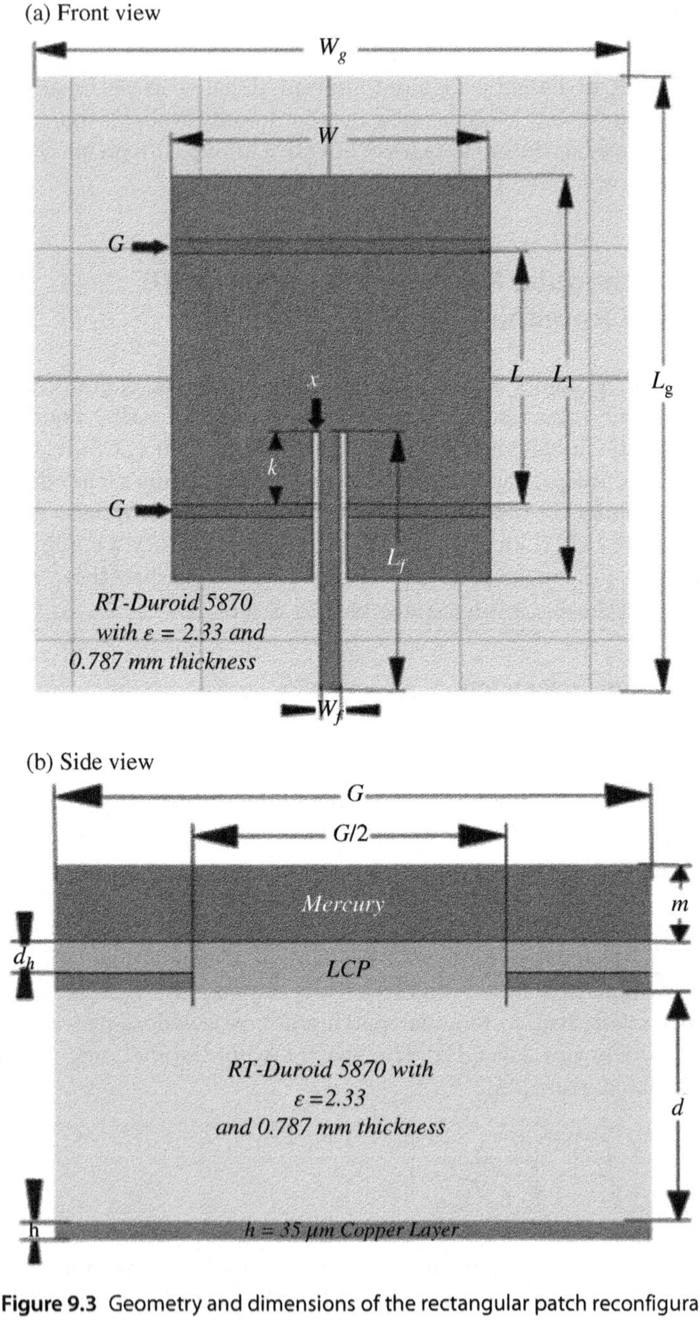

Figure 9.3 Geometry and dimensions of the rectangular patch reconfigurable LTE antenna (all units are in mm).

of free space ($1.257 \times 10(-6) H/m$) and ϵ_o is the permittivity of free space ($8.854 \times 10^{-12} F/m$). Substituting values of all parameters in Equation 9.1 results in $W = 47.41$ mm.

The length (L) of the radiating patch was calculated as follows [45],

$$L = \left(\frac{0.5}{f_r \sqrt{\epsilon_o \mu_o} \sqrt{\epsilon_{r_{eff}}}} \right) - (2 \times \Delta L) \tag{9.2}$$

where $\epsilon_{r_{eff}}$ and ΔL can be given by [12],

$$\epsilon_{r_{eff}} = \frac{\epsilon_r - 1}{2 \times \sqrt{1 + 12\left(\frac{d}{W}\right)}} + \left(\frac{\epsilon_r + 1}{2}\right) \tag{9.3}$$

and

$$\Delta L = \frac{0.412d\left(\frac{W}{d} + 0.264\right)\left(\epsilon_{r_{eff}} + 0.3\right)}{\left(\frac{W}{d} + 0.8\right)\left(\epsilon_{r_{eff}} - 0.258\right)} \tag{9.4}$$

Substituting values of ϵ_r and W in Equations 9.2 and 9.3 gave the value of $L = 39.764$ mm. Next, the increased length L_1 of the radiating patch was calculated for the antenna working at 1.575 GHz for the GPS operation following the steps stated before using Equations 9.1–9.4. In this way, two patches of lengths L and L_1 were created. Both the patches were then separated from each other by making gaps (G) of 2 mm.

This analytical model provided the basic dimensions of the 3D model of the antenna. The antenna structure modelled in CST Microwave Studio is illustrated in Figure 9.3. The antenna structure was then optimized to obtain the frequency reconfigurability using the liquid-metal switching mechanism as detailed in the next section.

9.3.2 Frequency Reconfiguration Mode

The side view of the gap (G) is shown is Figure 9.3(b). In this figure, d is the thickness of the dielectric substrate RT-Duroid 5870 and h is the thickness of copper. The two patches of lengths L and $L1$ are separated from each other when these gaps are not filled with any material. When both the patches are separated from each other then, L is the length of radiating element and antenna radiates at 2.45 GHz.

The switching mechanism for the frequency reconfiguration is implemented by filling all the three gaps with liquid crystal polymer (LCP) on the bottom and mercury on the top. LCP has gained much attention in recent years due to its amazing performance at high frequencies. It is cheaper than other dielectric substrates, has a lower dielectric constant value and has low loss tangent ($tan\sigma$), usually from 0.002 to 0.0045 for frequencies less than 105 GHz [46, 47]. It is extremely inert and this means it can resist reactions with strong acids [46]. It also has very low moisture absorption ratio that makes it suitable for extreme environmental conditions [46, 47]. Mercury is a metal and exists in a liquid state at room temperature. On filling the gaps with LCP and mercury, the length of the radiating patch extends from L to $L1$ making the antenna radiate at 1.575 GHz for GPS operation without any change required in the dimensions of microstrip feedline and inset width (x) or length (k).

A thorough optimization process was carried out in order to match the impedance and get the antenna to operate at the right centre frequencies using the liquid-metal switching mechanism. The final dimensions of the simulated antenna structure are summarized in Table 9.2.

Surface current distribution is given in Figures 9.4 and 9.5 for the filled and unfilled configurations, respectively. It is clear from these results that the filling and unfilling mechanism perturbs the surface current. The current is significantly reduced on the patch in the filled configuration at the higher frequency band due to an increase in the effective antenna length resulting in a shift in the frequency band to 1.575 GHz and reducing the radiation at 2.45 GHz.

A similar phenomenon has been observed for the unfilled configuration where the surface current is significantly reduced at the lower frequency due to a decrease in the effective antenna length. It results in an upwards frequency shift making the antenna operate at 2.45 GHz.

9.3.3 Antenna Performance Evaluation

The antenna was simulated in two configurations, gaps unfilled and gaps filled with LCP and mercury. The antenna performance was analysed in terms of simulated free space parameters of reflection

Table 9.2 Optimized dimensions of the proposed rectangular patch frequency reconfigurable antenna.

Parameter	Dimension (mm)
Width of patch (W)	49.1
Length corresponding to 2.45 GHz (L)	39.7
length corresponding to 1.575 GHz (L_1)	62.1
Width of substrate (W_g)	98
Length of substrate(L_g)	96
Width of feedline (W_f)	3
Length of feedline (L_f)	41
Inset width (x)	1.1
Inset length (k)	11.6
Thickness of substrate (d)	0.79
Thickness of copper (h)	0.035
Width of gap (G)	2
Thickness of LCP (d_h)	0.1
Thickness of mercury (m)	0.4

coefficient (S_{11}), impedance matching, VSWR, radiation pattern, efficiency and gain values.

The simulated reflection coefficient response of antenna before and after filling the gaps is plotted in Figure 9.6. The results show that the proposed antenna configurations of filled and unfilled gaps achieved a smooth switching of the frequency between 2.45 GHz and 1.575 GHz bands. In the unfilled configuration, the antenna was radiating at 2.45 GHz while the resonance frequency shifted to the GPS band of 1.575 GHz in the filled configuration due to increase in the size of the patch. The plot depicts that antenna has very good impedance matching at both the frequencies. The antenna has achieved a −10 dB impedance bandwidth of 23.57 MHz at 2.45 GHz and 10.8 MHz at 1.575 GHz.

The radiation characteristics of the proposed antenna in unfilled and filled configurations were analysed in terms of radiation patterns. The far-field radiation patterns of the unfilled antenna structure radiating at 2.45 GHz are depicted in Figure 9.7 while Figure 9.8

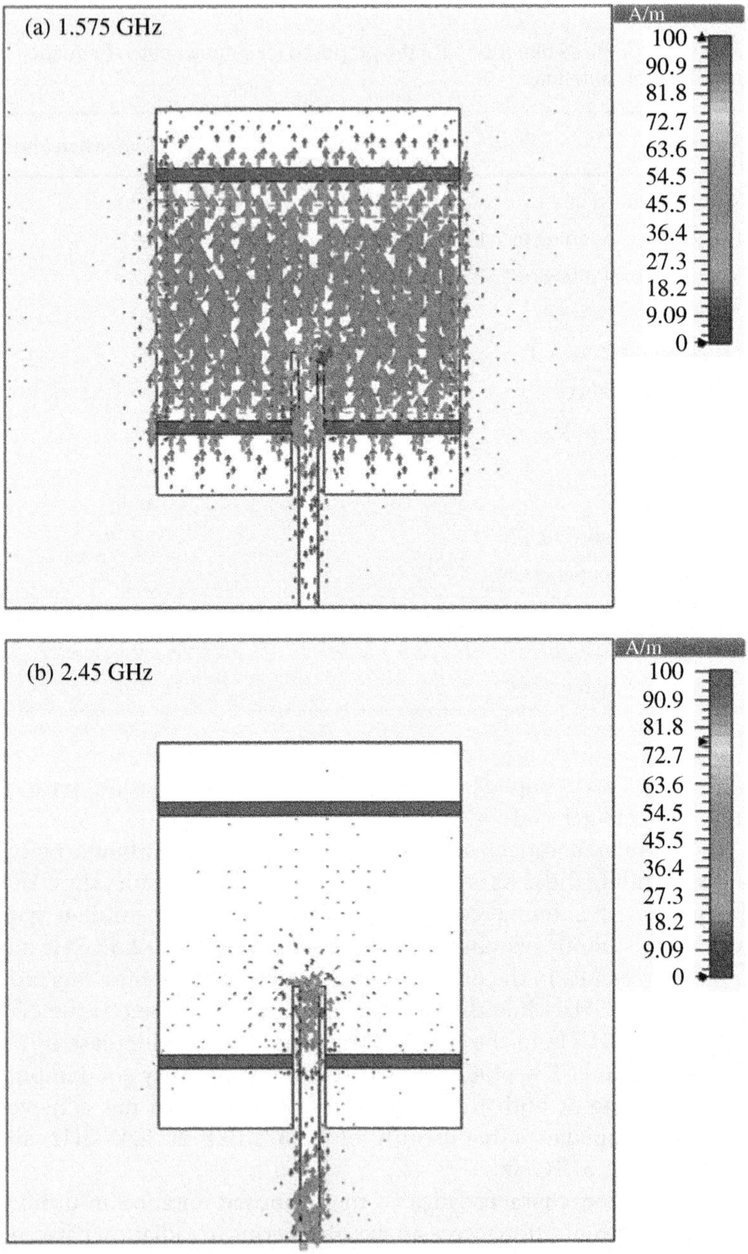

Figure 9.4 Surface current distribution of the rectangular patch reconfigurable antenna in the filled configuration at 1.575 and 2.45 GHz.

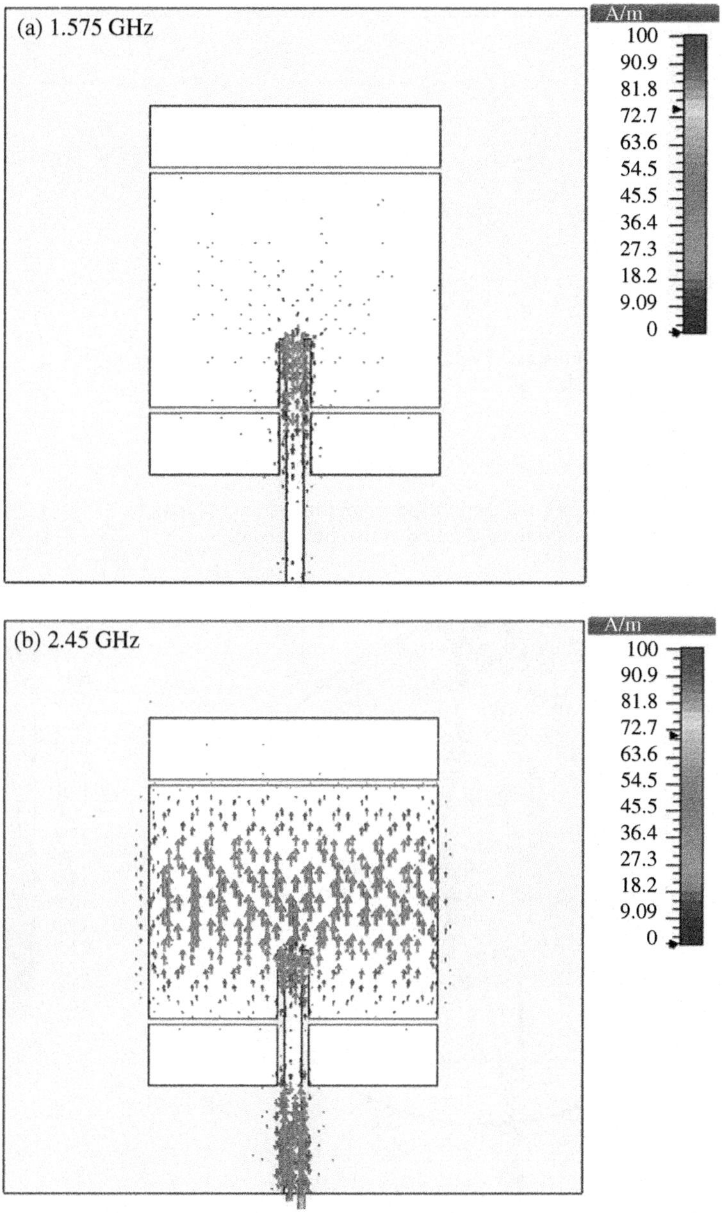

Figure 9.5 Surface current distribution of the rectangular patch reconfigurable antenna in the unfilled configuration at 1.575 and 2.45 GHz.

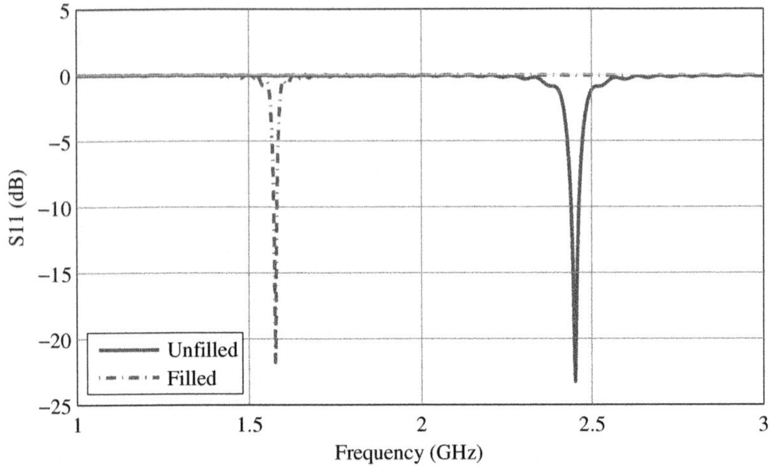

Figure 9.6 Reflection coefficient response of the rectangular patch reconfigurable LTE antenna operating at two frequencies.

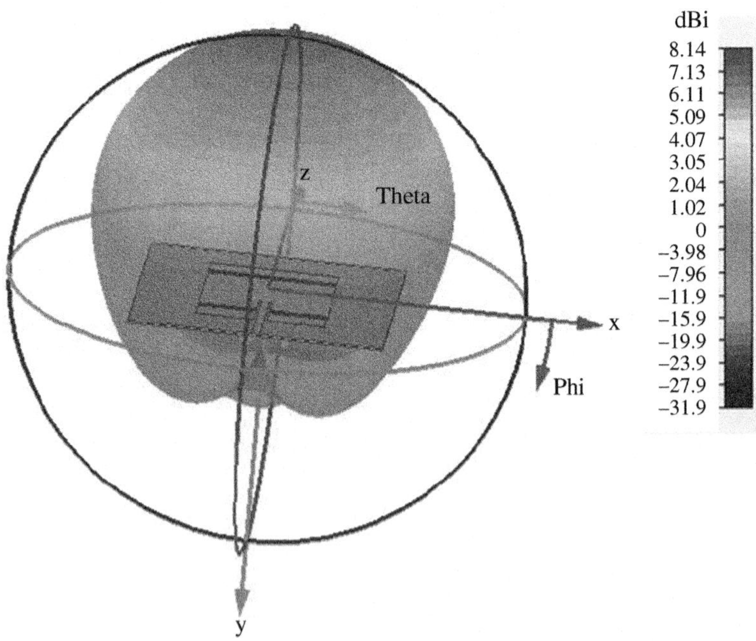

Figure 9.7 Three-dimensional radiation pattern of the rectangular patch reconfigurable antenna in unfilled configuration operating at 2.45 GHz.

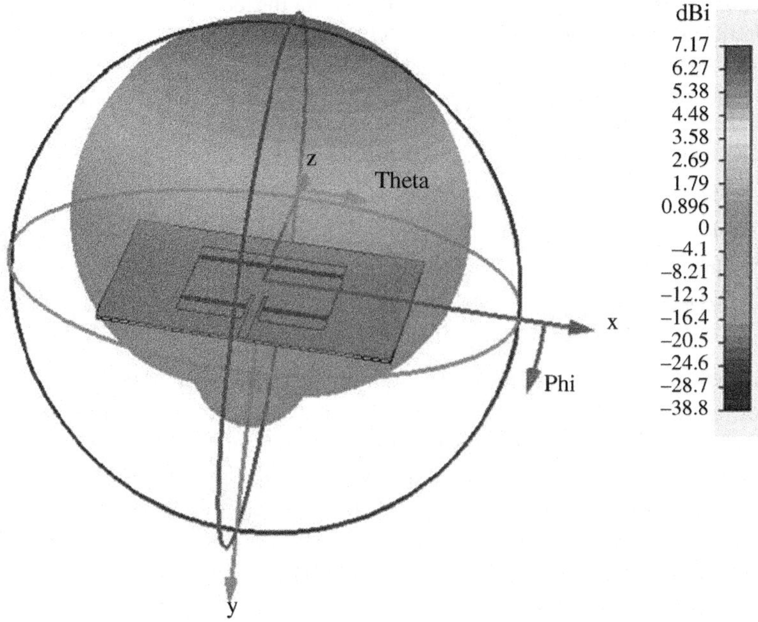

Figure 9.8 Three-dimensional radiation pattern of the rectangular patch reconfigurable antenna in filled configuration operating at 1.575 GHz.

illustrates the radiation patterns for the filled antenna configuration radiating at 1.575 GHz. These results show that the antenna has a similar radiation pattern profile at the two frequencies. It radiates omnidirectionally in XZ-plane and covers the whole upper hemisphere in the YZ-plane with a very high gain and a small back lobe. The value of the gain is noted to be 8.14 dBi for the unfilled structure radiating at 2.45 GHz, while the gain value is 7.17 dBi for the filled structure operating at 1.575 GHz.

The antenna is also observed to be good in terms of efficiency. The simulated values of antenna efficiency are observed to be 94.31% at 2.45 GHz and 88.56% at 1.575 GHz. Comparison of the proposed antenna performance at the two operating frequencies in terms of simulated free space parameters is given in Table 9.3.

These results have shown that the rectangular patch frequency reconfigurable antenna can support good performance at both frequency bands with a simple and clean switching method. The antenna also offers a simple design, is light weight, and has a low profile, good

Table 9.3 Comparison of performance metrics for the rectangular patch frequency reconfigurable LTE antenna at two operating frequencies.

Antenna parameter	Gaps not filled	Gaps filled with LCP and mercury
Resonant frequency	2.45 GHz	1.575 GHz
Reflection coefficient	−24.69 dB	−21.77 dB
Bandwidth	23.57 MHz	10.8 MHz
Gain	8.14 dBi	7.31 dBi
Efficiency	94.31%	88.56%

impedance matching, peak high gain at both the frequencies and good efficiency. This switching mechanism can also be extended to other frequency bands using the simple optimization steps for changing volume of the filling metals. Moreover, the liquid nature of the filling/unfilling mechanism helps to avoid mechanical failure of these parts due to material fatigue, creep or wear. Also, the displacement of liquid metals is more effective than solid materials. This antenna is, therefore, a good candidate for LTE systems.

9.4 Circular Patch Reconfigurable LTE Femtocell Antenna

The circular patch reconfigurable antenna structure is shown in Figure 9.9. The antenna is designed on the Rogers RT 5870 substrate with a permittivity (ϵ_r) of 2.33, thickness of 0.787 mm and tangential loss ($tan\sigma$) of 0.0012 [48].

The size of the antenna is $65 \times 50\,\text{mm}^2$. For performing frequency reconfiguration, two LTE frequencies of 3.6 GHz and 5 GHz were selected. Two circular patches were then constructed corresponding to these two frequencies. These two circular patches were then made isolated from each other by creating a circular gap 0.5 mm wide. The first circular patch has a diameter of 19.6 mm that corresponds to the 5 GHz frequency. The second circular patch has a 31.1 mm diameter that corresponds to a 3.6 GHz frequency. Both the patches are then excited by a common microstrip feedline with a feed length of 35.5 mm and feed width of 0.4 mm. The side view of the proposed antenna is shown in Figure 9.9(b). It can be seen that the substrate

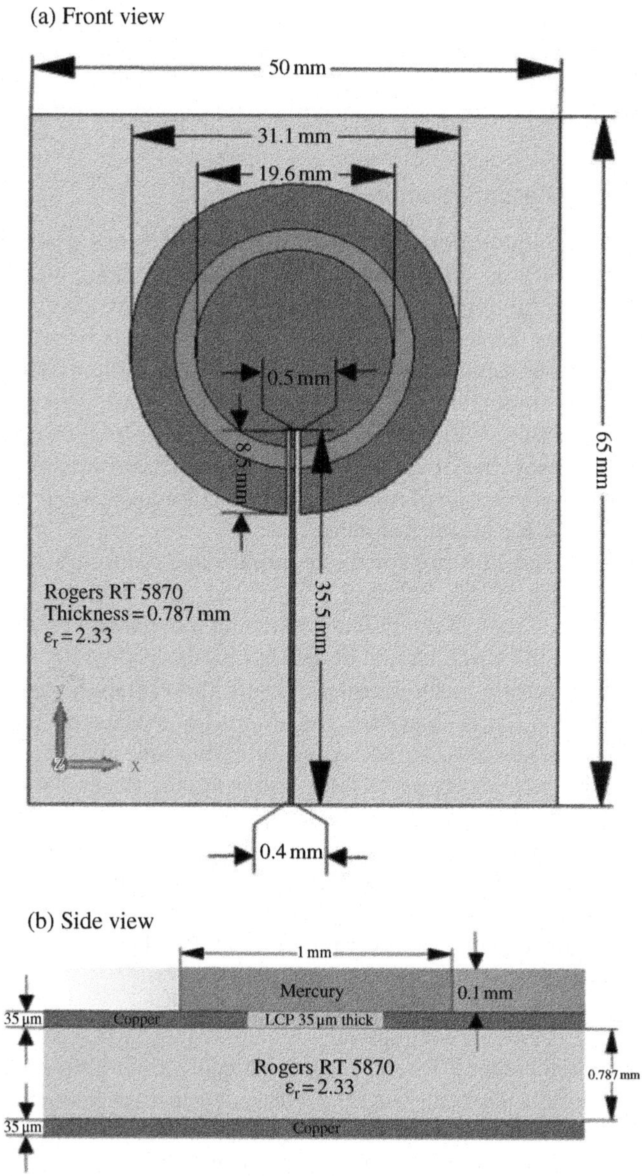

Figure 9.9 Geometry and dimensions of the circular patch reconfigurable LTE antenna (all units are in mm) [48].

is 0.787 mm thick and is coated with a 35 μm copper layer on both sides. The gap or isolation (0.5 mm wide) between both the patches is filled with LCP. The thickness of LCP layer is also μm. Finally, a 0.1-mm thick and 1-mm wide layer of mercury is placed on top of the LCP.

9.4.1 Frequency Reconfiguration Mode

The frequency reconfiguration operation is performed by using layers of LCP and mercury as discussed in the previous section. In the absence of LCP and mercury, the two circular patches become completely isolated from each other. In this way, an inner patch with a 19.6 mm diameter becomes the radiating patch and antenna starts radiating at 5 GHz frequency. For shifting the radiating frequency of antenna, the gap is filled with layers of LCP and mercury as shown in Figure 9.9. In this way, the diameter of radiating patch extends from 19.6 to 31.1 mm, which corresponds to the 3.6 GHz frequency and the antenna starts radiating at this frequency.

Surface current distributions for filled and unfilled configurations of the circular patch reconfigurable antenna are depicted in Figures 9.10 and 9.11, respectively. The switching mechanism affects the distribution of the surface current due to change in the effective antenna length resulting in the frequency shift and radiation variations. In the filled antenna structure, the amount of surface current on the patch increases due to increased effective antenna length making the antenna resonate at 3.6 GHz and reducing the radiation at 5 GHz.

Unfilled configuration of the antenna also exhibits the same trend. The surface current is now increased significantly at 5 GHz due to a decrease in the effective antenna length thus resulting in resonance shift to 5 GHz at the expense of lower radiation at 3.6 GHz.

9.4.2 Antenna Performance Evaluation

The antenna performance is analysed in terms of its reflection coefficient, radiation pattern, gain and efficiency in two frequency reconfigurable modes.

Figure 9.12 presents the reflection coefficient response of the antenna in an filled and unfilled configuration. It is evident that the antenna successfully achieves frequency reconfigurability at the two LTE frequencies with good impedance matching. The antenna

(a) 3.6 GHz

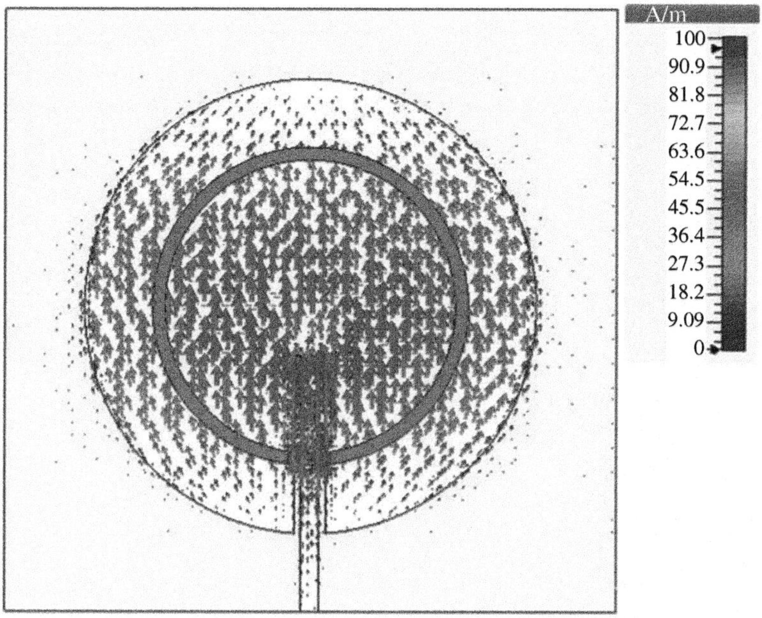

(b) 5 GHz

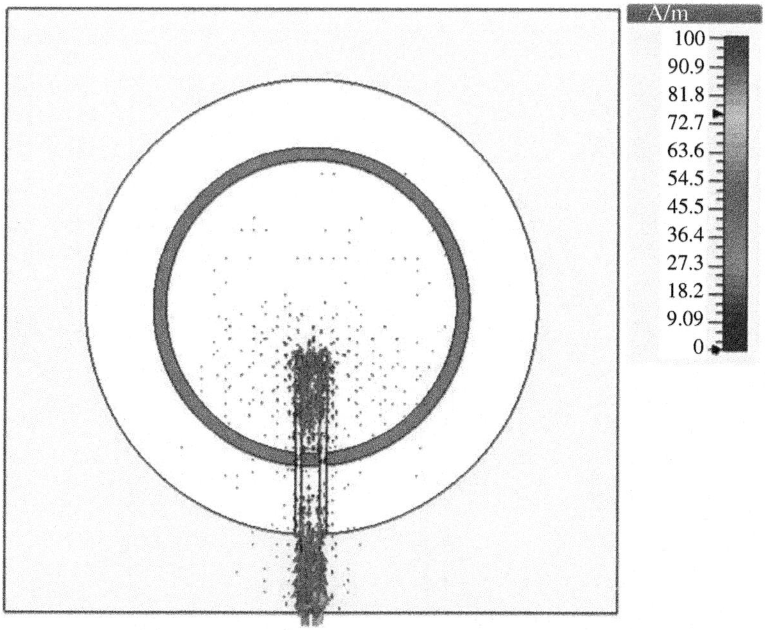

Figure 9.10 Surface current distribution on the circular patch reconfigurable antenna in a filled configuration at 3.6 and 5 GHz.

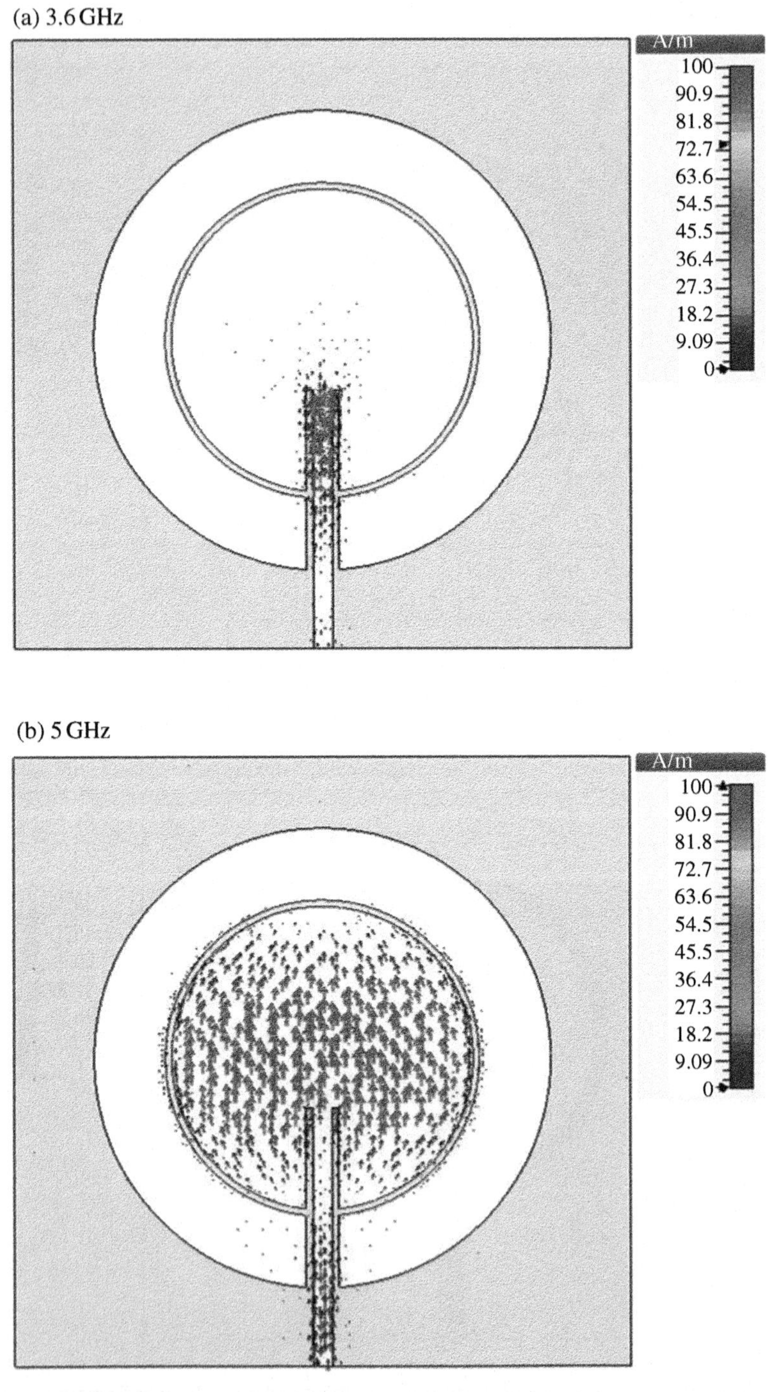

Figure 9.11 Surface current distribution on the circular patch reconfigurable antenna in an unfilled configuration at 3.6 and 5 GHz.

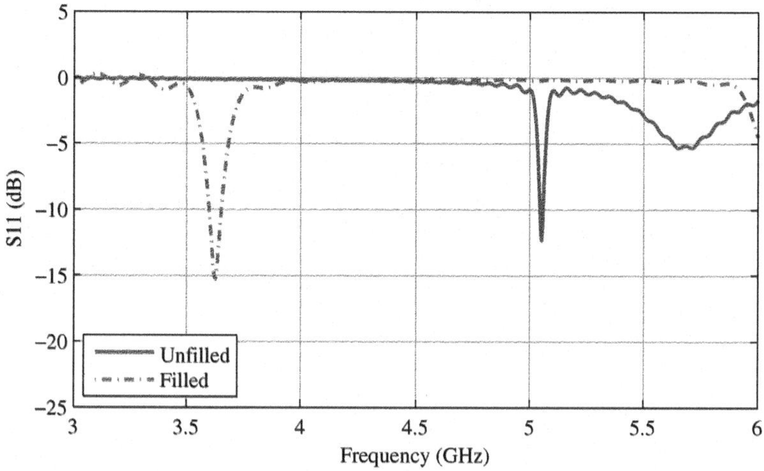

Figure 9.12 Reflection coefficient response of the circular patch reconfigurable LTE antenna operating at two frequencies.

exhibits a −10 dB impedance bandwidth of 25.47 MHz at 3.6 GHz and 12.5 MHz at 5 GHz. The bandwidths are sufficient for the antenna to operate well at the two frequencies.

Radiation characteristics of the antenna are evaluated by analysing 3D radiation patterns and efficiency. Figure 9.13 illustrates the antenna radiation pattern in a filled configuration operating at 3.6 GHz while the radiation pattern in an unfilled state making the antenna work at 5 GHz is presented in Figure 9.14. It is observed that the antenna provides good coverage in the upper hemisphere with wide beamwidth and low backside radiation at both frequencies. The antenna also offers high peak gain values at the two frequencies. The gain is noted to be 6.38 dBi at 5 GHz while it becomes 7.67 dBi at the lower frequency band of 3.6 GHz.

The antenna also has good efficiency values of 78 and 72% at 3.6 and 5 GHz, respectively, showing acceptable of radiation and ohmic losses. The summary of the antenna performance at the two frequencies is given in Table 9.4.

These results show that the circular patch frequency reconfigurable antenna performs well with excellent switching operation. The antenna offers good impedance matching, peak high gain on both the frequencies and good efficiency for operation in LTE systems.

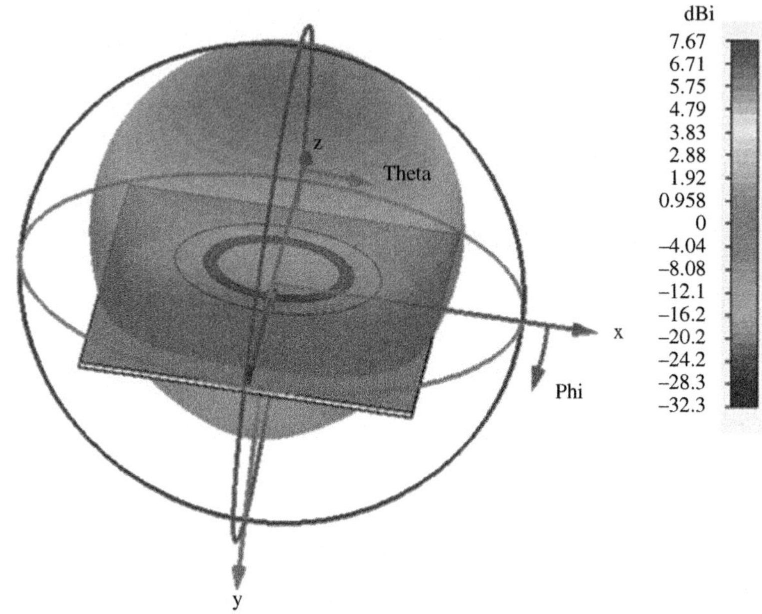

Figure 9.13 Three-dimensional radiation pattern of the circular patch reconfigurable antenna in a filled configuration operating at 3.6 GHz.

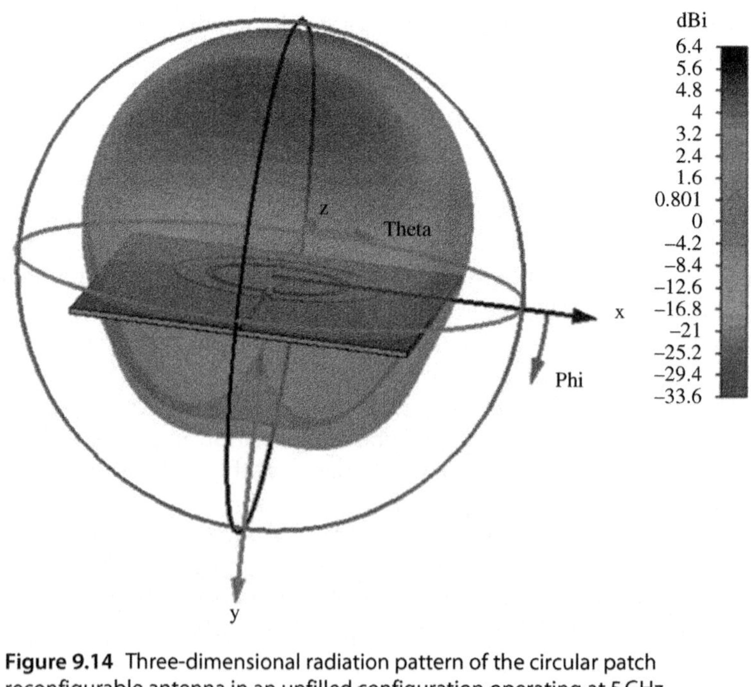

Figure 9.14 Three-dimensional radiation pattern of the circular patch reconfigurable antenna in an unfilled configuration operating at 5 GHz

Table 9.4 Comparison of circular patch frequency reconfigurable antenna performance at the two LTE frequency bands.

Antenna parameter	Gaps not filled	Gaps filled with LCP and mercury
Resonant frequency	5 GHz	3.6 GHz
Reflection coefficient	−12.30 dB	−15.24 dB
Bandwidth	12.5 MHz	25.47 MHz
Peak gain	6.38 dBi	7.67 dBi
Efficiency	72%	78%

The discussed antenna structures achieve the switching operation by creating/filling gaps between two lengths of the radiating patch using LCP and mercury. This technique adds a good option to the existing traditional switching mechanisms. The results are very encouraging in the sense that frequency reconfigurable patch antennas can be designed by modifying the conventional microstrip patch antennas. Since there exists an inverse relation between the resonance frequency and the length of radiating element, this method can be easily extended to reconfigure between more than two resonance frequencies by creating additional gaps along the length of the radiating element and filling them with LCP and mercury. The only limitation of this method is to the design of the containers of LCP and mercury that are effectively invisible to RF waves. It can, however, be dealt with easily by using the technique discussed in [43]. The problem of controlling the flow of mercury at normal operating conditions can also be solved by encapsulating the gaps with poly-dimethyl-siloxane (PDMS), which is silicon based organic polymer.

9.5 Open Research Issues

The advantages of using reconfigurable antennas in the context of overall system perspective is yet not fully established. There is also a lack of quantitative analysis based on a well-developed procedure to assess the feasibility of the reconfigurable antennas for new communication systems due to requirements of added material and components to support the switching mechanism. Moreover, the theoretical modelling and design principles for the reconfigurable antennas are not explored in detail to help the antenna designers devising novel solutions. These aspects are open issues and need efforts to enhance this very useful field of antenna technology. Furthermore, novel tunable materials and more practical solutions for switching assemblies with lower losses and cost would always be in demand for reconfigurable antennas.

References

1 J. Bernhard, *Reconfigurable Antennas*, Morgan and Claypool Publishers (USA), 2007.
2 F. Yang and Y. Rahmat-Samii, "Patch antennas with switchable slots (PASS) in wireless communications: concepts, designs, and applications," *IEEE Antennas and Propagation Magazine*, vol. 47, no. 2, pp. 13–29, April 2005.
3 S. Xiao, Z. Shao, M. Fujise, and B.-Z. Wang, "Pattern reconfigurable leaky-wave antenna design by FDTD method and Floquet's theorem," *IEEE Transactions on Antennas and Propagation*, vol. 53, no. 5, pp. 1845–1848, May 2005.
4 J. L. Freeman, B. J. Lamberty, and G. S. Andrews, "Optoelectronically reconfigurable monopole antenna," *Electronics Letters*, vol. 28, no. 16, pp. 1502–1503, July 1992.
5 Z. Jiajie, W. Anguo, and W. Peng, "A survey on reconfigurable antennas," in *2008 International Conference on Microwave and Millimeter Wave Technology*, vol. 3, April 2008, pp. 1156–1159.
6 M. T. Ali, M. Y. M. Zain, N. Ramli, A. L. Yusof, and M. N. M. Tan, "Frequency reconfigurable quasi-yagi microstrip antenna with beam shaping for WIMAX application," in *2011 IEEE International RF Microwave Conference*, Dec 2011, pp. 434–438.
7 X. Chen, C. G. Parini, B. Collins, Y. Yao, and M. Ur Rehman, *Antennas for Global Navigation Satellite Systems*, John Wiley and Sons, Inc. Ltd, (UK), 2012.
8 J. L. Freeman, B. J. Lamberty, and G. S. Andrews, "Optoelectronically reconfigurable monopole antenna," *Electronics Letters*, vol. 28, no. 16, pp. 1502–1503, July 1992.
9 C. J. Panagamuwa, A. Chauraya, and J. C. Vardaxoglou, "Frequency and beam reconfigurable antenna using photoconducting switches," *IEEE Transactions on Antennas and Propagation*, vol. 54, no. 2, pp. 449–454, Feb 2006.

10 A. Vasylchenko, X. Rottenberg, B. X. Broze, M. Nuytemans, W. D. Raedt, and G. A. E. Vandenbosch, "A frequency switchable antenna based on mems technology," in *Proceedings of the Fourth European Conference on Antennas and Propagation*, April 2010, pp. 1–3.

11 S. Shelley, J. Costantine, C. G. Christodoulou, D. E. Anagnostou, and J. C. Lyke, "FPGA-controlled switch-reconfigured antenna," *IEEE Antennas and Wireless Propagation Letters*, vol. 9, pp. 355–358, 2010.

12 V. K. Kunda and M. Ali, "Reconfigurable stacked patch antenna for satellite and terrestrial applications," in *2003 IEEE Topical Conference on Wireless Communication Technology*, Oct 2003, pp. 152–153.

13 S. W. Lee, H. S. Jung, and Y. J. Sung, "A reconfigurable antenna for LTE/WWAN mobile handset applications," *IEEE Antennas and Wireless Propagation Letters*, vol. 14, pp. 48–51, 2015.

14 Y. Li, Z. Zhang, J. Zheng, Z. Feng, and M. F. Iskander, "A compact hepta-band loop-inverted F reconfigurable antenna for mobile phone," *IEEE Transactions on Antennas and Propagation*, vol. 60, no. 1, pp. 389–392, Jan 2012.

15 H. B. Zhang, Y. L. Ban, Y. F. Qiang, J. Guo, and Z. F. Yu, "Reconfigurable loop antenna with two parasitic grounded strips for WWAN/LTE unbroken-metal-rimmed smartphones," *IEEE Access*, vol. 5, pp. 4853–4858, 2017.

16 Y. Xu, Y. W. Liang, and H. M. Zhou, "Small-size reconfigurable antenna for WWAN/LTE/GNSS smartphone applications," *IET Microwaves, Antennas Propagation*, vol. 11, no. 6, pp. 923–928, 2017.

17 B. Z. Wang, S. Xiao, and J. Wang, "Reconfigurable patch-antenna design for wideband wireless communication systems," *IET Microwaves, Antennas Propagation*, vol. 1, no. 2, pp. 414–419, April 2007.

18 H. A. Majid, M. K. A. Rahim, M. R. Hamid, N. A. Murad, and M. F. Ismail, "Frequency-reconfigurable microstrip patch-slot antenna," *IEEE Antennas and Wireless Propagation Letters*, vol. 12, pp. 218–220, 2013.

19 Keysight Technologies, "Understanding RF/microwave solid state switches and their applications," *Agilent Technologies Application Note*, pp. 14–30, 2010. Available online at: http://literature.cdn.keysight.com/litweb/pdf/5989-7618EN.pdf (accessed December 2017).

20 K. R. Boyle and P. G. Steeneken, "A five-band reconfigurable pifa for mobile phones," *IEEE Transactions on Antennas and Propagation*, vol. 55, no. 11, pp. 3300–3309, Nov. 2007.

21 T. Yamagajo and Y. Koga, "Frequency reconfigurable antenna with mems switches for mobile terminals," in *2011 IEEE-APS Topical Conference on Antennas and Propagation in Wireless Communications*, Sept. 2011, pp. 1213–1216.

22 A. Grau, J. Romeu, M. J. Lee, S. Blanch, L. Jofre, and F. D. Flaviis, "A dual-linearly-polarized mems-reconfigurable antenna for narrowband mimo communication systems," *IEEE Transactions on Antennas and Propagation*, vol. 58, no. 1, pp. 4–17, Jan 2010.

23 H. Rajagopalan, J. M. Kovitz, and Y. Rahmat-Samii, "Mems reconfigurable optimized e-shaped patch antenna design for cognitive radio," *IEEE Transactions on Antennas and Propagation*, vol. 62, no. 3, pp. 1056–1064, March 2014.

24 N. Nguyen-Trong, A. Piotrowski, and C. Fumeaux, "A frequency-reconfigurable dual-band low-profile monopolar antenna," *IEEE Transactions on Antennas and Propagation*, vol. 65, no. 7, pp. 3336–3343, July 2017.

25 S. J. Mazlouman, M. Soleimani, A. Mahanfar, C. Menon, and R. G. Vaughan, "Pattern reconfigurable square ring patch antenna actuated by hemispherical dielectric elastomer," *Electronics Letters*, vol. 47, no. 3, pp. 164–165, February 2011.

26 P. Tummas, P. Krachodnok, and R. Wongsan, "A frequency reconfigurable antenna design for UWB applications," in *2014 11th International Conference on Electrical Engineering/Electronics, Computer, Telecommunications and Information Technology (ECTI-CON)*, May 2014, pp. 1–4.

27 H. Li, J. Xiong, Y. Yu, and S. He, "A simple compact reconfigurable slot antenna with a very wide tuning range," *IEEE Transactions on Antennas and Propagation*, vol. 58, no. 11, pp. 3725–3728, Nov. 2010.

28 L. Pazin and Y. Leviatan, "Reconfigurable slot antenna for switchable multiband operation in a wide frequency range," *IEEE Antennas and Wireless Propagation Letters*, vol. 12, pp. 329–332, 2013.

29 L. Ge and K. M. Luk, "A band-reconfigurable antenna based on directed dipole," *IEEE Transactions on Antennas and Propagation*, vol. 62, no. 1, pp. 64–71, Jan 2014.

30 D. Peroulis, K. Sarabandi, and L. P. B. Katehi, "Design of reconfigurable slot antennas," *IEEE Transactions on Antennas and Propagation*, vol. 53, no. 2, pp. 645–654, Feb 2005.

31 K.-H. Chen, S.-J. Wu, C.-H. Kang, C.-K. Chan, and J.-H. Tarng, "A frequency reconfigurable slot antenna using pin diodes," *2009 Asia Pacific Microwave Conference*, pp. 1930–1933, Dec. 2009.

32 P. Bhartia and I. Bahl, "Frequency agile microstrip antennas," *Microwave Journal*, vol. 25, pp. 67–70, Oct. 1982.

33 E. Kiely, G. Washington, and J. Bernhard, "Design and development of smart microstrip patch antennas," *Smart Materials and Structures*, vol. 7, no. 6, p. 792, 1998.

34 J. T. Bernhard, E. Kiely, and G. Washington, "A smart mechanically actuated two-layer electromagnetically coupled microstrip antenna with variable frequency, bandwidth, and antenna gain," *IEEE Transactions on Antennas and Propagation*, vol. 49, no. 4, pp. 597–601, Apr 2001.

35 R. Al-Dahleh, L. Shafai, and C. Shafai, "A frequency-tunable mechanically actuated microstrip patch antenna," in *IEEE Antennas and Propagation Society International Symposium. Digest. Held in conjunction with: USNC/CNC/URSI North American Radio Sci. Meeting (Cat. No.03CH37450)*, vol. 4, pp. 548–551, June 2003.

36 D. M. Pozar and V. Sanchez, "Magnetic tuning of a microstrip antenna on a ferrite substrate," *Electronics Letters*, vol. 24, no. 12, pp. 729–731, June 1988.

37 P. E. Sieber and D. H. Werner, "A reconfigurable dual-band circularly polarizing miniturized conformal patch antenna exhibiting small frequency ratios," in *2014 IEEE Antennas and Propagation Society International Symposium (APSURSI)*, July 2014, pp. 1001–1002.

38 Y. Wang, Y. Liu, H. Du, C. Liu, Q. Xue, X. Gao, S. Li, and Y. Lu, "A frequency reconfigurable microstrip antenna based on $(rmBa, rmSr)rmTiO_3$ substrate," *IEEE Transactions on Antennas and Propagation*, vol. 63, no. 2, pp. 770–775, Feb 2015.

39 C. Y. Huang, C. W. Ling, and J. S. Kuo, "Dual-band microstrip antenna using capacitive loading," *IEE Proceedings – Microwaves, Antennas and Propagation*, vol. 150, no. 6, pp. 401–404, Dec 2003.

40 S. Pendharker, R. K. Shevgaonkar, and A. N. Chandorkar, "Optically controlled frequency-reconfigurable microstrip antenna with low photoconductivity," *IEEE Antennas and Wireless Propagation Letters*, vol. 13, pp. 99–102, 2014.

41 Y. Tawk, A. R. Albrecht, S. Hemmady, G. Balakrishnan, and C. G. Christodoulou, "Optically pumped frequency reconfigurable antenna design," *IEEE Antennas and Wireless Propagation Letters*, vol. 9, pp. 280–283, 2010.

42 A. Mehdipour, T. A. Denidni, A. r. Sebak, C. W. Trueman, I. D. Rosca, and S. V. Hoa, "Mechanically reconfigurable antennas using an anisotropic carbon-fibre composite ground," *IET Microwaves, Antennas Propagation*, vol. 7, no. 13, pp. 1055–1063, October 2013.

43 M. Kelley, C. Koo, H. Mcquilken, B. Lawrence, S. Li, A. Han, and G. Huff, "Frequency reconfigurable patch antenna using liquid metal as switching mechanism," *Electronics Letters*, vol. 49, no. 22, pp. 1370–1371, Oct 2013.

44 CST, "CST microwave studio user's manual," 2017.

45 C. A. Balanis, *Antenna Theory, Design and Analysis* (3rd edition), John Wiley & Sons, Inc. Hoboken, USA, pp. 816–826, 2005.

46 D. C. Thompson, O. Tantot, H. Jallageas, G. E. Ponchak, M. M. Tentzeris, and J. Papapolymerou, "Characterization of liquid crystal polymer (LCP) material and transmission lines on LCP substrates from 30 to 110 GHz," *IEEE Transactions on Microwave Theory and Techniques*, vol. 52, no. 4, pp. 1343–1352, April 2004.

47 Q. H. Abbasi, M. U. Rehman, X. Yang, A. Alomainy, K. Qaraqe, and E. Serpedin, "Ultrawideband band-notched flexible antenna for wearable applications," *IEEE Antennas and Wireless Propagation Letters*, vol. 12, pp. 1606–1609, 2013.

48 W. Farooq, M. Ur Rehman, Q. H. Abbasi, and K. Qaraqe, "A circular patch frequency reconfigurable antenna for wearable applications," in *2015 IEEE 11th International Conference on Wireless and Mobile Computing, Networking and Communications (WiMob)*, Oct 2015, pp. 103–106.

10

Multimode Antennas for LTE Femtocells

Oluyemi Peter Falade[1], *Xiaodong Chen*[2] *and*
Masood Ur Rehman[3]

[1] *Tempfad Limited, Leyton, London, UK*
[2] *School of Electronic Engineering and Computer Science, Queen Mary, University of London, London, UK*
[3] *School of Computer Science and Technology, University of Bedfordshire, Luton, UK*

The use of enabling technologies (such as highly innovative work in the electronics, sensors and photonics, robotics etc.) in recent years has brought a dramatic growth in wireless and satellite communications in radar systems and in terrestrial cellular and mobile satellite services. Multiple antennas are required to meet the diverse services provided by these devices. The small form factor of such devices makes it hard to integrate these several antennas as a result of space-limited handheld terminals. Mutual coupling between the various antennas is also a challenging factor. A single multifunctional antenna operating at various multi-frequency bands or broadband can overcome this problem efficiently. With the advent of the release 13 of LTE-A, evolution towards 5G will bring up effective and credible applications in public safety, vehicular communication, multi-cell broadcast and the internet-of-things. All these demand compact, cost effective, simple and highly efficient antennas.

In this chapter, requirements and challenges for multimode antennas supporting multifunctional compact handheld devices are explored. Design and analysis of two low profile multimode LTE antennas are studied as test cases.

LTE Communications and Networks: Femtocells and Antenna Design Challenges, First Edition. Edited by Masood Ur Rehman and Ghazanfar Ali Safdar.

10.1 Multimode Antennas: Fundamentals and Types

With the development of LTE technology, there is a need to provide multimode operation utilizing different frequency bands in order to improve the data rate and coverage accessibility[1]. Recently, attention has been drawn to variety of LTE antennas but with some different challenges. Multiple antennas are needed but the compact size of LTE antenna structure makes it hard to integrate these numerous antennas [2–4]. Several antennas designed to meet the LTE standards were presented in [5–8] for compact devices. In [5] a wideband antenna the covered the frequency bands 1.76–2.68 GHz. The large size of an antenna with a radius of 50 mm makes it awkward for compact handheld devices. Coupled-fed techniques are used to achieve lower and wide upper bands in [6–8]. The antennas are folded at the edges of the strip to make them more compact, which finally leads to complex fabrication. Introduction of the chip inductor in the design for miniaturization resulted in losses. Research work has shown that circularly polarized (CP) antennas are more resilient to multipath reflections, better in orientation between transmitter and receiver, suitable mobility and weather penetration than linearly polarized (LP) antennas, hence they enhance wireless system capacity [9]. A narrow impedance bandwidth and axial ratio have been the challenge for single feed CP antennas. Several works have been carried out to ameliorate this problem using a stacked patch [10, 11], a coupled loop with parasitic loop [12] and profile slot antenna [13]. In [14], an asymmetrical dipole and a slit in the ground plane are used to achieve an axial ratio bandwidth of 770 MHz. A slotted ground plane results in reduced antenna size compared to a conventional CP within the same resonant frequency context. More so, it improves the gain and axial ratio of the antenna [15–17].

To have effective coverage through the use of LTE frequency bands, there is a need to have a simple, low profile and cost-effective antenna. Antennas that will be able to serve multifunctional and simultaneous services either as broadband or multiband in a compact handheld terminal is needed. The benefits of a quality multimode or multiband antenna for commercial and industrial LTE applications cannot be over emphasized. This ensures constant coverage and extremely reliable data transfer services. Multimode antennas are a prerequisite requirement for present and future LTE terminal devices. A multimode antenna helps in support for roaming between different regions where different standards are used for communication services. Due to the limited size of mobile terminal devices, strong mutual coupling among antenna elements is inevitable, which affects antenna efficiency and also influences correlation.

Antennas that are commonly used in multimode design are the patch or planar antennas because of their efficiency due to their resonant nature. They contain ground planes that are an added advantage when deployed in terminals. Dipoles, PIFAs, stacked patches, slot antennas and so on can also be used to achieve multimode and multiband. With the compact nature of most devices, an antenna designer must avoid designing a larger size multimode antenna since size reduction is essential in order to meet miniaturization requirements. More so, low order (fundamental) mode design will guarantee a good antenna efficiency. A single feed antenna design should be the first choice, otherwise the antenna designer should avoid a complicated design that will not guarantee easy mass production [18–20]. Cost, complexity and power demand must be put into consideration when designing a broadband or multiband antenna.

The performance of multimode antennas is characterized by numbers of electrical parameters that the antenna designer must consider. These are impedance, voltage standing wave ratio (VSWR), bandwidth, gain, efficiency, polarization and isolation. Sometimes, this will require a trade-off based on what is more essential; gain or bandwidth, efficiency or mutual coupling and so on.

10.2 Design of a Compact Multimode LTE Femtocell Antenna for Handheld Devices

The proposed compact wideband CP slotted ground plane (CWSGP) antenna shown in Figure 10.1 was achieved to generate a broadband operation that covered multimode LTE frequency bands. The antenna is made up of two truncated annular ring slots with asymmetric stubs built on a grounded dielectric substrate of permittivity of $\varepsilon_r = 3.38$, tan $\delta = 0.0027$ and a thickness of 1.524 mm. Multiple rectangular slots positioned parallel to the x- and y-axes of the ground plane are embedded in the ground plane. The slots are used to lengthen the excited surface current paths in the ground plane thus lowering the operating

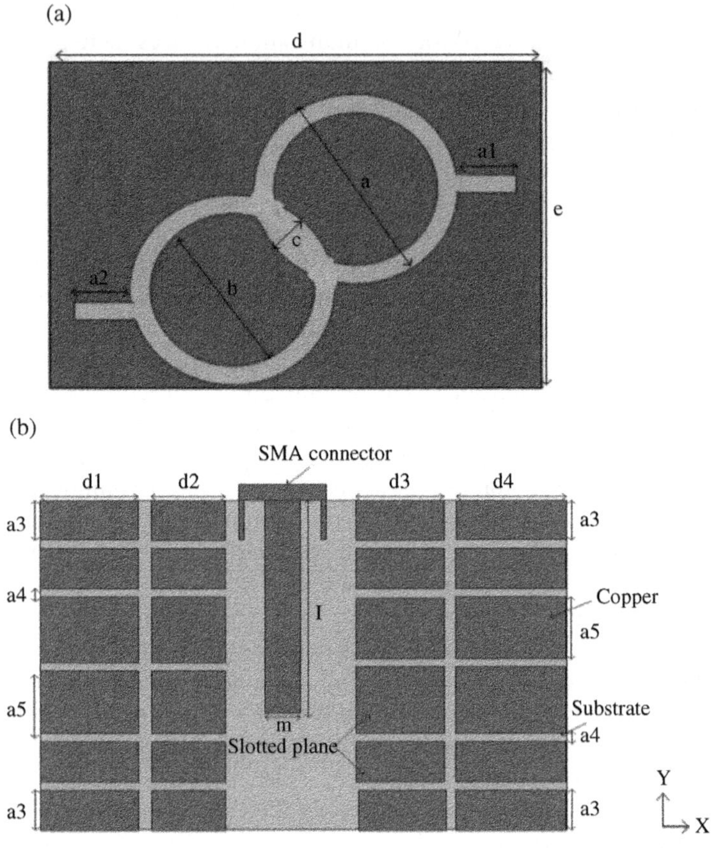

Figure 10.1 Geometry of the proposed CWSGP antenna showing the (a) top view and (b) back view of the antenna.

frequency. A dual orthogonal mode needed to achieve a CP radiation pattern is introduced through truncation of the inner patch coupled with the asymmetric stubs. The effect of different truncated inner patch, C, on the axial ratio is shown in Figure 10.2. The slotted ground plane contributes to the miniaturization of the antenna and improvement of the CP radiation pattern. In terms of size, the proposed antenna design is more compact than those presented in [22, 23]. Without the slot in the ground plane, the resonant frequency of the antenna increases and the axial ratio becomes very poor. The quality factor of the antenna has been effectively decreased by the slotted

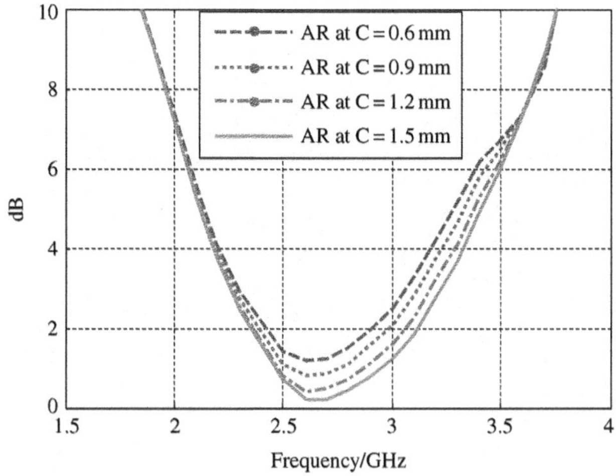

Figure 10.2 Simulated axial ratio of different truncated inner patch, C.

ground plane, thus improving the impedance bandwidth [16, 17]. The antenna is excited by a single microstrip line placed almost at the centre of the ground plane. By properly selecting the asymmetric stubs, the impedance matching of the antenna is improved.

10.2.1 Numerical Analysis

The antenna performance on the resonant frequency and axial ratio is investigated through the effect of the multiple rectangular slots in the ground planes. When there is no slot in the ground plane, the impedance matching is very poor and the bandwidth is very narrow. The axial ratio in this situation is greater than 12 dB, which portrays a linearly polarized antenna. With the introduction of four horizontal slots on the ground plane in the x-axes, as shown in Figure 10.3(a), the reflection coefficient and axial ratio curves are presented in Figure 10.4. A dual frequency band is achieved at both the reflection coefficient and the axial ratio. This results from the horizontal slots placed exactly at the edges of the truncated annular slots. The wide ground plane in between the slots causes the separation in the two frequency bands. In Figure 10.3(b) and 10.3(c), eight and ten horizontal slots in the x-axes have been introduced and this gives a wider impedance bandwidth and wider axial ratio. The increase in the slots in the horizontal x-axis

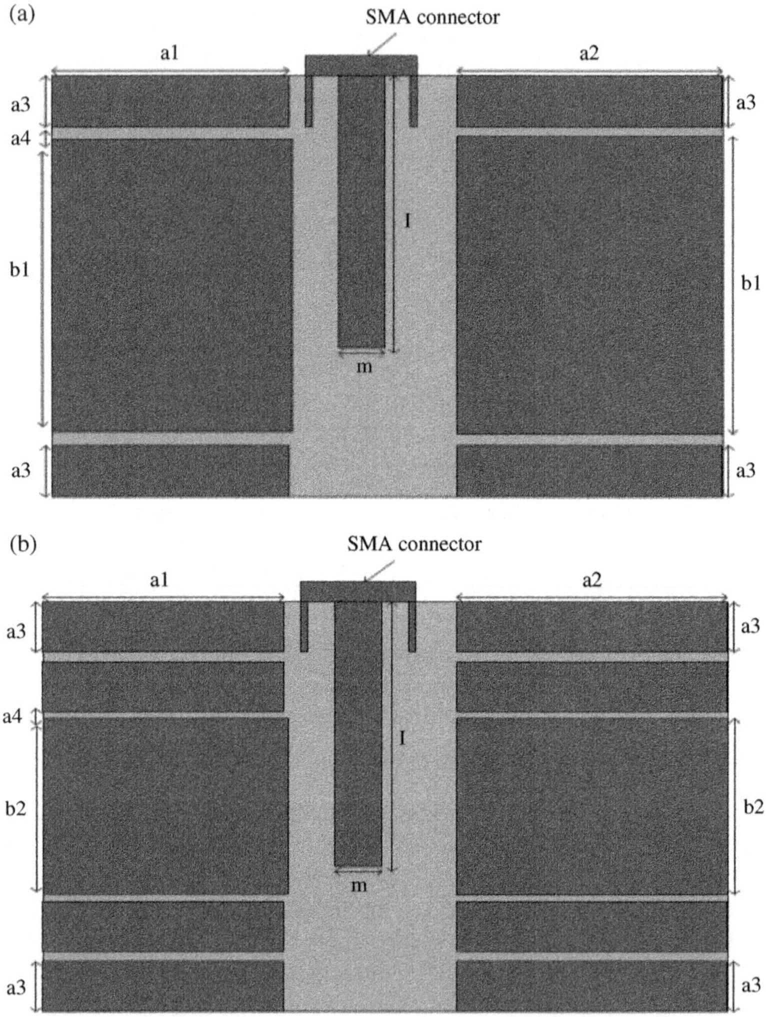

Figure 10.3 Different stages of the ground plane slots in the proposed CWSGP antenna. (a) Four horizontal slots. (b) Eight horizontal slots. (c) Ten horizontal slots. (d) Ten horizontal and two vertical slots. *Note: a1 = 17 mm, a2 = 21 mm, a3 = 4 mm, a4 = 1 mm, b1 = 24 mm, b2 = 14 mm, a5 = 6.5 mm, d1 = d4 = 10.5 mm, d2 = 5.5 mm and d4 = 9.5 mm, m = 3.1 mm and l = 23.5 mm*

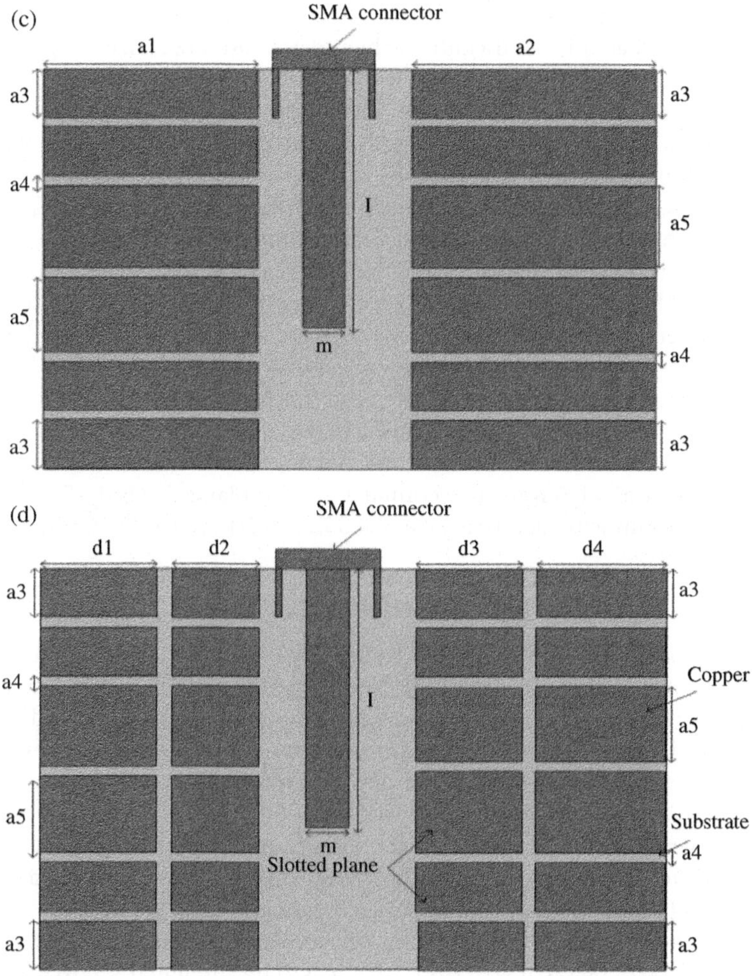

Figure 10.3 (Continued)

direction has improved the bandwidth of both the reflection coefficient and axial ratio. The slots succeed in creating an overlapping between the multiple bands and result in a broader bandwidth. To further improve the impedance and axial ratio bandwidth, a vertical slot is introduced in the y-axis direction, as shown in Figure 10.3(d). This causes a decrease in the quality factor of the antenna and a broader impedance bandwidth is achieved. More so, a better axial ratio and axial ratio bandwidth is achieved as a result of dual orthogonal perturbation. The introduction of the vertical slot extends the excitation of the two orthogonal modes in the antenna that produces a better 90° phase shift in the modes. A better equal amplitude and 90° phase difference in theta and phi is observed in broader frequency ranges when the vertical slot is included. In addition, from the e-field of the antenna, the vertical slots create better symmetry fields on the dual annular perturbation slot. This has resulted in a broader axial ratio of the proposed antenna. The combined results of the reflection coefficients and axial ratios for the various ground slots are presented in Figure 10.4(a) and 10.4(b), respectively.

Through extensive investigation, the optimized dimensions of the antenna obtained are: a = 11 mm, a1 = 4.8 mm, a2 = 5.2 mm, c = 1.5 mm, d1 = 10.5 mm, d2 = 6.5 mm, d3 = 4 mm, d4 = 10.5 mm, a3 = 4 mm, a4 = 1 mm, a5 = 6.5 mm. The compact ground plane size is d = 50 mm by e = 34 mm with electrical size of 0.3 λ_0 by 0.21 λ_0 at 1.842 GHz.

10.2.2 Experimental Investigation

CST Microwave Studio is used to investigate the performance of its reflection coefficient, radiation pattern performance, antenna gain and surface current distribution [24]. The antenna is fabricated as shown in Figure 10.5 using optimized dimensions and tested in the Antenna Measurement Laboratory at Queen Mary University of London. The return loss is measured using the Agilent vector network analyser N5230C PNA-L. There is good agreement between the measured and simulated curves of the reflection coefficient presented in Figure 10.6. A measured impedance bandwidth of 67% (1.858–3.145 GHz) is achieved while the simulated impedance bandwidth is 64% (1.842–3.023 GHz) at −10 dB return loss. The simulated ≤3 dB axial ratio bandwidth is 45% (2.243–3.253 GHz) while the measured is 2.35–3.32 GHz. The discrepancies are due to fabrication and measurement inaccuracies.

The measured radiation patterns at 2.6 GHz and 2.3 GHz are compared with the simulation results in Figures 10.7 and 10.8. Good agreement is obtained. The Right-Hand Circular Polarization (RHCP) pattern is stronger than the Left-Hand Circular Polarization (LHCP) pattern by more than 16 dB at the boresight direction. The measured gain and radiation efficiency of the antenna at 2.6 GHz and 2.3 GHz

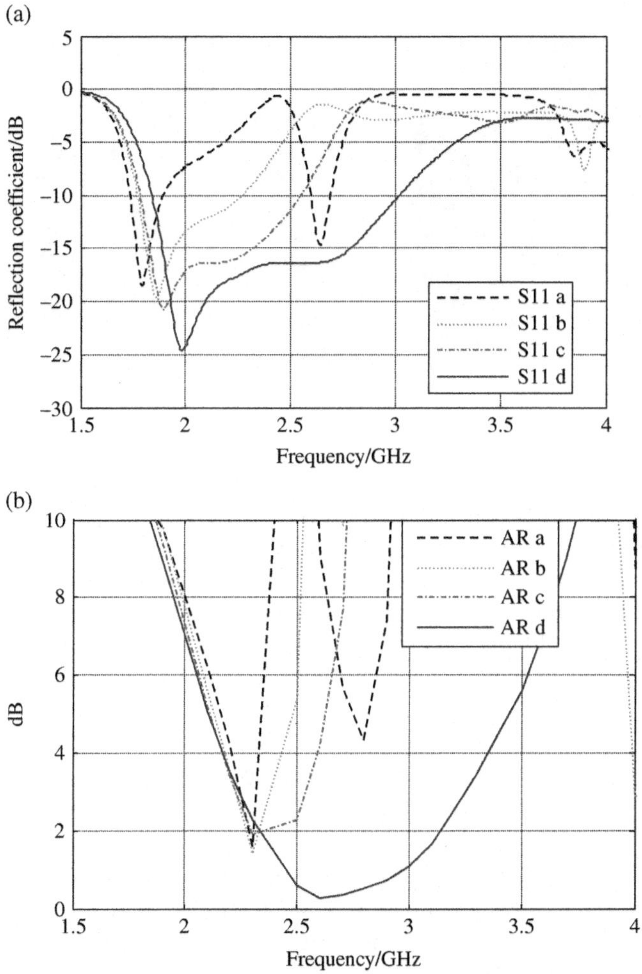

Figure 10.4 Simulated reflection coefficients and axial ratios of different stages of the grounded slots in the proposed CWSGP antenna. (a) Reflection coefficients and (b) axial ratios.

are 4.04 dBic and 6.95 dBic and about 78 and 80%, respectively. The ripples in the pattern occurred as a result of reflection from the cable.

Figures 10.9 and 10.10 show the simulated surface current distribution of the antenna. The current distribution reveals closely overlapping resonances that result in a wider impedance bandwidth. The

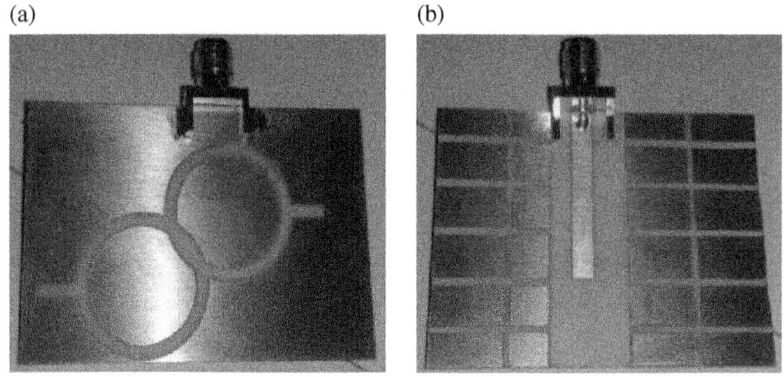

Figure 10.5 Fabricated prototyped CWSGP antenna: (a) top view and (b) back view.

current distribution is concentrated on the edges of the asymmetry stubs and the dual annular ring. As the resonant frequency increases, less current is observed starting from the feed line towards the ground plane and patch edge. In addition, symmetry current distribution is observed in the dual truncated ring slot coupled with the asymmetry stubs. This results in achieving a better CP pattern.

10.3 Design of a Multifunctional Compact Antenna for LTE Femtocells and GNSS Systems

This section focuses on the design of a stacked patch triple-band dual-polarized antenna for compact LTE and GNSS handheld terminals.

The growing demand for more accurate positioning using Global Positioning System (GPS) has increased research and development needs in this area. The GPS L1 frequency (1.575 GHz) has been used in most of the handheld devices for survey and geo-information applications. For example, the GPS L1 frequency band and GPS L2 (1.227 GHz) frequency band for Differential GPS (DGPS) are used to achieve a higher positioning accuracy with a less than 10 cm error margin [25]. The integration of GPS L1 and L2 bands with terrestrial wireless communication networks such as 4G Long Term Evolution (4G LTE 1800). 4G LTE 1800 can boost the dissemination of data from the information sources and the transaction stations. Prompt

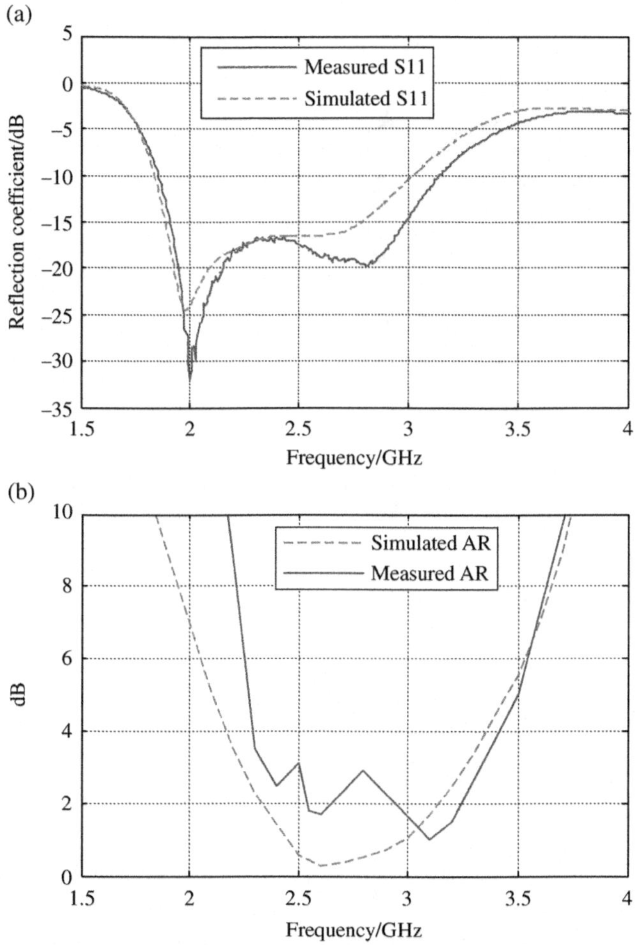

Figure 10.6 Measured and simulated reflection coefficient and axial ratio of the proposed CWSGP antenna: (a) measured and simulated reflection coefficient and [40] (b) simulated axial ratio.

evaluation with accuracy in timing and position can be obtained on such terminals. This will provide an all-in-one antenna for mobile communications and surveying data collection. 4G LTE 1800 has been selected since it is the backbone that underpins mobile telecommunications around the world. It is a robust and mature technology. It has the widest coverage and largest number of subscribers.

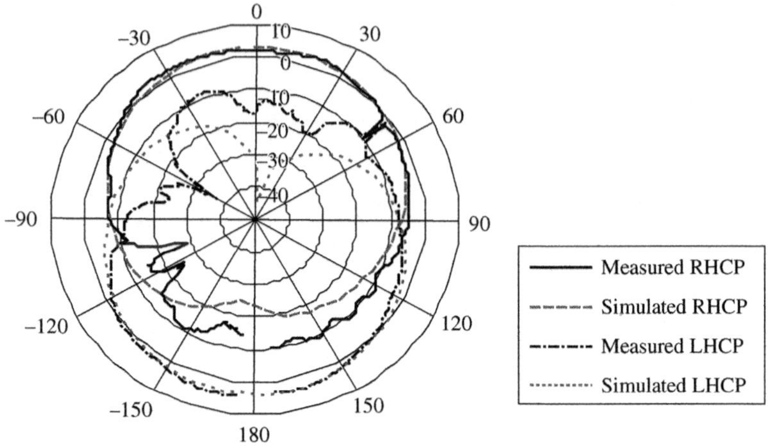

Figure 10.7 Measured and simulated radiation pattern at 2.6 GHz showing the RHCP and LHCP of the proposed CWSGP antenna.

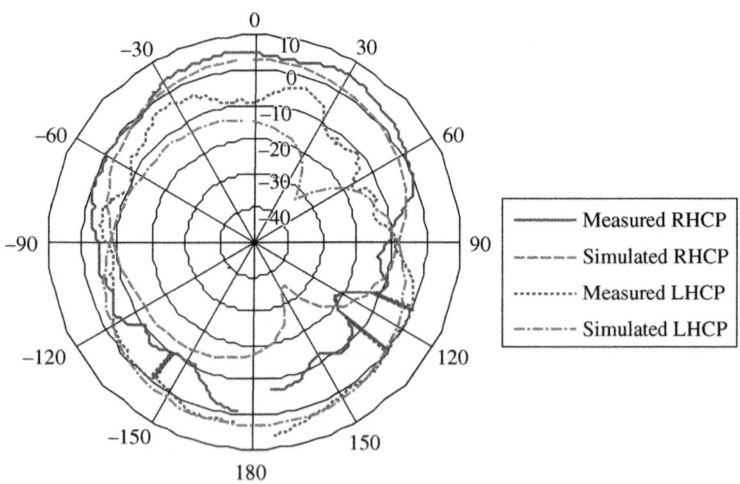

Figure 10.8 Measured and simulated radiation pattern at 2.3 GHz showing the RHCP and LHCP of the proposed CWSGP antenna.

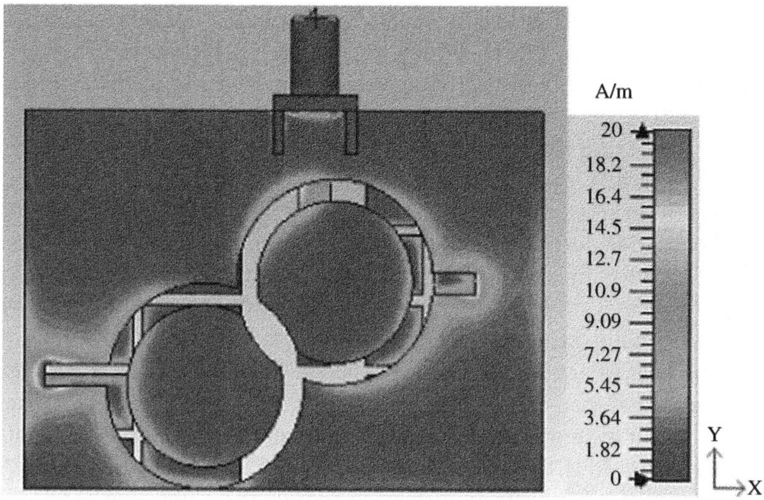

Figure 10.9 Simulated current distribution of the proposed CWSGP antenna at 2.6 GHz.

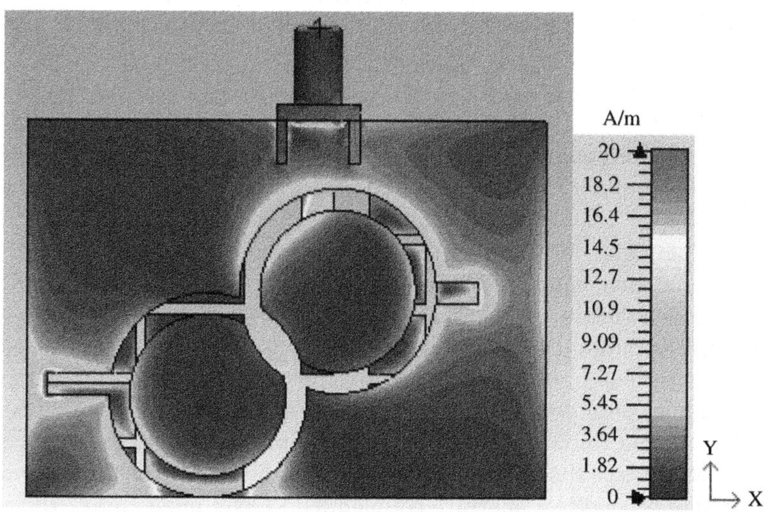

Figure 10.10 Simulated current distribution of the proposed CWSGP antenna at 2.3 GHz.

Presently, LTE1800, the band 3 spectrum is in commercial service in 104 countries on 226 LTE networks (LTE1800) representing over 45% of all LTE networks globally [26]. Combination of the GPS L1, L2 and 4G LTE 1800 will enable smooth running survey overflows while delivering the capabilities and convenience of a smart phone: voice, SMS text, data transfer and so on.

A number of studies have been reported in the literature that proposes antenna designs with dual-band operation covering GPS L1 and mobile communication frequencies [27–36]. For example, a low-cost radiating stacked wire-patch GSM antenna with an active corner truncated GPS printed element is presented in [30]. Two different ground planes are used for this proposed antenna. Kevin *et al.* presented a dual stacked patch antenna for GPS L1 and satellite digital audio radio systems (SDARS) in [31]. The antenna supports dual-polarization: RHCP at GPS L1 band while LHCP at SDARS band. A high permittivity substrate is used to reduce the patch size. An integration of monopoles and Archimedean spiral antenna has been proposed in [32] to achieve GPS, GSM/GPRS/COSPAS-SARSAT frequency bands. In [33], a combination of a corner truncated square ring patch antenna with an air substrate for GPS L1 band and a monopole for Universal Mobile Telecommunication Systems (UMTS) produces both circularly and linearly polarized radiation patterns. A stacked structure is used to realize the dual-band frequency for the applications of GPS/Digital Cellular Service (DCS) in [34]. The truncated square patch for the GPS operates in the fundamental mode while the annular ring patch used to generate the DCS operates in the TM_{21} mode. The antenna is connected to the dual feeds coupled with four slots embedded in the ground plane, which effectively increases the impedance bandwidth and lowers the resonant frequency. These studies only deal with dual-band antennas combining GPS L1 with a mobile communication frequency band. In [35], a triple-stacked microstrip patch antenna is built to achieve a triple-band operation for cellular phone, GPS and Personal Communications Service (PCS). The first and third patches produce the upper frequency bands (GPS and PCS) while the middle patch generates the lower frequency band (cellular phone). An aperture coupled feed technique is used with a thicker foam material (10.16 mm) to achieve a broader impedance bandwidth. A multiband stacked patch antenna employing five radiating patches is presented in [36]. The ground plane size is $220 \times 180\ mm^2$ while the largest patch element size is $80 \times 80\ mm^2$. Triple patches placed side by side are used to achieve multiband by utilizing a microstrip fed technique [37]. The first patch is directly connected to the microstrip feed while the other two patches are electromagnetically coupled. The complexity of the feeding systems and the use of high permittivity dielectric materials in these antennas reduces the gain. It affects the performance of these antennas and reduces the efficiency of handheld terminals. Also, the bulky size of these antennas makes them unsuitable for compact handheld terminals.

This section presents a novel design of a single feed, stacked patch, triple-band, dual polarized antenna for operation in the GPS L1 and L2 and LTE1800 band. The GPS performance in both L1 and L2 bands has a CP with a broadside radiation pattern while a LP is achieved in the LTE1800 band. The compactness, simplicity and distinct frequency bands make this proposed antenna quite suitable for handheld terminals.

10.3.1 Numerical Analysis

The configuration of the proposed dual polarized antenna is shown in Figure 10.11. This antenna is made up of three stacked radiating elements. The three square patches are etched on each side of an inexpensive FR4 substrates with thickness of h1 = h2 = h3 = 1.6 mm and relative permittivity of 4.4, where h1, h2 and h3 are the thickness of the lower, middle and upper substrates, respectively. The lower patch is used for producing the GPS L1 band, the middle patch for the GPS L2 band and the upper patch for the LTE1800 band. A slot is used in the lower and middle patches which serves two purposes. First, it lengthens the surface current path of the antenna and thus reduces the resonant frequency by increasing the physical length of the patch [38]. Second, it is used to improve axial ratio (AR) due to the resonant degradation of a singly fed patch antenna [39]. The lower and middle square patches are corner truncated to generate the CP broadside radiation patterns. The upper patch without corner truncation gives a linearly polarized pattern required for the LTE1800 system. A single probe (radius 0.65 mm) of 50Ω input impedance is connected to the upper patch through via holes in the lower and middle patches. The lower and middle patches are electromagnetically coupled to the probe.

In order to be able to predict the antenna performance, extensive optimizations are conducted using CST Microwave Studio package [24]. For operating frequencies 1.227 GHz (RHCP), 1.575 GHz

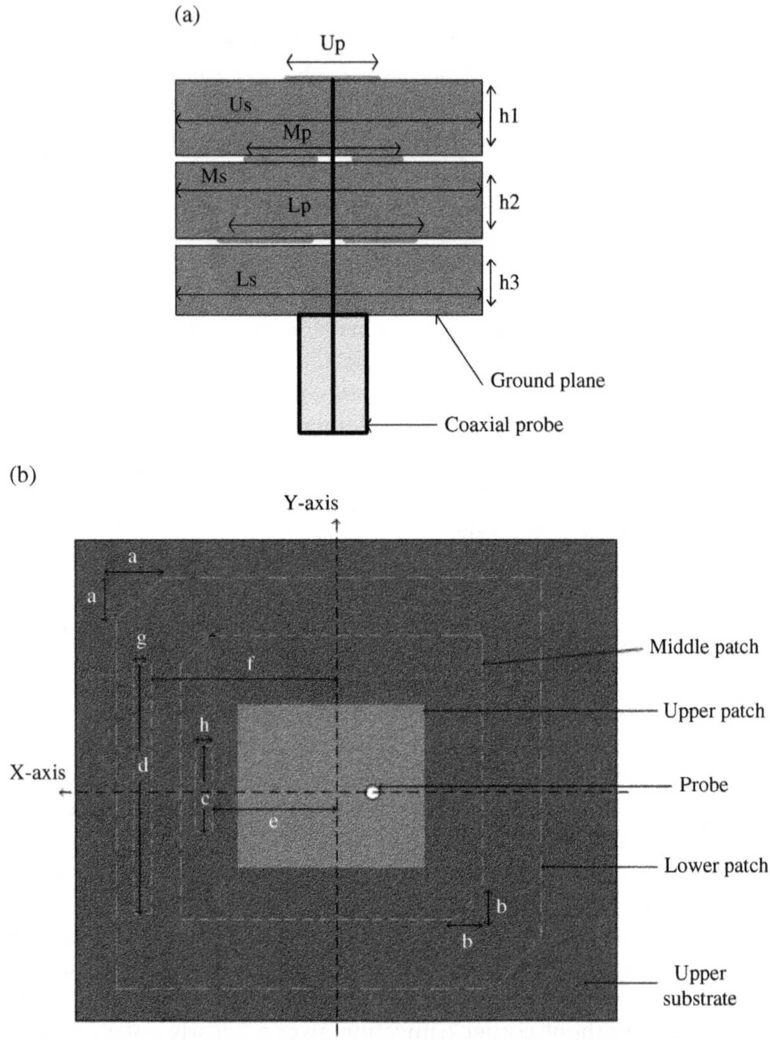

Figure 10.11 Geometry of the proposed stacked patch triple-band dual-polarized antenna: (a) configuration of the antenna and (b) top view of the antenna [41].

(RHCP) and 1.800 GHz (LP), the square patch sizes are 55.3 mm for the lower patch (Lp), 43.0 mm for the middle patch (Mp) and 39.6 mm for the upper patch (Up), respectively; while the via hole diameters for the lower patch and middle patch are 3.6 and 4.0 mm. The parametric sweep has been conducted on the probe position to achieve a better input impedance. The optimum position for the feeding probe is at the x-axis = 0 and y-axis = 13.73 mm. The sensitivity of the antenna performance on the resonant frequency and AR are investigated with the change of I-slot size and position in respect of the x- and y-axes in the lower and middle patch.

Figure 10.12 shows the effect of the I-slot breadth, g, of the lower patch on the reflection coefficient and the AR of the antenna. It can be observed that as the I-slot breadth decreases, the resonant frequency decreases while the bandwidth increases. This mainly affects the lower frequency band with little or no effect on the middle and upper frequency band. The AR improves as the breadth decreases and coincides with the centre frequency of the lower band frequency. No effect is observed on the AR of the middle frequency band. Similar effect was observed in the I-slot breadth, h, of the middle patch. The effect of the I-slot length, c, in the middle patch is shown in Figure 10.13. When the length, c, increases, the resonant frequency also increases with a reduction of the return loss by 1 dB. The AR improves as the length c decreases at the detriment of return loss. The optimum length c = 6 mm gives the best AR bandwidth for this middle frequency band. There is no significant effect on the lower and upper frequency bands. A similar effect is observed in the I-slot length, d, of the lower patch. Further study has been conducted on the effect of the I-slot position from the centre of the lower patch, f, and the plots are shown in Figure 10.14. As the I-slot position f increases, the impedance bandwidth and return loss of band 1 increases. In addition, the resonant frequency shifted forward with no traceable effect on the middle and upper frequency bands. This shows that the I-slot position f has shortened the surface current of the patch, therefore increasing the resonant frequency of the lower patch. The AR of band 1 improves as the I-slot position f increases coupled with better AR bandwidth. The optimum slot position, f, is 25 mm. A similar effect is observed in the I-slot position, e, of the upper patch. From these studies, it can be deduced that the lower and middle patch operate independently with the inclusion of the I-slot. The optimized dimension for the lower and middle I-slot are 'd × g' = 29.2 × 1 mm and 'c × h' = 6 × 1 mm. Other parameters used for the final design of this antenna are a = 5.7, b = 4.9, f = 25.0 and e = 16.0 mm, while the ground plane size is 70 × 70 mm^2. The size of the ground plane can be reduced but to the detriment of the antenna gain, efficiency and large back lobe in the radiation pattern.

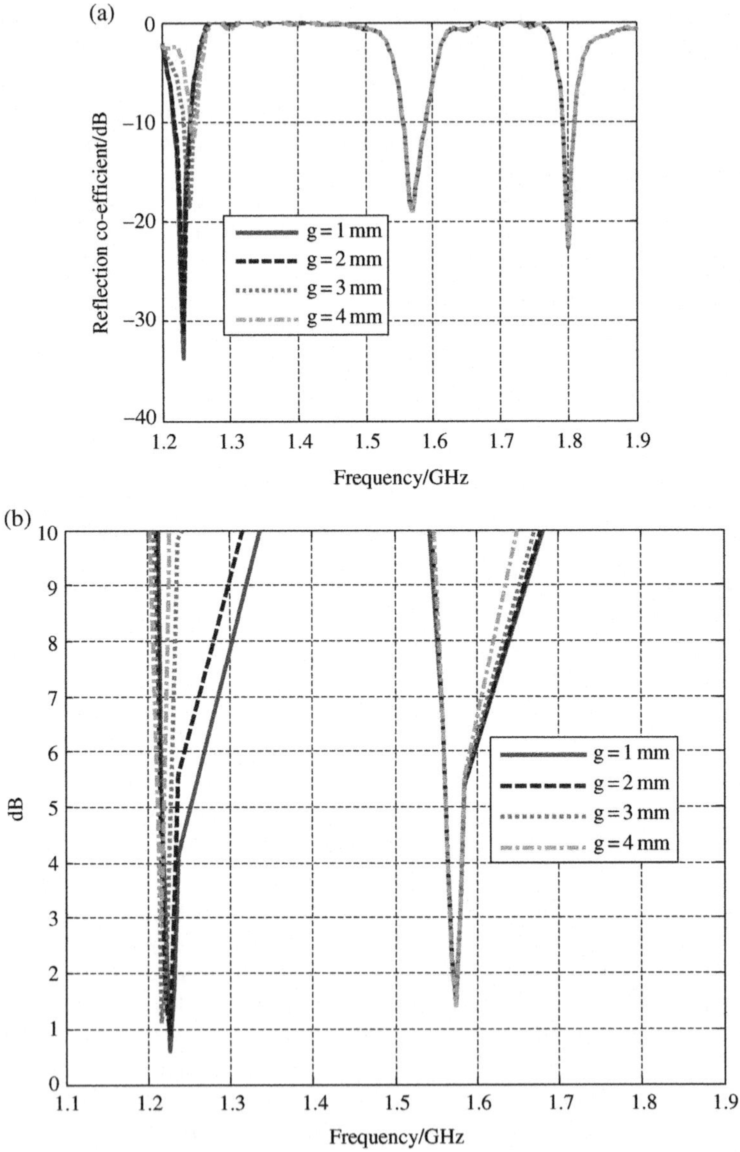

Figure 10.12 Variation of the antenna parameters for different values of breadth g of I-slot (lower patch): (a) reflection coefficient and (b) AR [41].

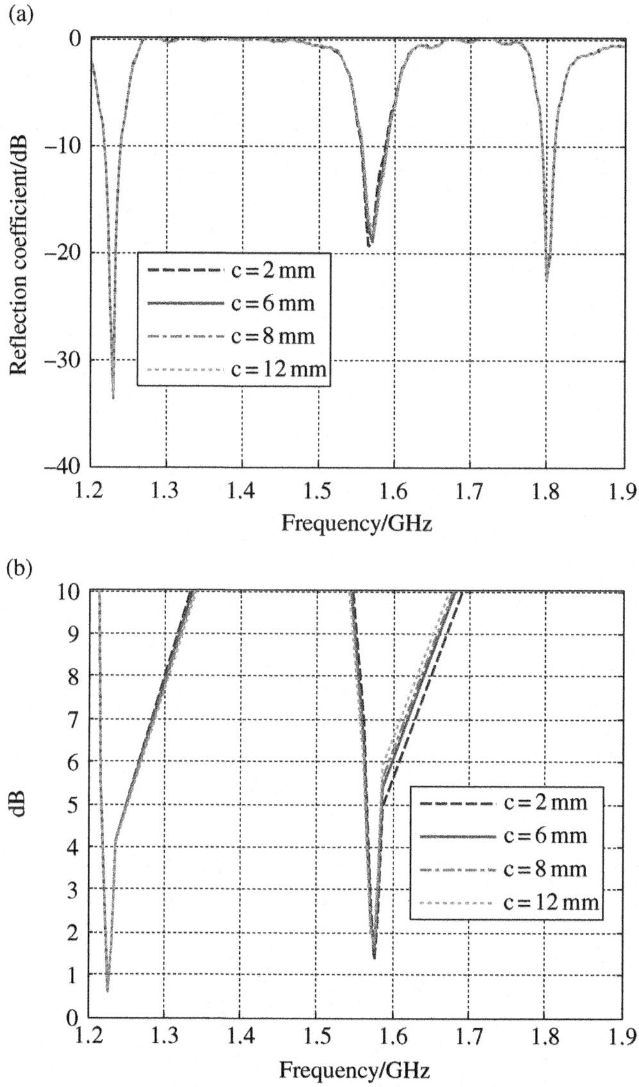

Figure 10.13 Variation of the antenna parameters for different values of length, c (middle patch I-slot): (a) reflection coefficient and (b) AR [41].

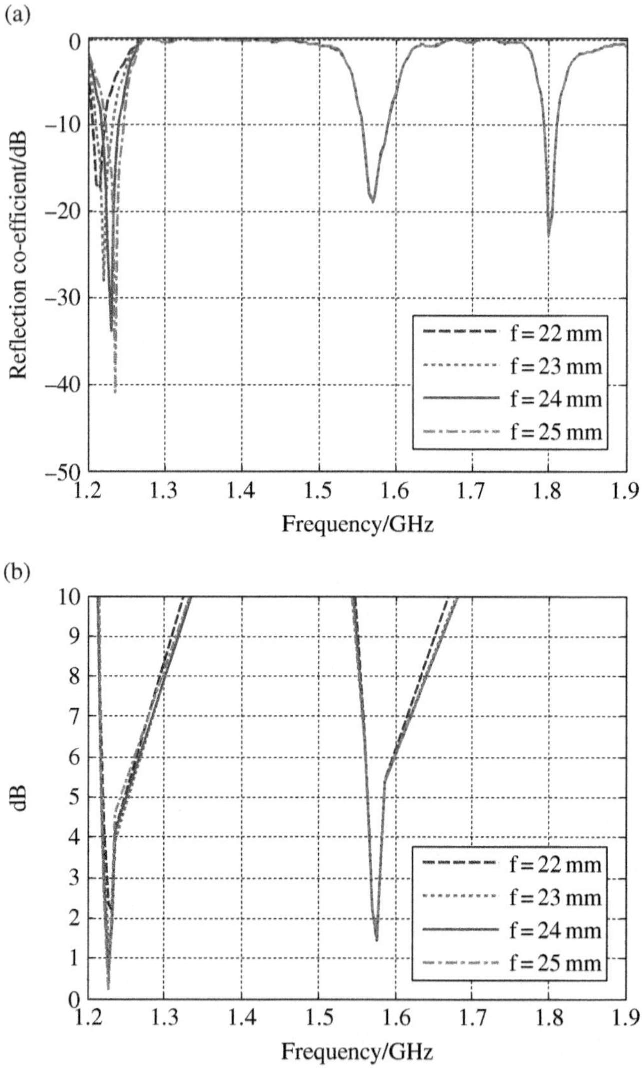

Figure 10.14 Variation of the antenna parameters for different values of breadth g of I-slot (lower patch): (a) reflection coefficient and (b) AR [41].

10.3.2 Experimental Investigation

The prototype of the proposed antenna is fabricated at QMUL, as shown in Figure 10.15, using optimized dimensions and tested in the Antenna Measurement Laboratory. The return loss is measured using an Agilent vector network analyser N5230C PNA-L while the radiation patterns are evaluated in the anechoic chamber. The simulated and measured reflection coefficient are shown in Figure 10.16(a). A good agreement between the two results is observed. Small discrepancies between the middle frequency band (GPS L2) and the upper frequency band (LTE1800) are attributed to fabrication errors. The measured minus 10 dB impedance bandwidth in the GPS L1 and L2 bands are 1.215–1.241 GHz (26 MHz) and 1.560–1.598 GHz (38 MHz) respectively; while the measured minus 6 dB impedance bandwidth at LTE1800 band is 1.771–1.846 GHz (75 MHz). The measured and simulated AR of the proposed antenna is presented in Figure 10.16(b). The ≤ 3 dB AR bandwidth at GPS L1 and L2 frequency band is 1.6% (1.220–1.240 GHz) and 0.8% (1568–1.580 GHz), respectively.

The measured and simulated radiation patterns in XZ-plane for the proposed antenna at 1.227 GHz and 1.575 GHz are shown in Figures 10.17 and 10.18, respectively. The RHCP pattern is stronger than the LHCP pattern by more than 20 dB and 18 dB at both frequency bands in the boresight direction. The right-hand back lobe is less than -20 dB while the front-to-back ratios of the RH circular

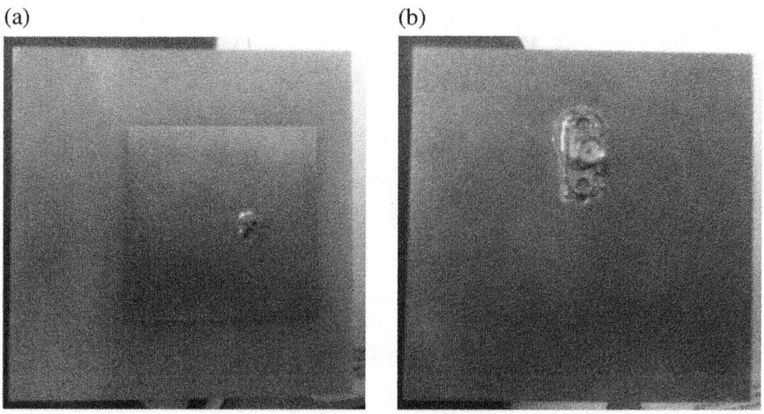

Figure 10.15 Fabricated prototype of the stacked patch triple-band dual-polarized antenna: (a) top view and (b) bottom view [41].

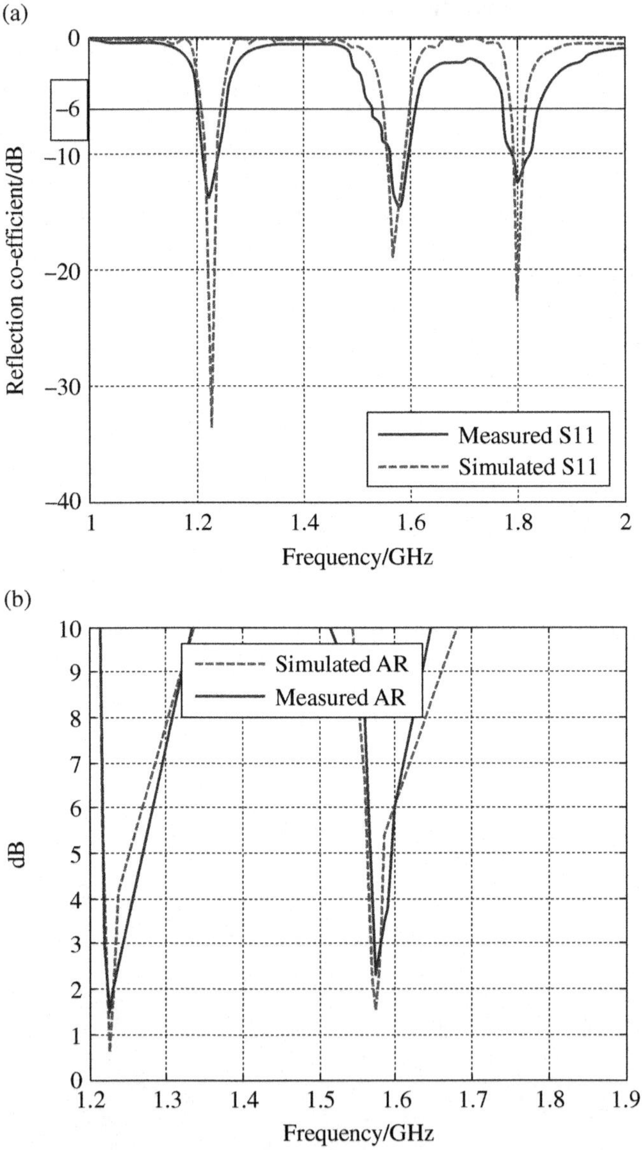

Figure 10.16 Measured and simulated reflection coefficient and AR of the stacked patch triple-band, dual-polarized antenna: (a) measured and simulated reflection coefficient and (b) measured and simulated AR [41].

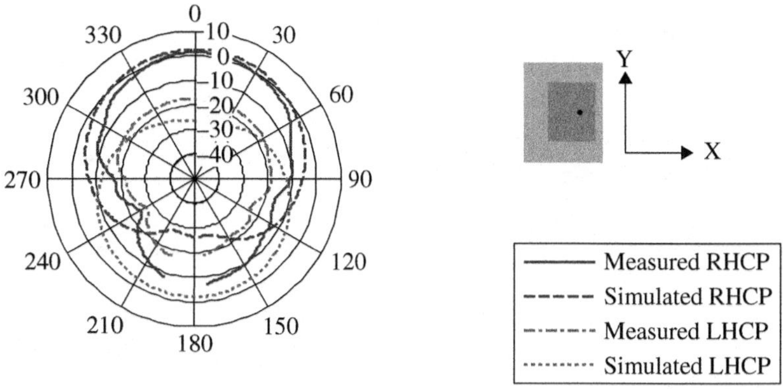

Figure 10.17 Measured and simulated RHCP and LHCP radiation patterns in XZ-plane of the stacked patch triple-band, dual-polarized antenna at 1.227 GHz [41].

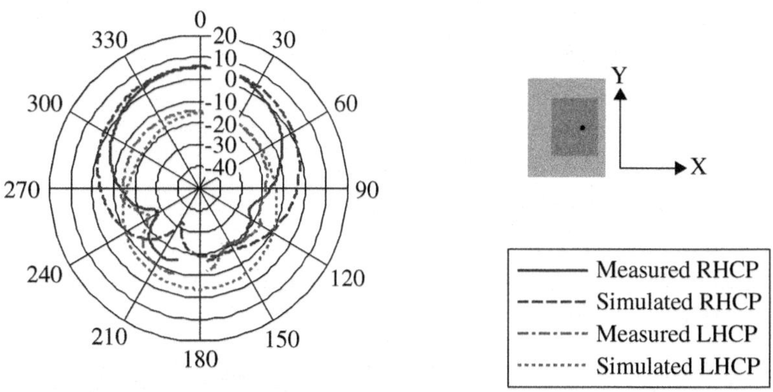

Figure 10.18 Measured and simulated RHCP and LHCP radiation patterns in XZ-plane of the stacked patch triple-band, dual-polarized antenna at 1.575 GHz [41].

polarization is more than 25 dB. The measured and simulated linear radiation pattern at 1.800 GHz in the XZ-plane is presented in Figure 10.19. The measurement agrees well with the simulation and the co-polarization is about 16 dB higher than the cross polarization. The measured and simulated radiation efficiencies of the antenna in the three bands are above 70 and 80%, respectively. The simulated 3D radiation patterns of the three frequency bands are presented in Figures 10.20—10.22. The Cartesian plot of the AR at 1.227 GHz and

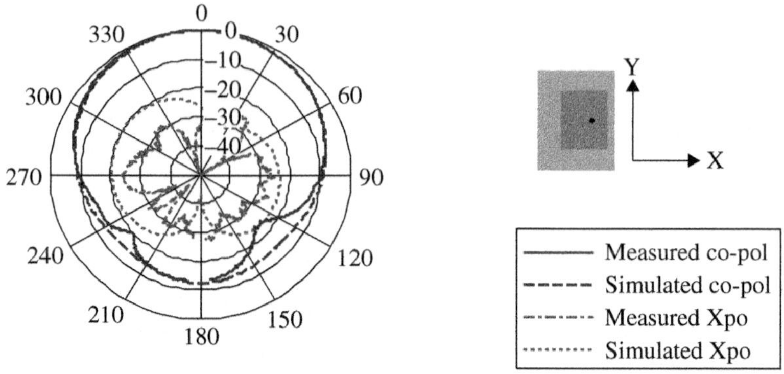

Figure 10.19 Linear polarized radiation pattern at 1.800 GHz showing both the measured and simulated co and cross polarization at XZ-plane [41].

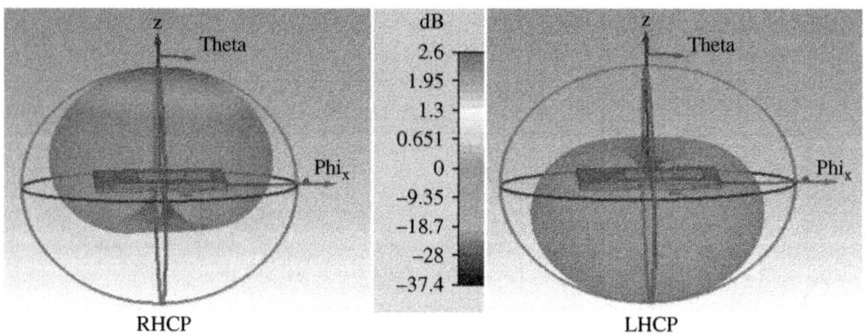

Figure 10.20 3D radiation pattern showing the RHCP and LHCP of the stacked patch triple-band, dual-polarized antenna at 1.227 GHz [41].

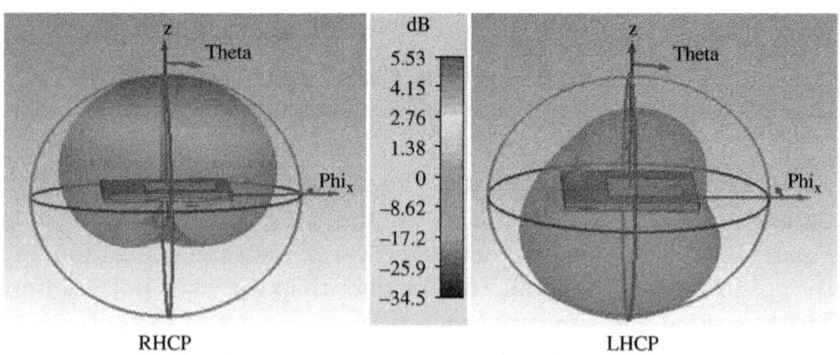

Figure 10.21 3D radiation pattern showing the RHCP and LHCP of the stacked patch triple-band, dual-polarized antenna at 1.575 GHz [41].

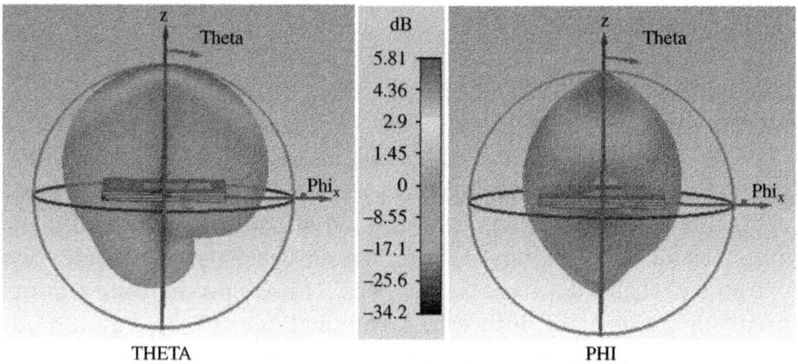

Figure 10.22 3D radiation pattern showing the THETA and PHI of the stacked patch triple-band, dual-polarized antenna at 1.800 GHz [41].

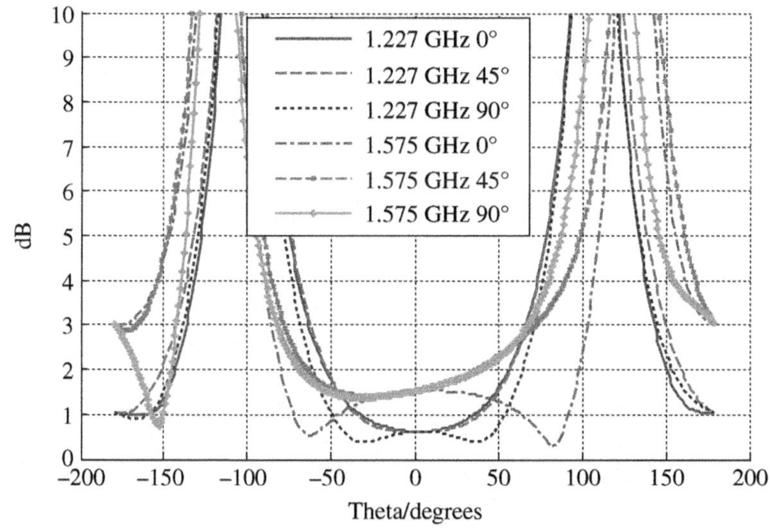

Figure 10.23 Simulated Cartesian plots of the stacked patch triple-band, dual-polarized antenna showing a wider beamwidth at the upper hemisphere for 1.227 GHz and 1.575 GHz at ∅ = 0°, 45° and 90° [41].

1.575 GHz depicts that at the boresight, minimum AR of 0.62 dB and 1.52 dB is achieved. At ≤3 dB AR, the beamwidth is about 126° for ∅ = 0°, 127° for ∅ = 45° and 190° for ∅ = 90° at 1.227 GHz. At 1.575 GHz, the AR beamwidth is 136° in ∅ = 0°, 148° for ∅ = 45° and 148° in ∅ = 90° plane, as shown in Figure 10.23.

10.4 Summary

The impasse trend to LTE technology has brought positive advancements to mobile handheld terminals. This necessitates the need for an effective compact and multimode antenna with several and simultaneous functions. This chapter has presented a blueprint for antenna engineers and researchers working on broadband and multiband antennas for LTE. Multimode antennas ensure constant coverage and extremely reliable data transfer services. This is a prerequisite requirement for present and future LTE terminal devices. The multimode antenna helps in support of roaming between different regions where different standards are used for communication services.

With the global coverage of the LTE1800 in over 104 countries, the need to integrate GNSS frequency bands for effective communication and positioning is essential. Good CP antenna with broad beamwidth and low cross polarization is achieved and presented. The compact antennas presented can work effectively in LTE mobile communication products coupled with satellite communications.

10.5 Open Challenges and Issues

Multimode antennas also face challenges similar to that discussed in Chapter 9 for reconfigurable antenna structures. Particularly, structured studies on the viability and feasibility of such antennas in the perspective of overall LTE system are not widely available. Moreover, performance of these antennas in a complete system needs to be evaluated in order to analyse their true potential and required possible changes. Moreover, design guidelines that can be used by the antenna designers to model structures at other than the already discussed frequencies also need serious effort.

References

1 5G Americas "Mobile broadband evolution towards 5G: Rel-12 & Rel-13 and beyond", available online at: www.5gamericas.org/files/6214/3569/1603/4G_Americas_Mobile_Broadband_Evolution_Toward_5G-Rel-12_Rel-13_June_2015.pdf (accessed December 2017), 2015.

2 J. Wannstrom, "LTE-Advanced" 3GPP A Global Initiative: The Mobile Broadband Standard, June 2013.
3 J. L. Burbank, J. Andrusenko, J. S. Everett and W. T. M. Kasch, *Wireless Networking: Understanding Internetworking Challenges*, IEEEE 2013, John Wiley & Sons, Inc., 2013.
4 S. Sesia, I. Toufik, and M. Baker, *LTE – The UMTS Long Term Evolution, From Theory to Practice*, John Wiley & Sons, Ltd, Chichester, 2011.
5 Y. Yu, F. Jolani and Z. Chen, "A wideband omnidirectional horizontally polarized antenna for 4G LTE applications", *IEEE Antennas and Wireless Propagation Letters*, vol. 12, pp. 686–689, 2013.
6 S. Zhang, K. Zhao, Z. Ying and S. He, "Adaptive quad-element multi wideband antenna array for user effective LTE MIMO mobile terminals", *IEEE Transactions on Antennas and Propagation*, vol. 61, no. 8, pp. 4275–4283, 2013
7 K. L. Wong, T. W. Kang and M. F. Tu, "Internal mobile phone antenna array for LTE/WWAN and LTE MIMO operations", *Microwave Opt. Tech. Lett.*, vol. 53, no. 7, pp. 1569–1573, 2011.
8 S. Yang, T. Jiang, J. L. W. Li and Y. L. Ban, "Internal low profile multi band MIMO antenna for WWAN/LTE mobile phone applications" *IEEE International Symposium Antennas and Propagation Society*, pp. 522–523, Orlando, FL, July 2013.
9 S. Gao, Q. Luo and F. Zhu, *Circularly Polarized Antennas*, John Wiley & Sons, Ltd, Chichester, 2014.
10 Z. Wang, S. Fang, S. Fu and S. Jia, Single-fed broadband circularly polarized stacked patch antenna with horizontally meandered strip for universal UHF RFID applications", *IEEE Trans. On Microwave Theory and Tech.*, vol. 59, no. 4, pp. 1066–1073, April 2011.
11 O. P. Falade, Y. Gao, X. Chen, and C. Parini, "Stacked patch dual polarized antenna for triple-band handheld terminals", *IEEE Antennas and Wireless Propag. Lett.*, vol. 12, pp. 202–205, Feb. 2013.
12 R. L. Li, J. Laskar, and M. M. Tentzeris, "Wideband probe-fed circularly polarised circular loop antenna", *IET Electron. Lett.*, vol. 41, no. 18, pp. 997–999, 2005.
13 J. Y. Sze, C. I. G. Hsu, Z. W. Chen, and C. C. Chang, "Broadband CPW-fed circularly polarized square slot antenna with lightening-shaped feedline and inverted-L grounded strips", *IEEE Trans. Antennas Propag.*, vol. 58, no. 3, pp. 973–977, March 2010.

14 X. L. Bao, M. J. Ammann and P. McEvoy, "Microstrip-fed wideband circularly polarized printed antenna", *IEEE Trans. Antennas Propag.*, vol. 58, no. 10, pp. 3150–3156, October 2010.

15 T. W. Chiou and K. L. Wong, "Designs of compact microstrip antennas with a slotted ground plane", in *IEEE Antenna and Propagation Society Symposium*, Boston, MA. vol. 2, pp. 732–735, July 2001.

16 H. D. Chen, "Compact circularly polarised microstrip antenna with slotted ground plane", *IET Electron. Lett.*, vol. 38 no. 13, pp. 616–617, June 2002.

17 X. L. Bao and M. J. Ammann, "Compact annular-ring embedded circular patch antenna with cross-slot ground plane for circular polarisation", *IET Electron. Lett.*, vol. 42, no. 4, pp. 192–193, February 2006.

18 S. Maci and G. Gentili, "Dual frequency patch antennas", *IEEE Antennas and Propagation Magazine*, vol. 39, pp. 13–20, December 1997.

19 R. Vaughan and J. Andersen, "A multiport patch antenna for mobile communications", in the *14th European Microwave Conference*, pp. 607–612, 1984.

20 K. L. Wong, *Compact and Broadband Microstrip Antennas*, John Wiley & Sons, Inc., Hoboken, 2002

21 LTE Frequency Bands. Website: available at: http://anisimoff.org/eng/lte_bands.html (accessed December 2017).

22 J. Liu and Q. Xue, "Broadband long rectangular patch antenna with high gain and vertical polarization", *IEEE Trans. Antennas Propag.*, vol. 61, no. 2 pp. 539–546, February 2013.

23 C. Y. Pan, J. Y. Jan and L.C. Wang, "Compact and broadband microstrip-line-fed modified rhombus slot antenna", *Radioengineering*, vol. 22, no. 3, pp. 694–699, September 2013.

24 CST-Microwave Studio, User's Manual 2015.

25 R. Leandro, H. Landau, M. Nitschke, M. Glocker, S. Seeger, X. Chen, et al., "RTX Positioning: The next generation of cm-accuracy Real-time GNSS positioning", *Proc., of the 24th Inter., Technical Meeting of the Satellite Division of the Institute of Navigation*, September 2010.

26 Global Mobile Supplier Association 2016, http://gsacom.com/paper/gsa-evolution-to-lte-report-494-lte-networks-commercially-launched-in-162-countries/.

27 G.Z. Rafi, M. Mohajer, A. Malarky, P. Mousavi, and S. Safavi-Naeini, "Low-profile integrated microstrip antenna for GPS-DSRC application", *IEEE Antennas and Wireless Propagation Lett.*, vol. 8, pp. 44–48, 2009.

28 C. Su, C. Su, and K.L. Wong, "Compact dual-band circularly polarized antenna for GPS/ETC operation on vehicles", *Microwave and Optical Technology Lett.*, vol. 40, no. 6, pp. 509–511, March 2004.

29 B. Lee, F. J. Harackiewicz, K.H. Kong, J. Byun, and S.H. Yang, "Dual-band dual-circularly polarized sector antenna for GPS and DAB system", *Microwave and Optical Technology Lett.*, vol. 46, no. 1, pp. 43–46, July 2005.

30 L. Zaid and R. Staraj, "Miniature combined GSM-GPS wire-patch antenna", *Microwave and Optical Technology Lett.*, vol. 48, pp. 970–972, 2006.

31 K. Geary, J.H. Schaffner, H. Hui-Pin, H.J. Song, J.S. Colburn and E. Yasan, "Single-feed dual-band stacked patch antenna for orthogonal circularly polarized GPS and SDARS applications", *Vehicular Technology Conference*, pp. 1–5, 2008.

32 R. Azaro, F.G.B. DeNatale, M. Donelli, A Massa and E. Zeni, "Optimized design of a multifunction/multiband antenna for automotive rescue systems", *IEEE Transaction on Antennas and Propagation*, vol. 54, no. 2 pp. 392–400, February 2006.

33 C. Y. D. Sim, J. S. Row, and S. H. Chen, "A dual-band antenna design for GPS and UMTS applications", *Microwave and Optical Technology Lett.*, vol. 49, pp. 1935–1939, August 2007.

34 S. Y. Lin and K. C. Huang, "A compact microstrip antenna for GPS and DCS application", *IEEE Transaction on Antennas Propagation*, vol. AP-53, pp. 1227–1229, March 2005.

35 H. C Ryu, H. R. Ahn, S. H. Lee and W. S. Park, "Triple-stacked microstrip patch antenna for multiband system", *Electronic Lett.*, vol.38, no. 24, pp. 1496–1497, November 2002.

36 J. Anguera, G. Font, C. Puente, C. Borja and J. Soler, "Multifrequency microstrip patch antenna using multiple stacked elements", *IEEE Microwave and Wireless Components Lett.*, vol. 13, no. 3, pp. 123–124, March 2003.

37 H. El Kamchouchi and Y. M. Madany, "Multiband triple-patch microstrip antenna using 3D-FDTD method", *Proceedings of the National Radio Science Conference 2002*, pp. 176–183, March. 2002.

38 K. P. Yanh, K. L. Wong, and J. H. Lu, "Compact circularly-polarized equilateral-triangular microstrip antenna with Y-shaped slot", *Microwave and Optical Technology Lett.*, vol. 20, pp. 31–34, January 1999.

39 G. Ramesh, P. Bhartia, I. Bahl, and A. Ittipiboon, *Microstrip Antenna Design Handbook*, Artech House Inc., London, UK, pp. 493–526, 2001.

40 O. P. Falade, Y. Gao, X. Chen and C. Parini, "Compact slotted ground plane circularly polarized antenna for mobile communication applications", *2013 IEEE Antennas and Propagation Society International Symposium (APSURSI), Orlando, FL*, pp. 426–427, 2013.

41 O. P. Falade, "Multifunctional antenna for global navigation satellite system and mobile communication", PhD thesis Queen Mary University of London.

11

Human Body Effects on LTE Femtocell Antennas

Masood Ur Rehman[1] *and Qammer Hussain Abbasi*[2]

[1] *School of Computer Science and Technology, University of Bedfordshire, Luton, UK*
[2] *School of Engineering, University of Glasgow, Glasgow, UK*

The development of portable devices including smart phones, smart watches, wireless health monitors, wireless sensor systems, Personal Navigation Devices (PNDs) and laptops has brought a revolution in the field of wireless communications. This rapid growth of research and development has been promoted by the wider approach, mobility and ease of use provided by these devices. It has necessitated the advances of several portable and wearable technologies within LTE femtocells utilizing Wireless Personal Area Networks (WPAN) and Wireless Body Area Networks (WBAN). Since the human body is an integral part of the WPAN/WBAN applications, interaction between such wearable devices and the human body is of paramount importance to establish a reliable communication link.

This chapter presents an investigative study to characterize the electromagnetic interaction between body-mounted LTE antennas and the human body. Various factors affecting this communication link including on-body antenna placement, antenna and body separation and body posture are considered. Thorough numerical modelling, supported by measurements, has been carried out to demonstrate the importance of numerical modelling and explore these body proximity effects.

LTE Communications and Networks: Femtocells and Antenna Design Challenges, First Edition. Edited by Masood Ur Rehman and Ghazanfar Ali Safdar.

11.1 Interaction of the Human Body with Antennas

The field of wireless communications is growing day by day and it is predicted by the Wireless World Research Forum (WWRF) that, by 2017, over 7 trillion wireless devices will serve 7 billion people [1]. This, along with continuous demand for convergent multiple technologies with ease of portability, mobility, flexibility, unobtrusiveness, low cost, low weight and miniaturization, has lead to a new wave of portable and wearable devices that now make up the largest share in this huge market. Applications of such devices span communications, health-care, navigation and positioning, entertainment, security and surveillance, space exploration and the military [2, 3]. LTE femtocells are one of the major fields that employ such devices.

The design of body-worn and handheld devices has many aspects to take into account including safety for the user, low power requirements, low profile, scalable data rates and low complexity to provide mobility with ease of use and low cost to be affordable for the masses [2, 4]. This area has undergone great research and development activities because of the huge scale of current usage and anticipated future increase in the use of WPAN/WBAN devices [1, 5–8].

The human body is an inherent part of WPAN/WBAN applications. Being wireless makes these applications vulnerable to electromagnetic distortions caused by lossy human body tissues. Efficient deployment of such systems necessitates thorough understanding of these effects. Therefore, the electromagnetic interaction between the human body and antennas has been an important research area [9–17].

It is now a well established fact that the performance of the antennas operating in close proximity of the human body is degraded due to losses caused by varying electric properties of human tissues. As a result, the WPAN/WBAN antenna suffers from distortion in radiation pattern, reduction in radiation efficiency and de-tuning of input impedance [2, 3, 18–21].

Fears about human safety resulted in standardization of maximum levels of exposure of human tissues to electromagnetic radiation emitted by body-worn antennas defined in terms of Specific Absorption Rate (SAR) [22–24]. Many studies have also been reported on human body interaction with WPAN/WBAN antennas and their compliance to SAR standards [25–28]. Human body phantoms have also been developed for experimental investigations into wearable antenna performance [29].

11.2 Numerical Modelling of the Human Body

Efficient deployment of the Body-centric Wireless Network (BCWN) systems makes it indispensable to evaluate the interaction of the human body and body-worn antennas. Measurement campaigns with human subjects have disadvantages of uncontrollable movements, longer experimental durations, lack of repeatability and pose health and safety hazards [12, 30]. Human body modelling is an alternative approach that can be accomplished either experimentally or numerically. Experimental modelling involves use of physical phantoms made from solid, liquid or gel materials while numerical modelling is based on body phantoms embedded in electromagnetic codes. Use of physical phantoms is limited due to greater cost and lower flexibility in terms of body shape and posture [31].

Numerical analysis is a powerful tool that provides useful information about such scenarios quicker than the actual measurements giving the user full control of the design environment. Modern software solutions of Maxwell's equations such as the finite difference time domain (FDTD) and method of moment (MoM) with efficient 3D modelling of large practical scenarios not only remove the drawbacks of measurements but also provide an overview of the expected on-body performance of antennas and a means to validate the results through comparison of simulation and measurement.

Size of the problem (i.e. voxel size) and definition of the input play important roles in governing the accuracy of the results obtained using any numerical method. It is a process where all inputs like geometry and electrical properties of the model are adjusted to get an output with acceptable precision. A good solution provides a balance between the precision of the outputs, required computational resources and simulation times. Therefore, it needs characterization of different types of human body models. A wide variety of numerical human body phantoms have been used in the literature ranging from simple canonical models to complex voxel models. Simple models include homogeneous or layered flat models [32] used to evaluate electromagnetic effects where energy is radiated from simple sources such as a plane wave or small dipole antenna, spherical models used to evaluate electromagnetic effects inside the human head [33, 34] and cylindrical models [35, 36] used to evaluate the electromagnetic effects of the whole-body presence. Advancement in medical imaging techniques like magnetic resonance imaging (MRI) along with greatly enhanced computing power and software approaches has made it possible to develop detailed realistic homogeneous and inhomogeneous head and whole-body models composed of many voxels [2, 4, 11, 12, 14]. Human-head models are usually used for the characterization of mobile terminal antennas when placed near to the head in a talking position [12, 25]. Whole-body voxel models with a spatial resolution as high as 1 mm^3 have been developed to represent humans of different physiques and classify more than 30 different types of human body tissues [31, 37]. CST Microwave Studio® has collaborated with Visible Human Project® and enables the import of Voxel Man, which is based on a layered model obtained using a dissected male corpse [38]. Different on-body scenarios with varying body postures and on-body antenna positions can be easily simulated by importing these models in electromagnetic numerical tools.

The low resolution models consisting of human body parts like a head, hand or arm cannot provide a full account of the complete human body effects while high resolution realistic body models require huge computational resources and are very time consuming. For efficient work flow, a compromise between resources, time and accuracy of results plays a crucial role.

The choice of the right model for the study of body effects is very critical as computational efficiency decreases with an increase in model resolution and its size. In this section, the on-body LTE communication link between two body mounted antennas at 2.45 GHz has been studied for three different human body models to assess the computational difficulties and impact on the accuracy of the results. Measurement results are used as a reference to benchmark the working of the three models and recommend a viable solution in terms of accuracy and computational efficiency.

CST Microwave Studio® has been used to develop the three numerical models of the human body. Due to the complexity of human body composition, these models are designed employing a single layer homogeneous structure by averaging the tissue properties estimating the weight of the main tissue contents. They are approximated to include 10% skin, 30% fat, 40% muscle and 20% bone (averages of bone concellous, bone cortical and bone marrow), which results in an weighted averaged relative permittivity of 28.81 and conductivity of 1.02 S/m at 2.45 GHz. The values of the four types of tissue at 2.45 GHz used in this study are summarized in Table 11.1.

The meshing scheme is made adaptive where finer cell sizes have been used around the vital parts of the body. This method reduces

Table 11.1 Electric properties of specific human tissues at 2.45 GHz used within the constructed homogeneous body models [39].

Tissue	*Electric Properties*	
	Dielectric Constant (ϵ_r)	**Tissue Conductivity (σ) (S/m)**
Average bone	11.74	0.43
Fat	5.28	0.11
Muscle	52.74	1.73
Skin	38.01	1.46

the required number of cell volumes (voxels) in the computational domain significantly, hence the computation and time requirements. The Perfectly Matched Layer (PML) absorbing boundary conditions are implemented [38] with a maximum mesh cell size of 10 mm near the boundaries of the computational domain and a minimum mesh cell size of 0.08 mm at the edges of the solids in the computational domain.

The high level discretization of the whole-body model represents an average built human with a height of 1755 mm, as shown in Figure 11.1(a). This very realistic model based on MRI scans was developed by the DSTL, UK [37]. The authors made necessary adjustments to make it compatible with CST Microwave Studio®. This model is rigid and replicates a standing human subject.

To obtain a medium resolution model, the high-resolution human body model is modified keeping the head as a high-resolution object while simplifying remaining organs, as illustrated in Figure 11.1(b). This medium-resolution model represents a nearly realistic shape of an average built human with a height of 1755 mm and thickness of 220 mm. This model has benefits of reduced complexity and flexibility in terms of re-positioning body parts (e.g., sitting, standing, talking and holding phone positions).

Finally, a low-resolution simpler homogeneous model based on canonical geometries is developed, as depicted in Figure 11.1(c). The complete height of this low-resolution body model is 1755 mm. The thickness of the torso for this low-resolution model is 120 mm, while the radius of the head is 86 mm.

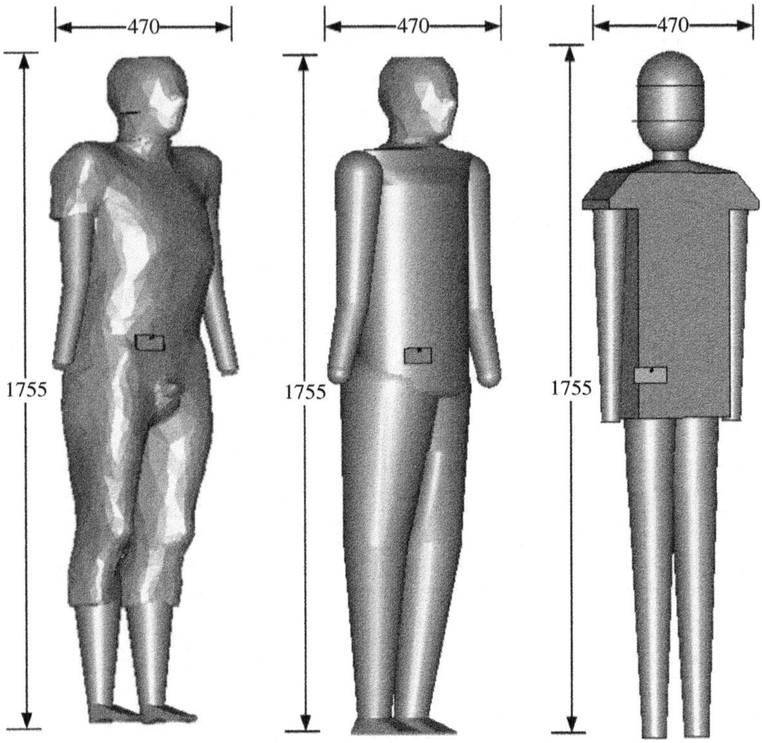

Figure 11.1 Structure of realistic high-resolution, simplified medium-resolution and basic low-resolution numerical models of the human body with on-body positioned headset and handset antennas (all lengths are in mm) [40].

11.2.1 Evaluation and Comparison of Numerical Models of Human Body

The performances of the three types of numerical models have been studied in terms of transmission coefficient, radiation patterns, electric field distribution and specific absorption rate (SAR) to evaluate their accuracy.

11.2.1.1 On-Body Transmission

Impact of different resolutions and shapes of the numerical models of the human body on the Bluetooth link between the body-worn handset and headset antennas is studied here. The Bluetooth link is characterized in terms of average path gain (S_{21}). The placement of the headset and handset antennas are kept similar for the three models to achieve a fair comparison. The separation between the body surface and the two antennas is 10 mm in order to incorporate the clearance for the cover assembly. The headset antenna is placed near the head at the approximate location of the ear. Its ground plane is longitudinal in the y-axis. The handset antenna is placed on right side of the body model at waist realizing a typical body worn position. The ground plane of the handset antenna is longitudinal in the x-axis with the PIFA at the top of the ground plane as depicted in Figure 11.2(a).

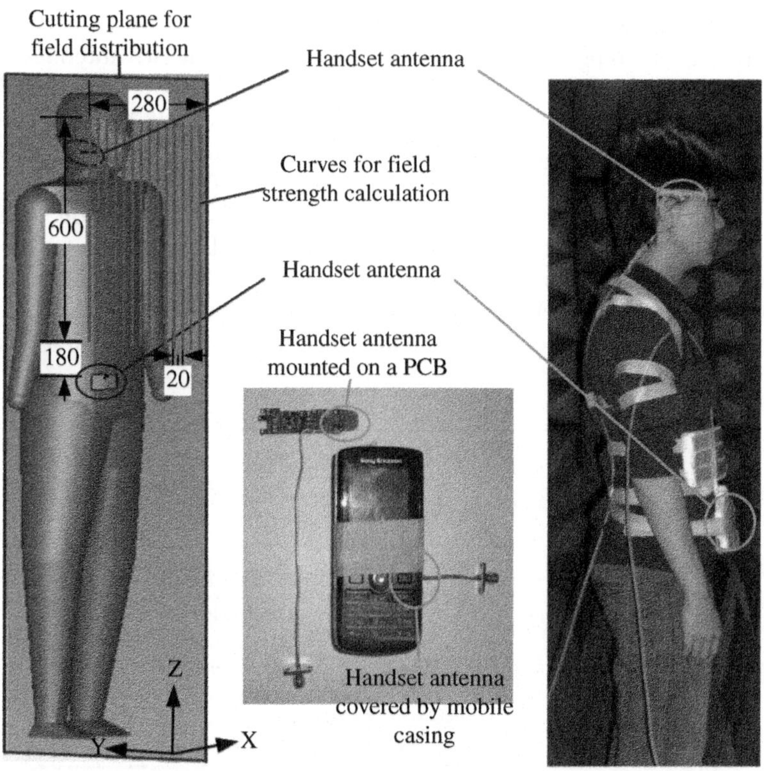

Figure 11.2 Simulation set-up (showing on-body test configurations with headset and handset antenna positions, location of cross-section plane and definition of curves for the observation of electric field distribution), antenna prototypes and measurement set-up in an anechoic chamber (all lengths are in mm) [40].

The numerical modelling of the human body is validated by confirming the simulated results through measurements of an on-body Bluetooth communication link. An Agilent HP8720ES Vector Network Analyser is used for the measurements. The prototypes of the two LTE antennas used in the measurements are shown in Figure 11.2(b). The measurement set-up in an anechoic chamber is illustrated in Figure 11.2(c). First, the LTE link is studied in the absence of the human body for the sake of bench-marking. In this scenario, the handset PIFA antenna and the headset meander line monopole antenna are placed at exactly the same locations where they would be in the presence of the human body. Therefore, the test set-up is essentially the same as described in Figure 11.2(a), but without the presence of the human body model. To locate these exact positions, a cube of polystyrene replicating the on-body gap between the two antennas is used in measurements. The two antennas are fed by low-loss coaxial cables of 5 m length. It is observed that the direct link between the handset and the headset antenna is −36.8 dB in simulation and −36.5 dB in measurement at 2.45 GHz, as illustrated in Figure 11.3.

The LTE link is then investigated for on-body transmission with body-worn handset and headset antennas. Simulated results for the three body models are compared and validated through measurements. In the experiment, the headset and handset antennas have been worn by the volunteer and arranged in the same configurations

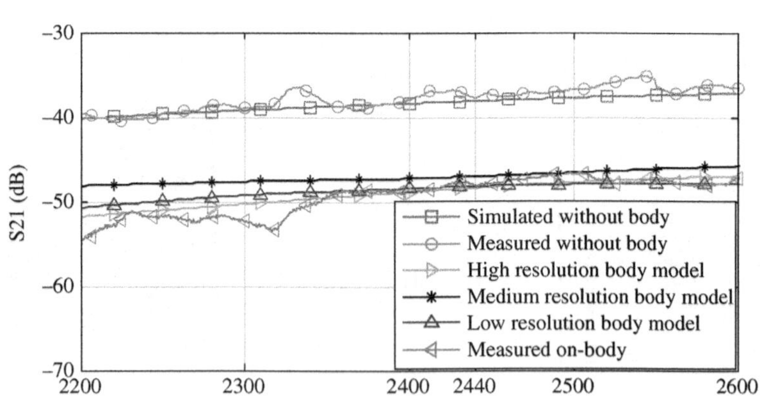

Figure 11.3 Comparison of simulated and measured average path gains for Bluetooth headset and handset antennas at 2.45 GHz [40].

as were shown in Figure 11.2. The coupling between the two low-loss feed cables is measured separately and a loss variation of 0.5 dB (maximum) is found on-body. The measured path gain between the body-worn, handset and the headset is found to be −48.1 dB at 2.45 GHz. This value is noted to be −48.3, −47.7 and −49.5 in simulation for the high, medium and low resolution models, respectively, as indicated in Figure 11.3. In the presence of the human body, the path gain value sees a substantial drop of 11.3 dB in measurement while 11.5, 10.9 and 12.7 dB in simulation using high, medium and low resolution body models, respectively. This drop is not surprising as the body blocks the line-of-sight between the headset and handset antennas.

It is evident from the close agreement of the measured and simulated path gain values that the three models are a good approximation of the real human subject. Moreover, from the good agreement of the measured path gain and simulated values for high and low resolution models, it can be deduced that for this communication link, a very simple homogeneous body model that utilizes fewer computing resources and offers higher time efficiency can provide results as accurate as those obtained using very detailed body representations.

Overall, the simulated path gains have shown a good agreement with the measured value. Few discrepancies are associated with small differences between the computer model and the experiment, such as the shape of the body, tissue properties, presence of clothes and feeding cables. This close agreement between the two values not only validates the numerical modelling approach and designed computer models but also serves as a benchmark for further investigation.

11.2.1.2 Effects on Antenna Radiation Pattern

The three designed models are used to investigate the impact of body shape and resolution on the radiation characteristics of the two antennas. Figure 11.4 highlights the effect of the three models on the radiation pattern of the headset antenna in XY and YZ planes while Figure 11.5 shows these effects on the radiation pattern of the handset antenna. These results depict that the human body deforms the radiation patterns of the two antennas due to electromagnetic absorptions in the body tissues and reflections from the body surface. It has caused up to 12 dB reduction in the back lobes and a 6 dB increase in frontal radiations for the handset antenna. The effect is more visible on the headset antenna due to the presence of body mass both at sides (i.e. head) and underneath (i.e. shoulder) decreasing its radiation in both directions

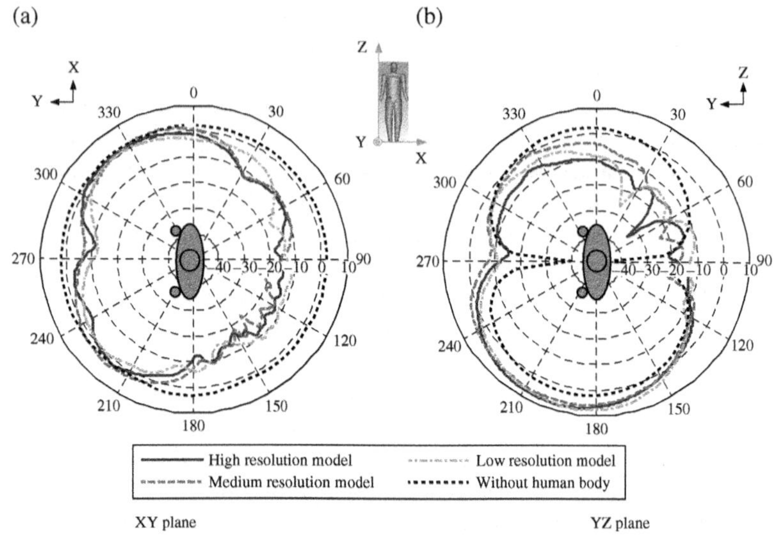

Figure 11.4 Comparison of simulated 2D radiation patterns of headset meander line monopole antenna at 2.45 GHz using different human body models [40].

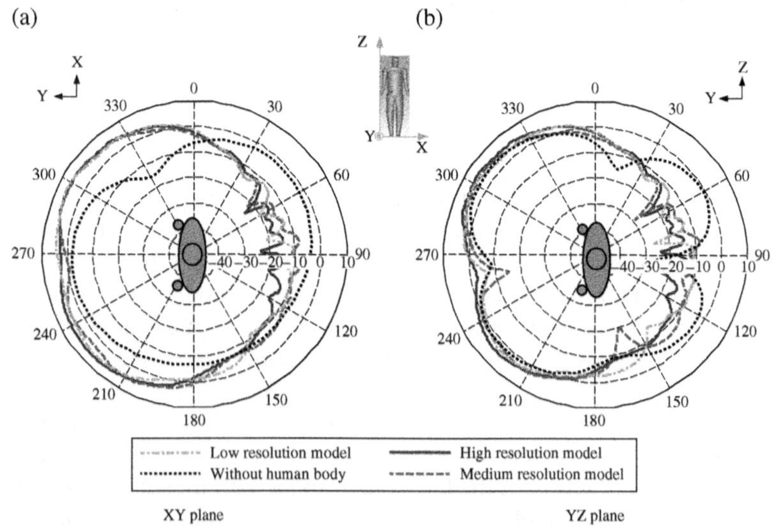

Figure 11.5 Comparison of simulated 2D radiation patterns of a handset PIFA antenna at 2.45 GHz using different human body models [40].

in the XY plane. The effect of the head is dominant in the YZ plane reducing its radiation between 0–90°. Overall, the radiation patterns of the two antenna are not greatly influenced by the type of the human body model as results are similar for the three models in terms of deformation trend and gain values.

11.2.1.3 Electric Field Distribution

The electric field distributions on the body surface are investigated and compared for the three models. Figure 11.6 shows distributions of normalized electric field magnitude, viewed on the cross-section plane (as illustrated in Figure 11.2) through the headset and handset LTE antennas. The fields are normalized to the value at the feeding

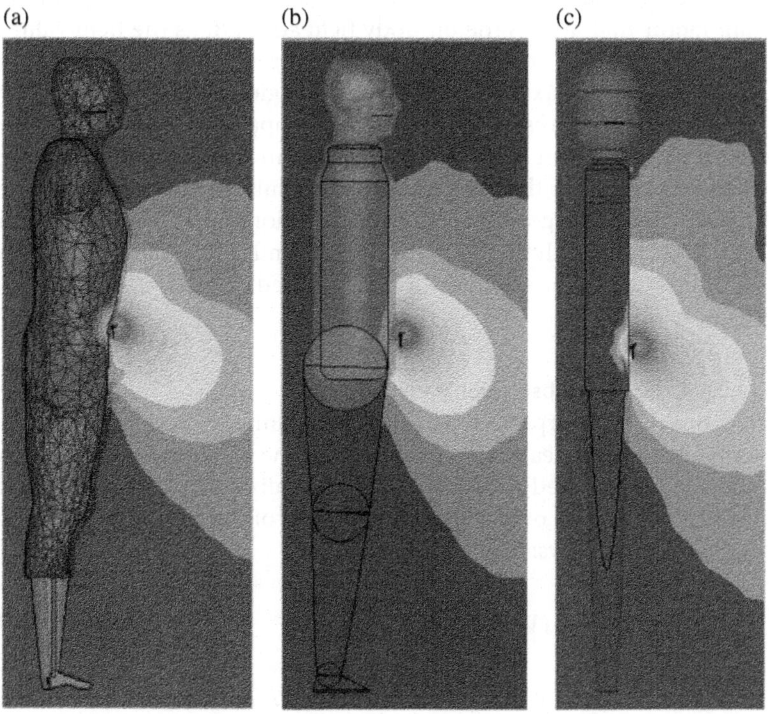

Figure 11.6 Comparison of normalized electric field distributions on the three human body models for handset LTE antenna on the cross section plane through a headset LTE antenna [40].

point of the handset antenna. From these results, it is obvious that the three models support a similar field pattern. In the three cases, the field strength at the headset antenna is in the order of −18 dB.

The electric field strength for the handset antenna is also plotted as a function of distance between the headset and handset antennas. The field strength is calculated on 14 curves defined to cover the computational domain between the two antennas. The origin of the three axes for this calculation is located on the right side of the body, above the handset antenna. The starting point in vertical direction (the origin of the *Z*-axis) for all the plots is chosen 180 mm away from the handset antenna to avoid the antenna near field region as shown in Figure 11.2(a).

The electric field strength in front of the human body, starting from the body surface and extending to a separation of 280 mm is presented in Figure 11.7. It is observed that body shape is an important factor that affects the on-body field strength as the field values are very low in the chest region of the high resolution model. On the first curve, the maximum value in this region is observed to be 9 V/m for the high resolution model as compared to 27 V/m for the medium resolution and low resolution models. However, going slightly away from the body makes the results similar for the three models. This, along with the field distributions shown in Figure 11.6 confirms that the three models generate similar electric field pattern with almost similar strengths for the studied on-body LTE communication link.

11.2.1.4 Specific Absorption Rate (SAR)

The Specific Absorption Rate (SAR) is commonly used as a measure to determine the health hazards posed by the RF signal to the human body. It is estimated as whole-body or localized SAR. The localized SAR is determined over 10 g or 1 g volumes of the tissues. The SAR is expressed as follows:

$$SAR = \frac{\sigma}{\rho}|E|^2 \left(W/kg\right) \tag{11.1}$$

Where, σ is conductivity of the tissue in S/m, ρ is mass density of the tissue in kg/m^3 and E is total RMS electric field strength in V/m.

Different standardization bodies use certain SAR limits for wireless devices to guarantee the safety of the user. These limits are different

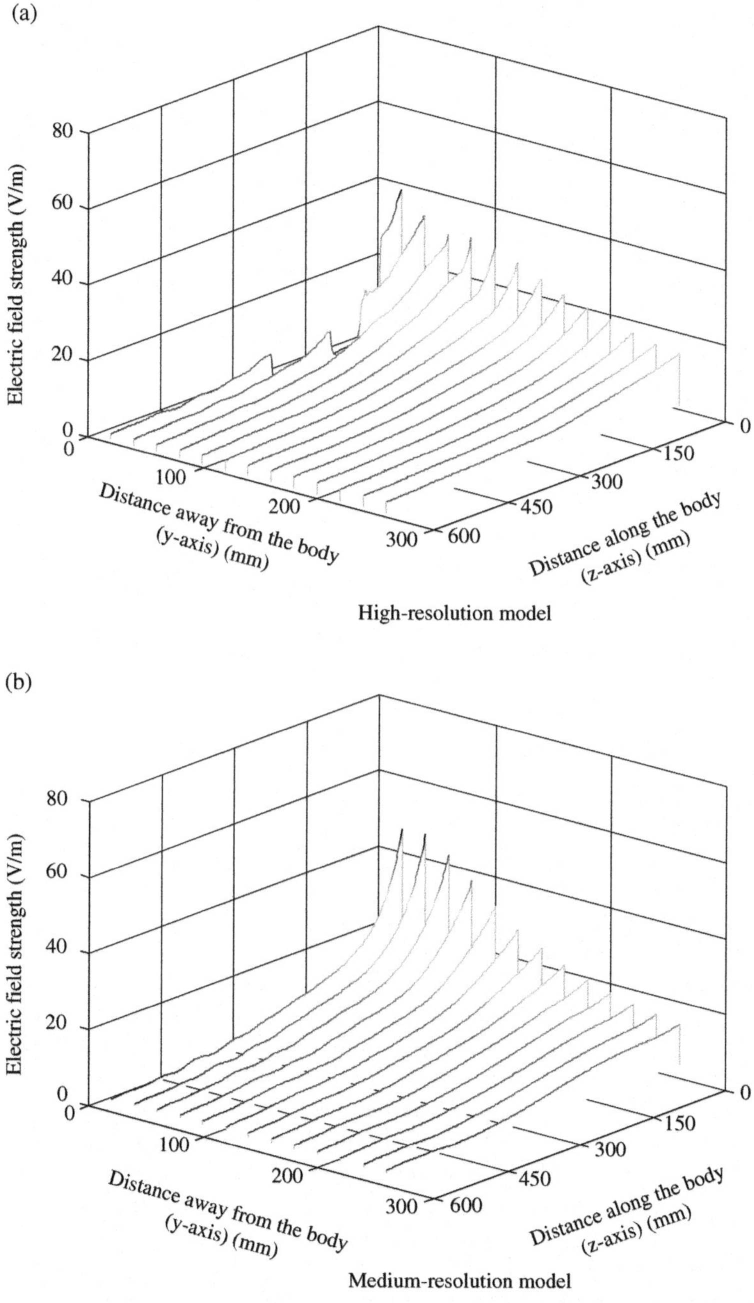

Figure 11.7 Electric field strength on the curves between the handset and the headset antennas in front of the three body models, corresponding to Figure 11.2(a) [40].

(c)

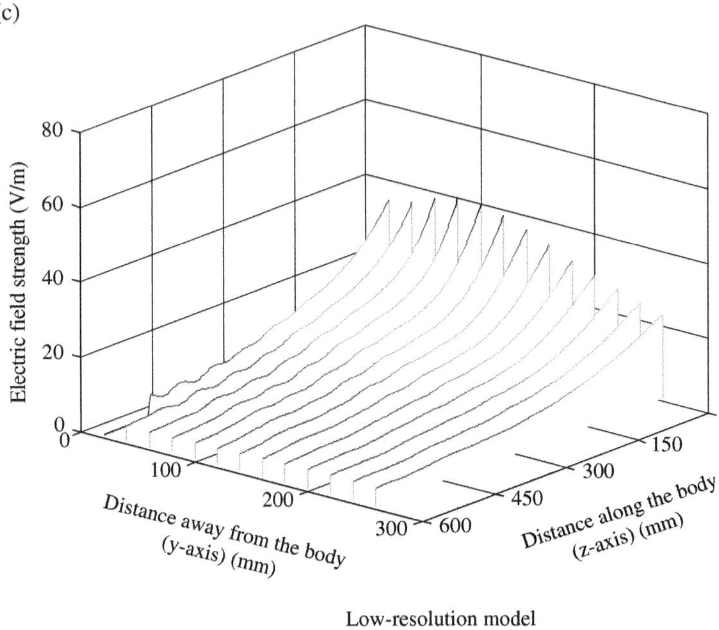

Figure 11.7 (Continued)

for controlled (occupational) and un-controlled (public) conditions of exposure to RF energy. The Federal Communications Commission (FCC) uses a 1.6 W/kg SAR value averaged over 1 g of tissue volume as a public safety standard while International Commission on Non-Ionising Radiation Protection (ICNIRP) and The Institute of Electrical and Electronics Engineers (IEEE) have adopted a 2 W/kg SAR value averaged over 10 g of tissue volume. The whole body averaged SAR limit is 0.08 W/kg for both the FCC and ICNIRP/IEEE [23, 41]. In this study, the European SAR standard of the ICNIRP is considered.

The whole-body, 10 g averaged and 1 g averaged SAR values for a handset antenna with a 1 W (peak) incident power at 2.45 GHz have been evaluated and summarized in Table 11.2. The maximum SAR distributions are compared for the three numerical models in Figure 11.8. It is evident from this figure that the three models support a similar pattern overall of SAR distribution. For the handset antenna, the SAR is more intense at the pocket position (where the antenna is placed) as this area is directly exposed to the feeding

Table 11.2 Comparison of the 10 g, 1 g and whole-body averaged SAR values at 2.45 GHz for a handset antenna using high, medium and low resolution numerical models of the human body.

	SAR (W/kg)		
Body Model	**10 g averaged**	**1 g averaged**	**Whole-body**
High resolution	1.7	2.7	0.0021
Medium resolution	1.2	2.1	0.001
Low resolution	0.8	1.8	0.0009

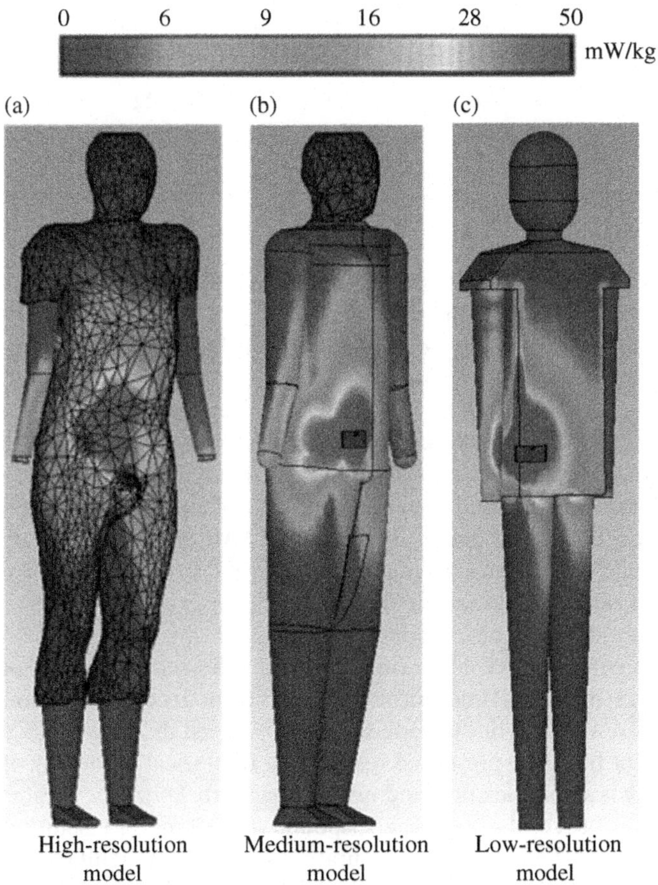

Figure 11.8 Comparison of maximum SAR distributions on the three human body models for handset antenna [40].

Table 11.3 Comparison of simulated parameters of on-body Bluetooth communication channel using high-, medium- and low-resolution human body models (with reference to high-resolution model).

Parameter		*Performance*		
		High	**Medium**	**Low**
S_{21}	Value (dB)	−48.3	−47.7	−49.5
	Rel. difference (%)	–	1.2	2.4
Electric field strength	Value (V/m)	38.4	36.2	42.1
	Rel. difference (%)	–	5.7	9.6
Maximum SAR	Value (W/kg)	2.7	2.1	1.8
	Rel. difference (%)	–	22	33
Computational efficiency	Mesh cells (millions)	186	154	114
	Simulation time (hours)	128	105	75

port of the antenna. The location for the maximum SAR value is observed to be similar for the three models.

The level of energy absorption is different in the three models due to varying body shapes. The 10 g averaged SAR values are observed to be 1.7 W/kg, 1.2 W/kg and 0.8 W/kg; the 1 g averaged SAR values are noted to be 2.7 W/kg, 2.1 W/kg and 1.8 W/kg while the whole-body averaged SAR is 0.0021 W/kg, 0.0001 W/kg and 0.0009 W/kg for the high resolution model, medium resolution model and low resolution model, respectively. The trend in change of SAR values with body shape is similar for the three averaging methods and in correspondence to the variation in the electric field strength.

The observed values of transmission coefficient (S_{21}), electric field strength, maximum SAR and computational efficiency in terms of mesh size and simulation time for the on-body Bluetooth communication link using three human body models are summarized in Table 11.3.

It is evident from the presented results that numerical modelling of human body is a complex task and needs great care. Different parameters can be affected by the body type, shape and size. Moreover, resolution of the body model affects the computational resources significantly. For on-body transmission coefficient and electric field strength, the three models give close values with little relative difference. The medium resolution model gives the closest value to the high resolution model with a maximum relative difference of 5.7% for electric field strength. The low resolution model values differ more with a maximum relative difference of 9.6%. The maximum SAR is affected the most by the type of model, as the medium resolution model has a relative difference of 22% while it becomes 33% for the low resolution model. Since the different models have different mass and shape, the absorption power and SAR have obvious differences. Location of the maximum SAR remains close to the feeding port of the handset antenna and, hence, near to the right side of the belly for the three models.

These results show that, for on-body LTE transmission studies, the medium resolution model provides significant accuracy compared with the high resolution model due to good approximation to real body physique, relatively fast computation and close agreement between the values observed for different parameters to the values offered by the high resolution model and measurements (S_{21}).

11.3 Evaluation of Human Body Effects on LTE Femtocell Antennas

Researchers have put a huge amount of effort into studying the performance of antennas in on/off body scenarios. The commercially used Planar Inverted F Antenna (PIFA) and meander line antenna have been used to study various factors affecting on-body communication links including handset-to-body separation and the presence of blocking objects has been studied in [42]. Two dual-band textile antennas with a higher conductivity-based textile material and a large-sized ground plane have been analysed for on-body performance by Boyes *et al.* [43]. The authors conclude that a larger ground plane makes the antennas more resilient when placed on-body undergoing a roughly 20% degradation in radiation efficiency as compared with free-space levels (which is much less than the lower conductivity textile material). A multiband textile-based body-worn antenna performance covering the GSM/PCS/WLAN frequency bands is discussed in [44]. The antenna is made of densely embroidered metal-coated polymer fibres of 15 μm thickness. Dierck *et al.* have proposed a wearable multiband circularly polarized active antenna for use in the GPS and Iridium satellite phone applications [45]. The antenna is constructed using flexible foam and fabric substrates and conductors etched on thin copper-on-polyimide films.

This section investigates the performance of a multiband antenna for on/off body applications using numerical studies and considering various body-mounting scenarios. The compact microstrip patch antenna employing a combination of two rings and semi-circular slot to generate multiband resonances (discussed in Chapter 8 and shown in Figure 11.9) is used to study the human body effects. The discussion in Chapter 8 shows that the antenna performance in free space is excellent at the four frequency bands. Good impedance matching (of more than −10 dB), good omni-directional coverage in both azimuth and elevation, and a high peak gain value of more than 5.1 dBi is observed for this multiband antenna. Good performance, simple configuration and low profile makes this antenna particularly attractive for portable/wearable LTE systems.

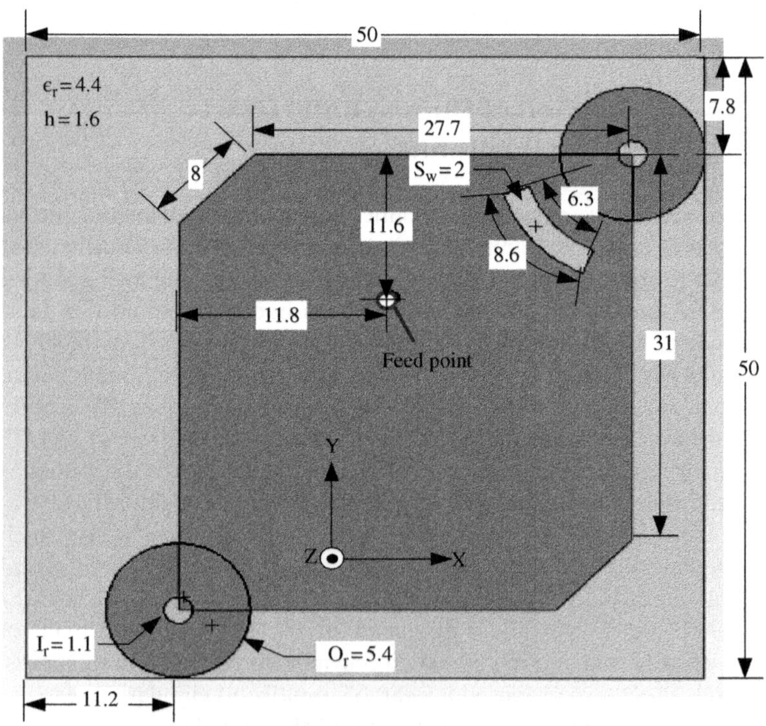

Figure 11.9 Geometry and dimensions of the slot-ring microstrip patch multiband antenna operating at GPS/4G/Wi-Fi frequencies (all units are in mm [46, 47]).

The high resolution realistic human body model discussed in the previous section is employed to study the human body effects on the antenna. The complexity of the human body composition compelled to design this model as a single layer homogeneous structure. To further reduce the overall number of cell volumes (voxels) in the computational domain that subsequently reduces the computation and time requirements, an adaptive meshing scheme with finer cell sizes around the vital parts of the body is used. The Perfectly Matched Layer (PML) absorbing boundary conditions are used [38]. This results in a maximum mesh cell size of 10 mm near the boundaries of the computational domain and a minimum mesh cell size of 0.08 mm at the edges of the solids in the computational domain. The human body model and on-body antenna set-up is shown in Figure 11.10.

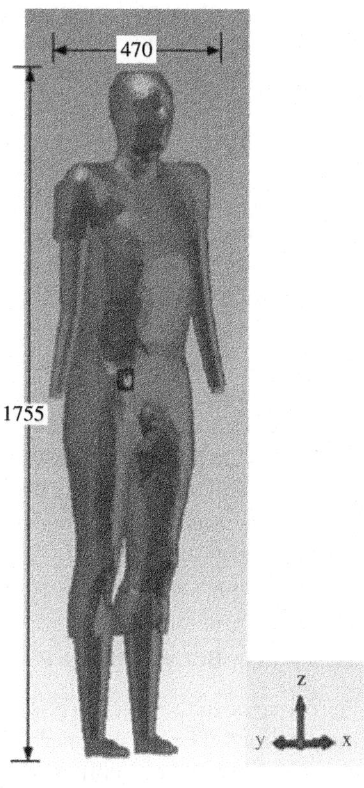

Figure 11.10 Structure of a realistic high-resolution numerical model of the human body (all lengths are in mm) [47].

Accurate modelling of the multiband antenna in the presence of the human body requires careful determination of the dielectric properties. For the single layer human body model, two-thirds muscle dielectric properties are considered that are well characterized up to 20 GHz [39]. The antenna operation at multiple frequencies in the range from 1.5 to 6 GHz lead to the adoption of the Debye dispersion equation to express the complex dielectric properties of the numerical human body [48]:

$$\varepsilon^* = \varepsilon_o\left(\varepsilon' - j\varepsilon''\right) = \varepsilon_o\left(\varepsilon_\infty + \frac{\varepsilon_s - \varepsilon_\infty}{1 + j\omega\tau}\right) \tag{11.2}$$

where ε_0 is the free space permittivity, ε_s represents the static permittivity, ε_∞ is the optical permittivity, ω depicts the angular frequency and τ is the relaxation time. The values used in this paper for these parameters to obtain best fit are $\varepsilon_s = 37.1$, $\varepsilon_\infty = 12.2$ and $\tau = 12.5 \times 10^{-12}$ [49].

For the on-body scenarios, the antenna is mounted on-body keeping a separation of 10 mm between the antenna and the body to replicate the space for the covering assembly of the wireless device. The antenna performance is analysed in terms of reflection coefficient response and Specific Absorption Rate (SAR) distributions.

The position of the antenna as the access point is selected to be on the right side of the belly, which is the common location if the device is placed in the user's pocket (as shown in Figure 11.10). The reflection coefficient (S_{11}) performance is compared to that observed in the absence of the human body model and illustrated in Figure 11.11. The results show that the human body detunes the antenna and brings a shift in the resonance frequency. However, the antenna still manages to give good performance at 2.1 GHz and 5.3 GHz frequency bands. The two bands can be used simultaneously for transmission/reception of body monitored data.

The Specific Absorption Rate (SAR) is used as a measure to determine the health hazards posed by the RF signal to the human body. The 1 g averaged SAR values for the multiband antenna at different operating frequencies are summarized in Table 11.4. The results show that the level of energy absorption is different at different frequencies. Overall, the SAR values are well below the required threshold.

11.3.1 On-Body Antenna Placement

Performance of the antenna is also evaluated for different on-body mounting positions. The most logical placements at the left belly, right belly, left chest, right chest and centre chest are studied as shown in Figure 11.12.

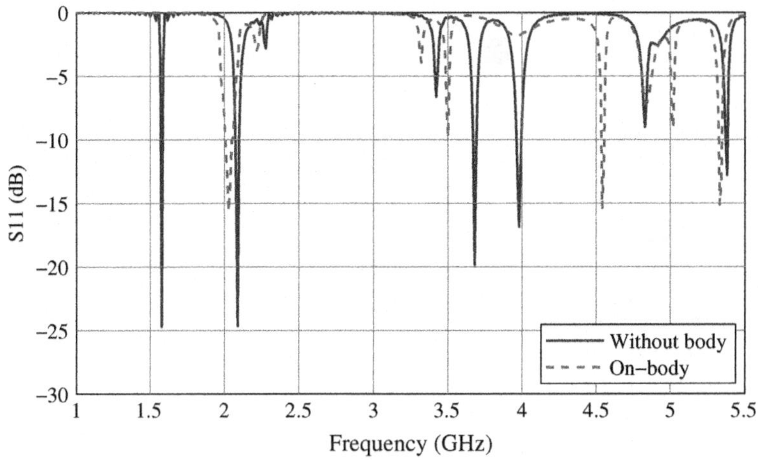

Figure 11.11 Reflection coefficient response of the multiband microstrip patch antenna with and without human body presence [47].

Table 11.4 Comparison of the 1 g averaged SAR values at different frequencies of operation for the on-body multiband antenna.

Frequency (GHz)	SAR (W/kg)
1.575	1.65×10^{-12}
2.1	5.7×10^{-11}
3.68	2.02×10^{-11}
5.37	2.65×10^{-11}

Reflection coefficient responses (S_{11}) are compared and presented in Figure 11.13. The results show that varying position changes the antenna performance in terms of impedance matching due to change in the underlying body physique. However, the antenna maintains its operation successfully at the 2.1 and 5.3 GHz bands. Left and right placements at the similar position (i.e. belly and chest) brings no significant variation in the antenna performance.

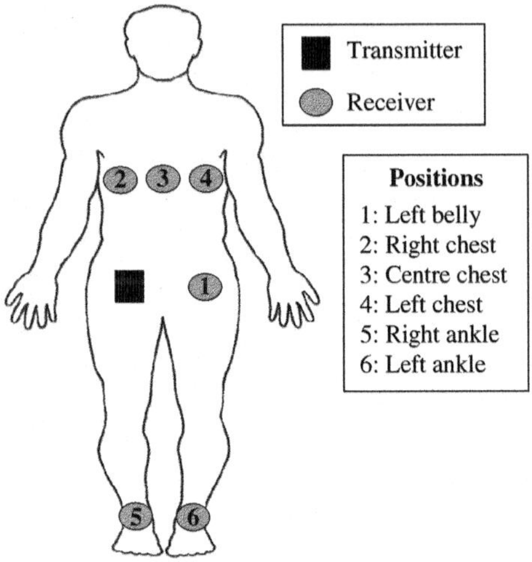

Figure 11.12 On-body placement of the antenna at different body locations [47].

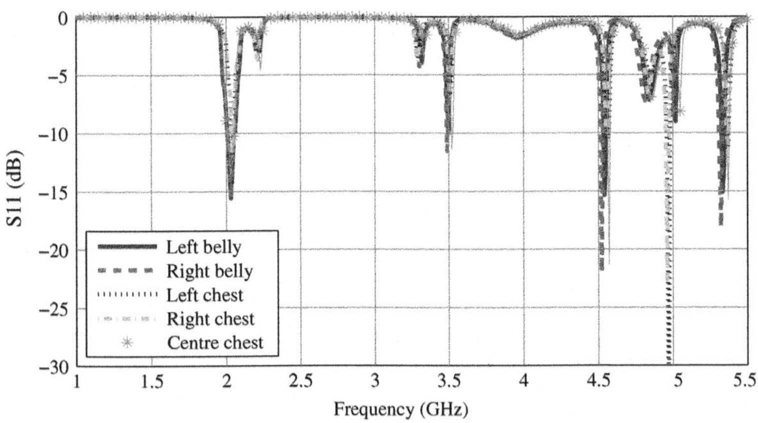

Figure 11.13 Reflection coefficient response of the multiband microstrip patch antenna while placed at different on-body positions.

11.3.2 Antenna-Body Separation

The excitation condition of the electromagnetic waves depends upon the separation of the electromagnetic device from the lossy human body. This has been examined by analysing the effects of varying separations between the antenna and the human body. The reflection coefficient response (S_{11}) has been studied for antenna-body separations (*d*) of 0, 5, 10, 15 and 20 mm as illustrated in Figure 11.14. Here, d = 0 mm represents a situation where the antenna is placed directly on the body surface without preserving any clearance for the covering assembly.

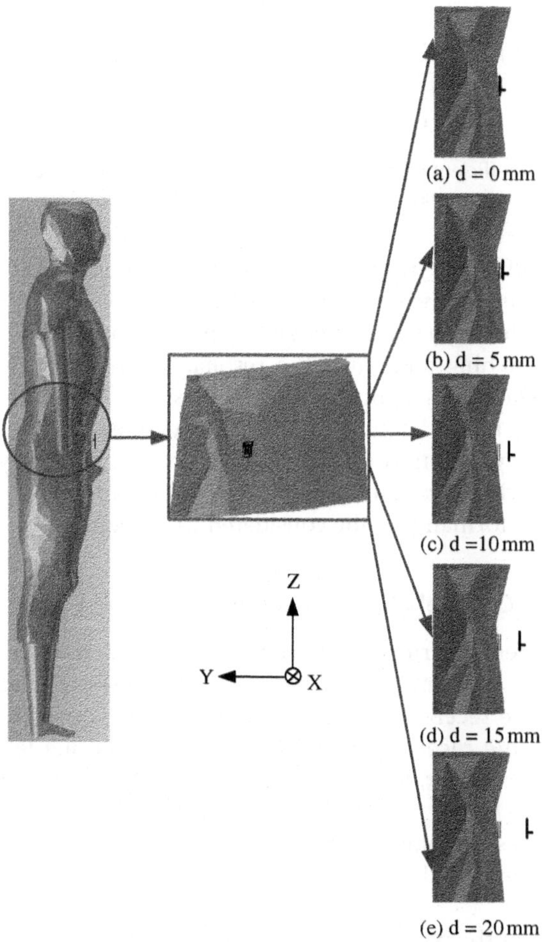

Figure 11.14 Antenna placement at different separations from the human body.

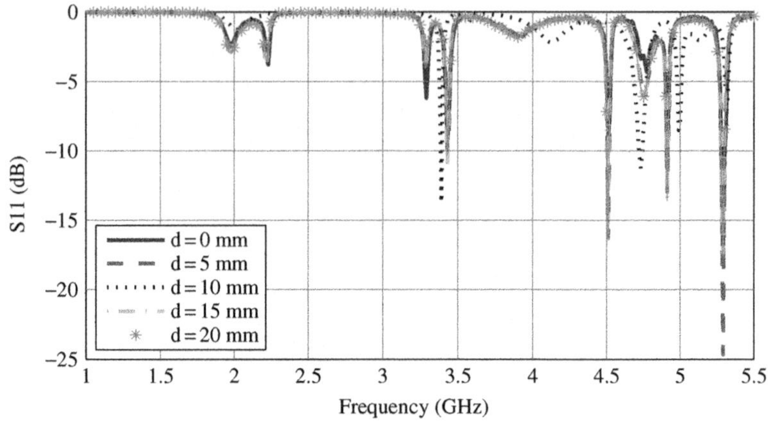

Figure 11.15 Reflection coefficient response of the multiband microstrip patch antenna with varying separations between the human body and the antenna.

The reflection coefficient response for varying antenna-body separations is plotted in Figure 11.15. It is evident from these results that decreasing the gap between the antenna and the human body deteriorates the antenna matching and shifts the resonances towards the lower end due to increased absorptions in the body tissues. This effect is most visible for d = 0 mm. Matching improves with increasing gaps with d = 15 mm and beyond offering fewer reflections from the body surface. The optimized separation appears to be d = 10 mm, which gives an acceptable performance at the considered LTE frequencies.

11.3.3 On-Body LTE Channel Characterization

In this section, the antenna performance is evaluated considering various on-body channels. The similar multiband antenna is used both as the transmitter and the receiver. The transmitter position is fixed at the right belly while the receiving antennas are rotated among left belly, right chest, centre chest, left chest, right knee and left knee as illustrated in Figure 11.12.

The path gain curves are plotted and compared by observing S_{21} values. Figure 11.16 illustrates the path gain values for different on-body channels.

The results show that the path gain values lie between −42 and −50 dB for different on-body channels when the antenna is operating at 2.1 GHz. Path gain values are between −50 and −55 dB for antenna

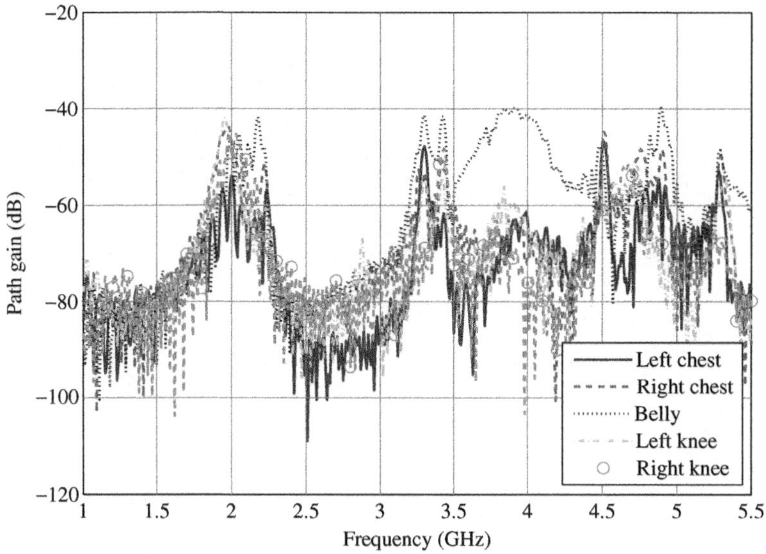

Figure 11.16 Path gain response of the multiband microstrip patch antenna in different on-body channels.

operation at 5.3 GHz. It is evident from these results that the antenna establishes quite a strong link at the two frequencies guaranteeing continuous and robust working of the overall monitoring system.

11.3.4 On-Off Body LTE Channel Characterization

Patient health monitoring systems require efficient channel performance between on-body and off-body antennas. Characterization of the on-off body channel is considered in this section. Both line-of-sight (LOS) and non-line-of-sight (NLOS) scenarios are studied, which could occur if a person moves. In the LOS configuration, the two antennas are in a straight line, one transmitting while mounted on-body at right belly while the other receiving at a distance of d = 50 cm in front of the human body to maintain a distance larger than twice the wavelength at the lowest operating frequency of 2.1 GHz. In the NLOS configuration, the receiving antenna is located in front of the centre chest. Distance between the body-mounted transmitter and the receiver is again 50 cm. The configuration of the tested scenarios is shown in Figure 11.17(a).

The path gain values for the two scenarios are plotted in Figure 11.18.

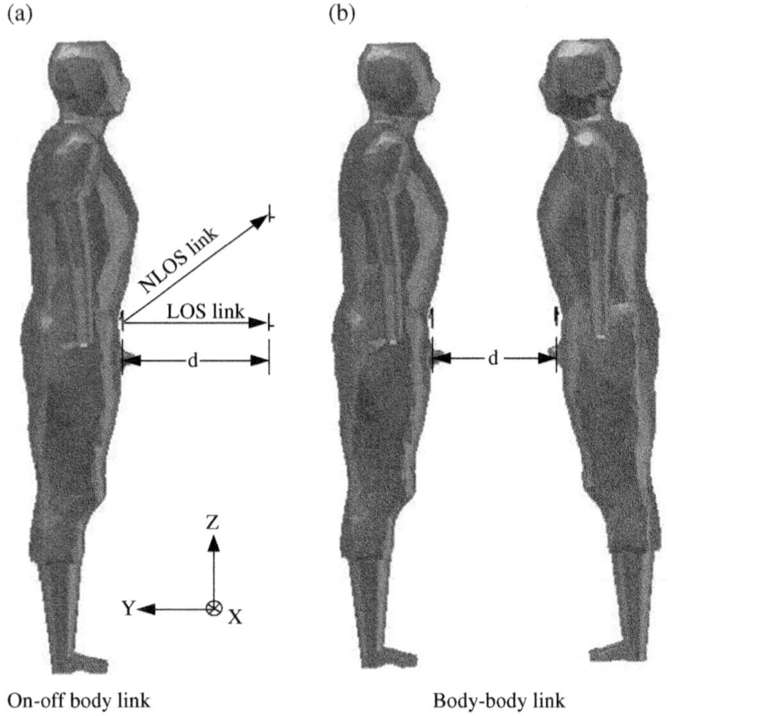

Figure 11.17 Considered scenarios for evaluation of the antenna performance in on-off body and body-body channels.

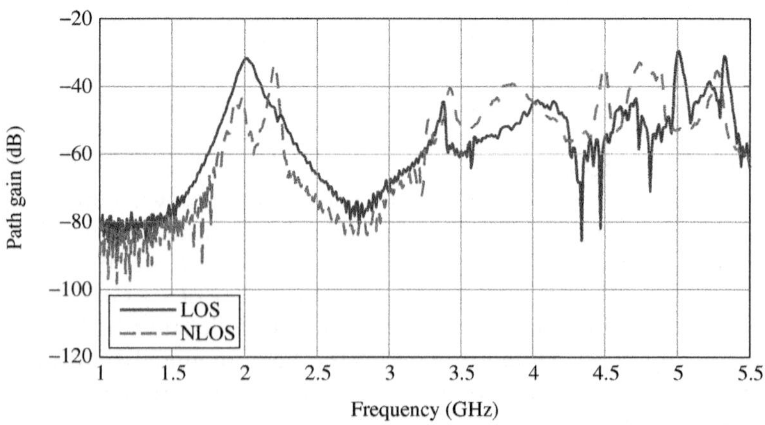

Figure 11.18 Path gain performance of the multiband microstrip patch antenna in on-off body LOS and NLOS configurations [47].

The presented results indicate that the performance of the antenna is good in both LOS and NLOS on-off body channels. For the LOS link, path gain value is observed to be −31 dB while this value is −42 dB for the NLOS link at 2.1 GHz. At 5.3 GHz, path gain value for LOS link appears to be −32 dB while for NLOS, this value is −37 dB. It shows that, for the on-off body channel, the antenna has the potential to establish an excellent link, much stronger than the on-body links.

11.3.5 Body-to-Body LTE Channel Characterization

Figure 11.17(a) shows that Body-to-Body communications is one of the key application scenarios for futuristic wearable technologies. LTE will also play a key role in these applications and, therefore, the potential use of LTE antennas in body-to-body communication necessitates an investigation of its response in a body-to-body channel. The antenna performance in terms of path gain (S_{21}) is, therefore, evaluated considering a representative body-to-body link. Both LOS and NLOS conditions are considered. The length of the channel is kept at 50 cm in continuation of the previous discussion. The test scenario is depicted in Figure 11.17(b).

The path gain values for the two body-to-body scenarios are plotted in Figure 11.19.

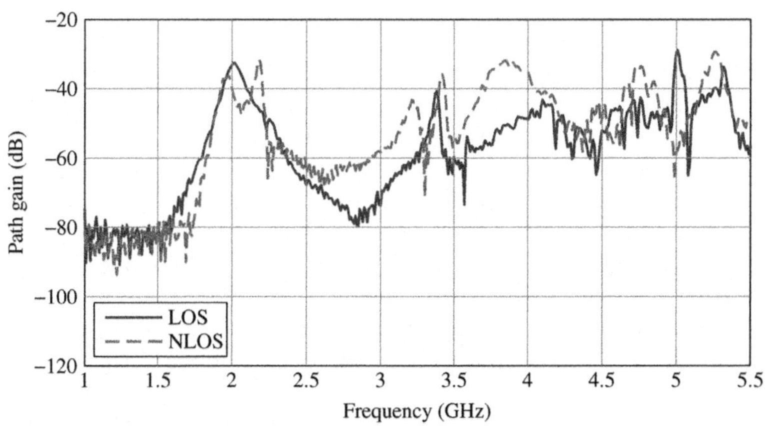

Figure 11.19 Path gain performance of the multiband microstrip patch LTE antenna in body-to-body LOS and NLOS configurations.

The body-to-body channel has a path gain profile with a higher average than the one observed for on-off body channel due to increased radiations away from the two body surfaces. The path gain value for the LOS body-to-body link is noted to be −40 dB while the NLOS body-to-body channel has achieved a path gain of −43 dB at the 2.1 GHz band. At the higher frequency of 5.3 GHz, the LOS link offers a path gain value of −39 dB while the NLOS link has achieved a path gain of −33 dB. Again, reflections from the body surface and level of absorbed power in the tissues due to varying shape and physique of the underlying body parts play an important role to make the NLOS link stronger than the LOS at 5.3 GHz. These results show that the LTE antenna establishes a good communication link at both frequencies guaranteeing continuous and robust working of wearable LTE systems. It is, however, pertinent that a free space performance evaluation has strict limitations to define antenna performance in wearable scenarios. Therefore, detailed subject specific investigations need to be carried out in order to assess the working of antennas in realistic working conditions.

11.4 Open Research Issues

Wearable LTE systems in femtocells work in a very volatile environment. Channel losses are high because of body absorption, surface reflections and difficult operational conditions. Inherent limitations of the antenna design, on-body antenna placement, human body postures and human body movements could lead to electromagnetic shielding resulting in incomplete data reception or complete link droppage in worst cases. Strict regulations on levels of power emission and Specific Absorption Rate (SAR) along with size and shape limitations of the device increases the complexity of the antenna design. Channel characterization is also very complex due to multipath and human movements.

Solutions need to be explored to improve performance of the LTE antennas in environments such as improved antenna design, use of novel materials in the antenna design, use of various types of antennas and optimal positioning of antennas on-body. Energy efficiency is also an open research issue due to anticipated enhanced battery life of wireless LTE devices.

References

1 L. Sorensen and K. Skouby, "User scenarios 2020 – a worldwide wireless future," *WWRF*, July 2009, URL: www.wwrf.ch/files/publications.

2 P. S. Hall and Y. Hao, Antennas and propagation for body-centric wireless networks (2nd Edition), Artech House, 2012.

3 X. Chen, C. G. Parini, B. Collins, Y. Yao, and M. U. Rehman, Antennas for global navigation satellite systems, John Wiley and Sons, Inc., (UK), 2012.

4 Z. N. Chen, Antennas for portable devices, John Wiley & Sons, Inc., (UK), 2007.

5 Tractica, "Wearable device market forecasts," Available online at https://www.tractica.com/research/wearable-device-market-forecasts/ (accessed December 2017), 2017.

6 E. Jovanov, A. Milenkovic, C. Otto, P. de Groen, and B. Johnson, "A wireless body area network of intelligent motion sensors for computer assisted physical rehabilitation," Journal of NeuroEngineering and Rehabilitation, vol. 2, no. 6, 2005.

7 J. Bernardhard, P. Nagel, J. Hupp, W. Strauss, and T. Von der Grun, "BAN-body area network for wearable computing," *9th Wireless World Research Forum Meeting*, July 2003.

8 S. Matsushita, "A headset-based minimized wearable computer," *IEEE Intelligent Systems Archive*, vol. 16, no. 3, pp. 28–32, May 2001.

9 J. Toftgard, S. Hornsleth, and J. B. Anderson, "Effects on portable antennas of the presence of a person," *IEEE Transactions on Antennas and Propagation*, vol. 41, no. 6, pp. 739–746, June 1993.

10 H. R. Chuang, "Human operator coupling effects on radiation characteristics of a portable communication dipole antenna," *IEEE Transactions on Antennas and Propagation*, vol. 42, no. 4, pp. 556–560, April 1994.

11 M. A. Jensen and Y. Rahmat-Samii, "EM interaction of handset antennas and human in personal communications," *IEEE Transactions on Antennas and Propagation*, vol. 83, no. 1, pp. 7–17, January 1995.

12 M. Okoniewski and M. A. Stuchly, "A study of the handset antenna and human body interaction," *IEEE Transactions on Microwave Theory and Techniques*, vol. 44, no. 10, pp. 1855–1864, October 1996.

13 J. S. Colburn and Y. Rahmat-Samii, "Human proximity effects on circular polarized handset antennas in personal satellite communications," *IEEE Transactions on Antennas and Propagation*, vol. 46, no. 6, pp. 813–820, Junuary 1998.
14 Z. Wang, X. Chen, and C. G. Parini, "Effects of the ground and the human body on the performance of a handset antenna," *IEE Proceedings on Microwave and Antenna Propagation*, vol. 151, no. 2, pp. 131–134, April 2004.
15 J. Wang and O. Fujiwara, "EM interaction between a 5GHz band antenna mounted PC and a realistic human body model," *IEICE Transactions on Communications*, vol. E88-B, no. 6, pp. 2604–260, 2005.
16 J. Guterman, A. A. Moreira, C. Peixeiro, and Y. Rehmat-Samii, "Electro-magnetic human interaction with ISM 2.4GHz laptop antennas," *European Conference on Antennas and Propagation (EuCAP)*, November 2007.
17 C. C. Y. Poon, B. P. L. Lo, M. R. Yuce, A. Alomainy, and Y. Hao, "Body sensor networks: In the era of big data and beyond," *IEEE Reviews in Biomedical Engineering*, vol. 8, pp. 4–16, 2015.
18 S. Curto and M. J. Ammann, "Electromagnetic interaction between resonant loop antenna and simulated biological tissue," *Microwave and Optical Technology Letters*, vol. 48, no. 12, pp. 2418–2421, September 2006.
19 Z. H. Hu, M. Gallo, Q. Bai, Y. I. Nechayev, P. S. Hall, and M. Bozzetti, "Measurement and simulations for on-body antenna design and propagation studies," *European Conference on Antennas and Propagation (Eu-CAP)*, November 2007.
20 T. W. Koo, Y. J. Hong, G. k. Park, K. Shin, and J. G. Yook, "Extremely low-profile antenna for attachable bio-sensors," *IEEE Transactions on Antennas and Propagation*, vol. 63, no. 4, pp. 1537–1545, 2015.
21 Q. H. Abbasi, M. Ur-Rehman, K. Qaraqe, and A. Alomainy, Advances in body-centric wireless communication: Applications and state-of-the-art, the IET, (UK), 2016.
22 "IEEE standard for safety levels with respect to human exposure to radiofrequency electromagnetic fields, 3kHz to 300GHz," *IEEE Standards, C95.1*, 1999.
23 "ICNIRP guidelines for limiting exposure to time-varying electric, magnetic and electromagnetic fields (up to 300 GHz)," ICNIRP website, http://www.icnirp.org/cms/upload/publications/ICNIRPemfgdl.pdf).

24 "Considerations for the evaluation of human exposure to electromagnetic fields (EMFs) from mobile telecommunication equipment (MTE) in the frequency range from 30MHz-6GHz," *CENELEC, European Specification, Ref. No. ES-59005:1998 E*, 1998.

25 V. Hombach, K. Meier, M. Burkhardt, E. Kuhn, and N. Kuster, "The dependence of EM energy absorption upon human head modelling," *IEEE Transactions on Microwave Theory and Techniques*, vol. 44, no. 10, pp. 1865–1873, October 1996.

26 D. Nashaat, H. Alsadek, and H. Ghali, "Investigation of the mutual effect between human head and new shapes of PIFAs used in mobile communication systems," *Microwave and Optical Technology Letters*, vol. 46, no. 3, pp. 243–248, Junuary 2005.

27 S. E. Troulis, W. G. Scanlon, and N. E. Evans, "Effect of hands-free leads and spectacles on SAR for a 1.8GHz cellular handset," *Physics in Medicine and Biology*, vol. 48, pp. 1675–1684, June 2003.

28 A. D. Ball, N. E. Evans, S. J. Burgess, and W. G. Scanlon, "Head-SAR dependence on spectacle frame shape for operator of 450MHz personal radio," *IET Electronics Letters*, vol. 43, no. 20, pp. 1063–1065, September 2007.

29 K. Ito, "Human body phantoms for evaluation of wearable and implantable antennas," *European Conference on Antennas and Propagation (EuCAP)*, November 2007.

30 S. Drude, "Requirements and application scenarios for body area networks," *Mobile and Wireless Communications Summit*, 2007.

31 T. Nagaoka, S. Watanabe, K. Sakurai, E. Kunieda, M. Taki, and Y. Yamanak, "Development of realistic high resolution whole-body voxel models of Japanese adult males and females of average height and weight, and application of models to radio frequency electromagnetic-field dosimetry height," *Physics in Medicine and Biology*, vol. 49, pp. 1–15, January 2004.

32 R. A. Tell, "Microwave energy absorption in tissue," *Twinbrook Research Laboratory Technical Report*, 1972.

33 C. M. Weil, "Absorption characteristics of multilayered sphere models exposed to uhf/microwave radiation," *IEEE Transactions on Biomedical Engineering*, vol. 22, pp. 468–476, 1975.

34 C. W. G. 3, "Proposal numerical canonical models in mobile communications," *Proceedings of COST244*, pp. 1–7, November 1994.

35 S. Nishizawa and O. Hashimoto, "Effective shielding analysis for three layered human model," *IEEE Transactions on Microwave Theory and Techniques*, vol. 47, pp. 277–283, 1999.

36 G. A. Conway, W. G. Scanlon, and D. Linton, "Low-profile microstrip patch antenna for over-body surface communication at 2.45GHz," *IEEE Vehicular Technology Conference*, April 2007.

37 S. J. Holden, R. D. Sheridan, T. Coffey, R. Scaramuzza, and P. Diamantopoulos, "Electromagnetic modelling of current flow in the heart from TASER devices and the risk of cardiac dysrhythmias," *Physics in Medicine and Biology*, vol. 52, pp. 7193–7209, 2007.

38 *CST Microwave Studio® 2017 User Manual.*

39 S. Gabriel, R. W. Lau, and C. Gabriel, "The dielectric properties of biological tissues: II. measurements in the frequency range 10 Hz to 20 GHz," *Physics in Medicine and Biology*, vol. 41, no. 11, pp. 2251–2269, 1996.

40 M. Ur-Rehman, Q. H. Abbasi, X. Chen, and Z. Ying, "Numerical modelling of human body for bluetooth body-worn applications," *Progress In Electromagnetics Research*, vol. 143, p. 1623–1639, 2013.

41 "Evaluating compliance with FCC guidelines for human exposure to radiofrequency electromagnetic fields," *FCC*, 1997.

42 M. Ur Rehman, Y. Gao, Z. Wang, J. Zhang, Y. Alfadhl, X. Chen, et al., "Investigation of on-body bluetooth transmission," *IET Microwaves, Antennas and Propagation*, vol. 4, no. 7, pp. 871–880, July 2010.

43 S. Boyes, P. Soh, Y. Huang, G. Vandenbosch, and N. Khiabani, "On-body performance of dual-band textile antennas," *IET Microwaves, Antennas and Propagation*, vol. 6, no. 15, pp. 1696–1703, December 2012.

44 Z. Wang, L. Lee, D. Psychoudakis, and J. Volakis, "Embroidered multiband body-worn antenna for GSM/PCS/WLAN communications," *IEEE Transactions on Antennas and Propagation*, vol. 62, no. 6, pp. 3321–3329, June 2014.

45 A. Dierck, H. Rogier, and F. Declercq, "A wearable active antenna for global positioning system and satellite phone," *IEEE Transactions on Antennas and Propagation*, vol. 61, no. 2, pp. 532–538, Feb 2013.

46 M. Ur-Rehman and B. Allen, "A compact multi-band slot-ring microstrip patch antenna for wireless applications," in *Loughborough Antennas and Propagation Conference (LAPC)*, Nov 2013, pp. 595–598.

47 M. Ur-Rehman, X. Yang, and S. Haxha, "On-off body performance of a multiband antenna for wireless health monitoring," in *2014 4th International Conference on Wireless Mobile Communication and Healthcare - Transforming Healthcare Through Innovations in Mobile and Wireless Technologies (MOBIHEALTH)*, Nov 2014, pp. 1–4.

48 O. G. Martinsen, S. Grimmes, and H. P. Schwan, Encyclopedia of surface and collied science, Marcel Dekker, 2002.

49 N. Chahat, M. Zhadobov, R. Sauleau, and K. Ito, "A compact UWB antenna for on-body applications," *IEEE Transactions on Antennas and Propagation*, vol. 59, no. 4, pp. 1123–1131, April 2011.

12

The Road Ahead for LTE Femtocells

Masood Ur Rehman and Ghazanfar Ali Safdar

School of Computer Science and Technology, University of Bedfordshire, Luton, UK

In July 2015, more than 6.2 billion subscribers that made up of 85% of the world's 7.3 billion population were using GSM-HSPA [1]. Cellular data revenues were estimated at $132 billion in 2015 alone in the US. The global mobile market is expected to include nearly 8.45 billion subscribers by 2019. An estimated 8.1 billion users representing about 97% of the market share would use 3GPP technologies [2]. LTE has become a global wireless foundation supporting continual enhancements. Competition in the mobile broadband market due to persistent growth in the demand for capacity has made LTE the favoured 4G solution because of its high data throughput, low latency and high spectral efficiency. Because of this, LTE has experienced faster deployment than any mobile technology ever and continues to see an increased demand.

Femtocells provide the solution to poor indoor coverage in cellular communication that has hugely attracted network operators and stakeholders. Femtocells are low powered, low cost and subscriber controlled units that provide a dedicated BS to indoor subscribers. Femtocell units, known as Femtocell Access Points (FAPs), connect standard mobile devices also known as Femtocell User Equipment (FUE) to the network of a mobile operator through residential Digital Subscriber Line (DSL), optical fibres, cable broadband connections or wireless last-mile technologies. By installing FAPs indoors, the cell

LTE Communications and Networks: Femtocells and Antenna Design Challenges, First Edition. Edited by Masood Ur Rehman and Ghazanfar Ali Safdar.

sites are reduced, thereby bringing the transmitter and receiver closer to each other. The use of the subscriber's broadband network to backhaul data offer improved indoor mobile phone coverage for both voice and data because of improved connectivity compared to the Macrocell Base Station (MBS). Femtocells are similar to WiFi as both are connected to a wired backhaul but unlike WiFi, femtocells make use of an existing cellular standard for their operation [3, 4].

The benefits of femtocells now and in the long term cannot be overemphasized as it has advantages for both network operators and subscribers. As far as network operators are concerned, the reduction in macrocells due to the deployment of femtocells will result in a huge saving in CAPEX (Capital Expenditure) of network operators and the reduction in traffic will also yield a saving in the OPEX (Operational Expenditure) through advanced self-management and optimization techniques. The subscribers also benefit as the close proximity of the transmitter and receiver offered by femtocells enables subscribers to have high speed services such as voice, video and multimedia. The close proximity greatly lowers transmission power and increases the battery life of mobile devices. With a dedicated FAP in their homes, it offers subscribers a single billing address for mobile phone, broadband and land line as they are all channelled through the same backhaul [5]. Femtocells also act as a solution towards convergence of landline and mobile [6].

Even though femtocells are discovering an important role in cellular mobile communication, the issue of interference when deployed in a co-channel fashion with the macrocell needs to be addressed. The integration of cognitive radio in femtocells introduces the ability of femtocells to dynamically adapt to varying network conditions through learning and reasoning. This network intelligence by cognitive femtocells have the capability of retrieving operating parameters that can be manipulated to mitigate the inherent mutual interference between femtocells and macrocells to enhance network performance.

12.1 Future Prospects and Challenges

This book has introduced new practical approaches to mitigate interference of cognitive femtocells in two-tier networks by means of novel algorithms and system level simulations in LTE networks. In order to achieve this, a structural analysis of different state-of-the-art interference mitigation schemes has been presented. In this analysis, the schemes were classified into categories based on their characteristic interference mitigating features such as power control, resource/spectrum allocation as well as antenna schemes. Also, hybrid schemes were introduced as a mix of two or more of the aforementioned categories.

It can be seen that the importance of cognitive radio (CR) in femtocells cannot be overemphasized. CR provides a wide range of channel statistics that are suitable for interference mitigation schemes in femtocells. The idea of an efficient CR interference mitigation scheme is how these individual channel statistics are utilized taking into consideration a number of parameters such as air interface technology, density of femtocells and femtocell deployment (e.g. co-channel/dedicated, open access, CSG or hybrid). Although cognitive femtocells can retrieve operating parameters, success of current and future LTE applications depends on efficient solutions to the technological factors it faces today. These factors will inhibit growth and drive innovation in the LTE femtocells implementation. Some of the key future prospects and challenges are discussed next.

12.1.1 Spectrum Sharing

Current LTE systems assume dedicated spectrum. However, spectrum scarcity continues to challenge the communications systems. A number of methods are being employed to make the best use of this limited resource. Some key schemes include deploying technologies with higher spectral efficiency, enabling operation of LTE in all available bands, using both FDD and TDD variants to take advantage of both paired and unpaired bands, designing carrier aggregation methods and deploying a larger number of new small cells.

Although, these initiatives have expanded capacity, they do not fully deal with the problem and do not remove the need for additional spectrum. Spectrum sensing using CR methods is deemed an efficient way out. This has been discussed in Chapter 3 in detail but still has potential of innovative solutions.

12.1.2 Intelligent/Efficient Spectrum Sensing Schemes

The implementation or improvement of the CR enabled mitigation schemes depends entirely on how efficient spectrum sensing is conducted to retrieve channel statistics utilizing some of the standard spectrum sensing techniques. Although spectrum sensing techniques can be conducted by FAPs either on a channel by channel basis (single channel sensing) across the whole spectrum or over a group of channels (group channel sensing) simultaneously, there are issues that arise with both types of sensing techniques and need to be addressed. Single channel sensing introduces accuracy in sensing as each individual channel is sensed thoroughly but introduces sensing overhead and delays. Channel degradation is also imminent when a group of FAPs sense each individual channel concurrently. Group channel sensing on the other hand is less accurate as available channels might be confused with interference channels. Thus, in practice, an intelligent and efficient spectrum sensing mechanism is of paramount importance in CR enabled interference mitigation schemes.

12.1.3 Primary/Secondary User Issue

Most spectrum schemes reflect a scenario where the MBS and MUEs are the licensed users and regarded as the PUs and the FAPs and FUEs as SUs. This approach sounds fair enough for SUs who opportunistically access resources of the PUs with priorities given to PUs. However, in reality, FAPs and FUEs require and utilize high data and traffic since they are mainly used indoors as compared with MUEs. Therefore, more priorities should be dedicated to SUs with constraints not to affect MBS and MUE transmission. This is an important research issue to be looked in as it may reshape the way how concept of PU and SU is addressed in CR enabled interference mitigation schemes.

12.1.4 Energy Saving

Cognition in CR enabled interference mitigation schemes is usually performed by the FAP and sometimes assisted by a cognitive enabled FUEs that can help to improve the interference mitigation technique and the overall capacity of the network. However, this puts a strain on the cognitive FUE with its limited energy capacity and makes it susceptible to battery drain. This is an important research challenge as efficient energy saving schemes need to be employed to cater for this added strain on cognitive FUEs.

12.1.5 Security

An open access mode significantly reduces interference in femtocells, but unlike CSG it also introduces an increased threat to the privacy of the owner as well the network itself. The ability of any UE to gain access into a FAP makes it vulnerable as a hacker with malicious intent can ultimately take control of it to access the restricted information, such as private key and authentication procedure with the core network. Although a secure gateway provides security between the femtocell and the core network, open access makes it exposed to advanced hacking techniques that could disrupt the entire network. Therefore, adequate security measures are required in open access mode to protect the subscriber and network operator and some issues such as criteria for joining the FAP should be addressed.

12.1.6 Pilot Power/Coverage Radius Issue

Most power control schemes utilize an adaptive power control mode where the pilot power levels of an FAP is controlled effectively to adjust coverage radius not only to mitigate interference but also to reduce the need for handover of close by UEs in open access mode. Power control schemes that are based on pilot power of an FAP become less significant in densely deployed femtocells if a FUE is located close to a neighbouring co-located FAP. This may result in lower co-tier SINR values and it leads to an important research issue to be further investigated.

12.1.7 Signalling Overhead

Femtocells employing CR to mitigate interference would need sophisticated cognition schemes. There will be lots of information sensing and gathering involved and some signalling channels would be needed to achieve the required coordination among nodes. This is an important research area to be investigated as signalling overhead can result in increased delays, energy consumption reduced bandwidth and sometimes in a total loss of quality of service.

12.1.8 Proximity Services

Proximity services, such as Device-to-Device (D2D), Vehicle-to-Vehicle (V2V) and Human-to-Human (H2H) are supported by

LTE technology that offers spectral efficiency and reduces end-to-end latency. D2D devices can communicate directly with one another when they are in close proximity through autonomous discovery. They will offload data from the network resulting in reduced cost of data processing and energy requirements related to signalling [7].

An H2H network employs the same concept and consists of several wearable/portable nodes each communicating with its neighbour. It is envisioned as a mesh network using the human subjects to transmit data within a limited geographic region [8]. This communication scenario will be the most widely used type of D2D communication since the majority of future communication devices could be body-worn [9].

Such communications mode can be used for crime alerts, public-safety for disaster management and home automation. The LTE network performs resource allocation and authentication while communication can take place via the network or directly between devices. Efficient realization of these proximity services still requires much work especially in the areas of energy efficiency and QoS considerations to support different data types (i.e. data, audio and video), mobility of the participating LTE devices, reliability, privacy and security of the data exchanges, efficient resource allocation along with novel antenna solutions, mitigation of the human body effects (which would be high due to increased penetration depth) and channel characterization [10].

12.1.9 The Internet-of-Things (IoT)

The IoT is the next technological marvel that has a potential to create over 1 trillion US dollars of economic value by 2020 through the technology development and corresponding market opportunities [11].

It refers to a web of ubiquitously connected objects, devices, sensors and processes providing applications ranging from wearable health trackers to connected cars encompassing fields of transportation, healthcare, consumer electronics, construction to name few. Presently, we have 10 billion connected devices in the form of Machine-to-Machine (M2M) communications. This number is expected to rise to 24–50 billion total connections in the next 5 years [11, 12]. M2M communications require reliability and availability as two basic needs. LTE systems offer support for this mode of communication through very high reliability and availability. However, issues such as scalability without risking the operation of data/voice connections and hardware costs require innovative device solutions.

12.1.10 The Age of Big Data

Rapid innovation in telecommunication has enabled us to flood ourselves with data of all types, sizes and speeds. The extent of data production is such that a value of 10.7 Exabytes per month in 2017 is estimated to reach 24.3 Exabytes per month by 2019 [13]. This massive amount of data will seriously test conventional processing and data mining paradigms. High data rate solutions along with novel access techniques would be required to deal with this issue.

12.1.11 5G and Femtocells

Evolution of 5G, where deployment is targeted to commence by 2020, is another venue that would require extensive research and development of LTE femtocells. 5G is designed to integrate with the existing LTE systems at the initial stages [14, 15].

A key challenge is the harmonious integration of the two technologies while exploiting their respective benefits. Particular issues that need to be addressed include combination of existing LTE bands with new 5G frequencies to provide interworking with access to both, reduction in high latency required for resource allocation before the effective data transmission and energy-aware radio access (possibly through transmission of alarm messages). Moreover, an arrangement of femtocell networks in IoT scenarios with centralized control in which the FAPs are organized by a single entity providing an energy efficient method where the redundant femtocells can be switching off as and when required [16, 17].

12.1.12 Antenna Design and Channel Modelling

In order to achieve unobtrusive communication in LTE femtocells, efficient implementation of wireless modules is vital. Despite extensive research and development in this area, design of LTE wireless modules is still very challenging due to limitations of form factor requirements, reliable data transmission and antenna design. Antenna design is particularly important as it dictates the radio link performance and physical size of the LTE user equipment. Ever-changing requirements of LTE devices thanks to the emergence of new operational scenarios along with new materials, methods and technologies further necessitates new and improved solutions.

Antennas and associated electronics are required to be conformal to the body shape for on-body or Human-to-Human communications environments. They also need to show immunity from detuning in frequency and polarization. Use in cluttered environments brings further challenges of multipath fading and multi-frequency operation on top of conventional complexities of physical dimensions, directivity, efficiency and SAR as discussed in Chapter 7. Techniques like reconfigurability, multi-band operation, beamforming and phased-controlled arrays also need extensive study in LTE scenarios to support high data rate applications as highlighted in Chapters 8, 9 and 10.

LTE technology considers MIMO while its integration with 5G will require massive MIMO antenna arrays (with more than 10 elements) with beamforming at the mobile receivers as one of the solutions to support high data rate, enhance spectral efficiency and mitigate interference [18, 19]. Although these techniques can offer multiple advantages, several major challenges need to be addressed before their practical deployment. First, overhead of acquiring the channel state information (CSI) to exploit full capacity gain of massive MIMO systems increases significantly. Second, the complexity of precoding and detection rises with the number of antennas. Finally, integrating a large number of antennas into a limited device volume while still maintaining low mutual coupling between the antenna elements remains very challenging. Moreover, the multiple access control (MAC) layer design would also need improvements in order to accommodate the massive MIMO antennas [20].

Accurate propagation prediction through channel modelling, which can account for complex propagation conditions in urban environment, is also necessary for efficient deployment of LTE systems. These estimations need to take into account wide range of antenna types, reflection and absorption from a wide range of materials such as glass, concrete, wood and of course vegetation. Moreover, frequency dispersion caused by mobility results in Doppler effect that will cause variation in the channel, which is likely to impact the requirement of CSI needed for a particular antenna scheme.

Furthermore, a clear understanding of wireless links between various LTE devices, either body-worn or operating in the vicinity of the human body is also essential for the efficient and reliable communication. It requires in-depth analysis of wave attenuation in lossy human body tissues and propagation in, on and around the body through channel characterization and modelling. Chapter 11 of this book offers an insight into this topic, however, much more work needs to be done in this direction.

References

1 US Census Bureau, "U.S. and World Population Clock," http://www.census.gov/popclock/, accessed on June 25, 2017.
2 Beyond LTE: Enabling the Mobile Broadband Explosion, Rysavy Research/4G Americas, August 2015.
3 J.G Andrews, H. Claussen, M. Dohler, S. Rangan, and M.C Reed, "Femtocells: Past, Present, and Future," *IEEE Journal on Selected Areas in Communications*, vol. 30, no. 3, April 2012, pp. 497–508.
4 S.F. Hasan, N.H. Siddique and S. Chakraborty, "Femtocell versus WiFi – A survey and comparison of architecture and performance," *1st International Conference on Wireless Communication, Vehicular Technology, Information Theory and Aerospace & Electronic Systems Technology*, May 2009, pp. 916–920.
5 H.A. Mahmoud and I. Guvenc, "A comparative study of different deployment modes for femtocell networks," *IEEE 20th International Symposium on Personal, Indoor and Mobile Radio Communications*, Sept. 2009, pp. 1–5, 13–16.
6 N. Saquib, E. Hossain, Long Bao Le and D. I. Kim, "Interference management in OFDMA femtocell networks: issues and approaches," *IEEE Wireless Communications*, vol. 19, no. 3, June 2012, pp. 86–95.
7 A. Meharouech, J. Elias, and A. Mehaoua, "Future body-to-body networks for ubiquitous healthcare: a survey, taxonomy and challenges," *2nd International Symposium on Future Information and Communication Technologies for Ubiquitous HealthCare (Ubi-HealthTech)*, May 2015.
8 "5G vision white paper," *DMCR &D Center at Samsung Electronics Co., Ltd.*, February 2015.
9 Q.H. Abbasi, M. Ur Rehman, A. Alomainy and K. Qaraqe (Ed.), *Advances in Body-Centric Wireless Communication: Applications and State-of-the-Art*, The IET (UK), 2016.

10 T. Bai, A. Alkhateeb, and R. Heath, "Coverage and capacity of millimeter-wave cellular networks," *IEEE Communications Magazine*, vol. 52, no. 9, pp. 70–77, September 2014.
11 S. Andreevy, O. Galinina, A. Pyattaev, M. Gerasimenko, T. Tirronen, et al., "Understanding the IoT connectivity landscape: a contemporary M2M radio technology roadmap," *IEEE Communications Magazine*, vol. 53, no. 9, pp. 32–40, September 2015.
12 5G innovation opportunities – A discussion paper, Future Technologies Network, UK, August 2015.
13 "Cisco Visual Networking Index: Forecast and Methodology, 2014–2019", Cisco Visual Networking Index, July 2014.
14 ITU, "ITU towards 'IMT for 2020 and beyond'," http://www.itu.int/en/ITU-R/study-groups/rsg5/rwp5d/imt-2020/Pages/default.aspx, retrieved on July 5, 2017.
15 3GPP, "Access System for Ultra Low Complexity and Low Throughput Internet of Things based on Cellular", GP-140301, May 2014.
16 C. Mavromoustakis, G. Mastorakis and J. Batalla (eds) Internet of Things (IoT) in *5G Mobile Technologies. Modeling and Optimization in Science and Technologies*, vol. 8. Springer, Chamonix.
17 M. Condoluci, M. Dohler, G. Araniti, A. Molinaro and K. Zheng, "Toward 5G densenets: architectural advances for effective machine-type communications over femtocells," in *IEEE Communications Magazine*, vol. 53, no. 1, pp. 134–141, January 2015.
18 S. Chen and J. Zhao, "The requirements, challenges, and technologies for 5G of terrestrial mobile telecommunication," *IEEE Communications Magazine*, vol. 52, no. 5, pp. 36–43, May 2014.
19 F. Rusek, D. Persson, B.K. Lau, "Scaling up MIMO: opportunities and challenges with very large arrays". *IEEE Signal Processing Magazine*, vol. 30, pp. 40–60, 2012.
20 D. Muirhead, M. A. Imran and K. Arshad, "A survey of the challenges, opportunities and use of multiple antennas in current and future 5G small cell base stations," in *IEEE Access*, vol. 4, pp. 2952–2964, 2016.

Index

a
Acknowledgement (ACK) 56
Analogue mobile phone systems (AMPS) 3
Antenna
  bandwidth 190
  directivity 191, 201
  efficiency 193, 201, 219, 220, 222, 245, 251, 266–267, 281
  gain 192, 201, 219, 222, 245, 251, 266–267
  impedance matching 201
  input impedance 188
  polarization 195, 203
  radiation pattern 189, 222
Autocorrelation 159
Axial ratio (AR) 194, 260, 262, 263, 265, 266, 273, 275, 281
Axial ratio bandwidth 266

b
Bandwidth 189
Bandwidth allocation 119
Barium strontium titanate (BST) 236
Base station (BS) 1
Beamforming 329
Block error rate 112
Bluetooth 294, 296
Body-centric Wireless Network (BCWN) 291
Body phantoms 291
Body-to-body communication 315
Boresight 283

c
Capacity limitation 138
Capital expenditure (CAPEX) 22, 323
Carrier aggregation 121, 122, 196
CDMA *see* Code division multiple access (CDMA)
Cell edges 132
Cell specific reference signals 129
Cellular networks 139
Centralised 52, 54, 61, 64
Channel
  additive Gaussian noise 151
  band limited 153

LTE Communications and Networks: Femtocells and Antenna Design Challenges, First Edition. Edited by Masood Ur Rehman and Ghazanfar Ali Safdar.

linear filter 152
quality 120
Channel state information (CSI) 329
Circularly polarized (CP) antennas 260
Close service group–close access (CSG) 28, 324, 326
Clusters 124
Co-channel 86
Code division multiple access (CDMA) 4, 220
Cognitive femtocell base station (CFBS) 50, 66, 67, 74
Cognitive femtocell network (CFN) 61
Cognitive femtocells (CF) 34
Cognitive femtocell users (CFUs) 66
Cognitive hybrid division duplex (CHDD) 59
Cognitive interference mitigation 51
Cognitive macrocell users (CoMUs) 67
Cognitive Manager for Resource Management (CM-RM) 62
Cognitive Manager for Spectrum Management (CM-SM) 62
Cognitive radio (CR) 21, 324–326
Cognitive radio networks (CRNs) 112, 128
Collaborative detection 163
Common macrocell users (CMUs) 67
Component carriers (CC) 58
Conductivity 292
Control channel signalling 134
Cooperative cells 118
Co-Tier 24
Coverage area bounds 99
CP radiation pattern 262, 273
CRNs *see* Cognitive radio networks (CRNs)
Cross polarization 203
Cross-tier 22
CR resource management (CRRM) 56
Crucial places (CP) 65
CSG *see* Close service group–close access (CSG)
CST Microwave Studio 215, 220, 237, 239, 266, 273, 292, 293
Cumulative distribution function (CDF) 93
Cyclic Redundancy check (CRC) 27
Cyclostationary feature detection
cyclic prefix 154, 159
SINR 160
spectral correlation function 160

d

Data rate 196
Debye dispersion equation 308
Decentralised 52, 61
Dedicated channel 86
Device-to-Device (D2D) 326, 327
Digital Cellular Service (DCS) 272
Digital Subscriber Line (DSL) 85, 322
Direction-of-arrival (DoA) 65, 202
Discontinuous transmission 109
Distance aware 133
Distance Sense Multiple Access (DSMA) 59

Distributed antenna systems (DAS) 2
Doppler effect 329
Dormant cells 118
Downlink (DL) 24, 200
Dynamic power 109
Dynamic spectrum access (DSA) 50

e

Efficiency
  mismatch 193
  radiation 193
  reflection 193
Eigen-based 66
Electric field strength 300
Electromagnetic absorptions 297
Electromagnetic waves
  polarization 193
  reflection 202
  refraction 202
  scattering 202
Energy conservation 113
Energy detector
  noise variance 156
  square law device 157
Energy efficient 115
Enhanced Data rates for GSM Evolution (EDGE) 4
Enhanced system 138
Environmental aspects 139
Error probability
  probability of error 155
  probability of false alarm 155
  probability of missed detection 155
ETSI (European Telecommunications Standards Institute) 49
Eutectic Gallium Indium (EGaIn) 237
Evolution data Optimised (EV-DO) 4
Evolved Packet Core (EPC) 8, 105, 106
Evolved Packet System 107
Evolved-UTRAN (E-UTRAN) 8
Extended sector 125, 126

f

Far-field radiation pattern 241, 251
The Federal Communications Commission (FCC) 49, 302
Femtocell access point (FAP) 24, 322, 325, 326
Femtocell User Equipment (FUE) 322, 325, 326
FETs 232–233
Finite difference time domain (FDTD) 291
Finite Integration Technique (FIT) 215, 237
FR4, 220, 273
Fractional Frequency Reuse (FFR) 55
Fractional power control (FPC) 68
Frequency assignment 46
Frequency Division Duplex (FDD) 11, 200, 324
Frequency reconfigurability 204
Front-to-back ratios 279
Fusion node 163

g

5G 328
Gale Shapely spectrum sharing (GSOIA) 55
Game-based Radio Resource Management (SGRRM) 56

Game based Resource Allocation in Cognitive Environments (GRACE) 68
General packet radio service (GPRS) 3
Generation (G)
  1G, 2G, 2.5G, 2.75G, 3G, 3.5G, 4G / LTE 3–5
Global Positioning System (GPS) 214, 220, 231, 234–235, 237, 241, 268, 272, 273, 279, 305
Global system for mobile communication (GSM) 3, 305
Global warming 107
Group channel sensing 325
GSM-HSPA 322

h

Half power beamwidth 191
Hand-held device 219, 290
Hidden node 163, 164
Higher order resonance 210
High rate packet data (HRPD) 4
High Speed packet access (HSPA) 5
Home eNodeB (HeNB) 85
Home NodeB (HNB) 85
Human-to-Human (H2H) 326, 327, 329
Hybrid Access 45
Hybrid Automatic Repeat Request (HARQ) 27
Hybrid Sensing 162
Hysteresis 130

i

Impedance bandwidth 251, 263, 265, 266
Information and Communication Technology 108
Institute of Electrical and Electronics Engineers (IEEE) 1, 302
Interference 43
International Telecommunication Union (ITU) 4, 197
Internet of Things (IoT) 13, 327, 328
Interoperability testing (IOT) 9
ISM 49

l

Left-Hand Circular Polarization (LHCP) 266, 279
Lightly loaded network 133
Linear filter channel
  linear time variant 152
  linear time invariant 152
Line-of-sight (LOS) 203, 297, 312, 315, 316
Link adaptation 114
Liquid Crystal Polymer (LCP) 237–240, 248, 253
Load balancing 120
Long Term Evolution (LTE) 1, 104
Loss tangent 240, 246
LTE architecture 105
LTE frequency bands 198–200
LTE Rel-8, 196

m

Machine-to-Machine (M2M) 327
MAC layer 115
Macrocell Base Station (MBS) 10, 323, 325
Macrocell User Equipments (MUEs) 23, 325
Magnetic Resonance Imaging (MRI) 291, 293

Matched Filter 158
MBSFN 117
Mechanically reconfigurable antenna 236
Mercury 237–238, 240
Method of moment (MoM) 291
Mobile Station (MS) 2
Mobile Switching Centre (MSC) 3
Mobile Telecommunication Switching Office (MTSO) 3
Mobility Management Entity (MME) 9, 107
Modulation
  FFT 153, 157, 160
  IFFT 153
  inter-symbol interference 154
  multipath fading 154, 163, 168
  OFDM 153
Monopole 211
Multifunctional antenna 259
Multimedia Messaging Service (MMS) 3
Multiple access control (MAC) 329
Multiple input Multiple output (MIMO) 11, 187, 197, 205–206, 329
Multiple resonant structures 211
Mutual coupling 259
Mutual Information based Exponential SNR Mapping (MIESM) 26

n
Non-line-of-sight (NLOS) 312, 315, 316
Numerical analysis 291
Numerical model 292

o
Offset 130
Open service group–open access (OSG) 28
Operational expenditure (OPEX) 10, 104, 323
Optically reconfigurable antenna 236
Optical permittivity 308
Optical switches 232–235
Orthogonal frequency division multiple access (OFDMA) 5, 105
Overheads 135

p
Path gain 295, 297, 312, 315, 316
Path loss (PL) 25
Peak to average power ratio (PAPR) 9
The Perfectly Matched Layer (PML) 293, 307
Performance key indicator (PKI) 28
Permeability 234, 237, 239
Permittivity 234, 237, 239, 246, 261, 292
Personal Communications Service (PCS) 272, 305
Personal navigation devices (PNDs) 227, 289
Physical channels 113
Physical Clusters (PCs) 69
Physical layer (PHY) 27
PIFA 201, 211, 213, 232, 261, 295, 296, 305
Pilot symbols 159
PIN diodes 232–235
Polarization 193–196

circular 194
elliptical 194
LHP 194
linear 194
RHP 194
Polymer fibres 305
Power control 46, 111
Primary user (PU) 33, 325
Probability density 154, 155
PU *see* Primary user (PU)
Public Switched Telephony Network (PSTN) 3

q

Q-Learning 53
Quadrature Amplitude Modulation (QAM) 27
Quadrature phase shift keying (QPSK) 27
Qualified Channels (Qc) 64
Quality factor 265
Quality of service (QoS) 21, 327
constraints 123

r

Radiation intensity 192
Radiation pattern
farfield 190
major lobe 190
minor lobe 191
null 191
Radio frequency (RF) 33
Radio frequency micro-electromechanical system (RF-MEMS) 232–235
Radio link failure 130
Radio Resource Management 123
Radio Resource Unit 108–109
RBs allocation 114
Received Signal Strength (RSS) 2
Received Signal Strength Indicator (RSSI) 34
Receiver uncertainty 163
Reconfiguration 203
Reduced early handover 129
Reference Signal Received Power (RSRP) 53, 128
Reflection coefficient (S11) 189, 241, 248, 265, 266, 308, 309, 312
Reflections 297
Regions of Interest (ROI) 33
Reinforced continuous carbon fibre composite (RCCFC) 236
Relaxation time 308
Relay Nodes 196
Remote antenna Unites (RAU) 2
Return loss (RL) 189
Right-Hand Circular Polarization (RHCP) 266, 274, 279
RSRP *see* Reference Signal Received Power (RSRP)
RT-Duroid 237–239

s

Satellite digital audio radio systems (SDARS) 272
Secondary user (SU) 33, 325
Sectors 111
Selfish Frequency Access (SFA) 68
Serving Gateway (S.GW) 9
Shadowing 163
Shared relay 127
Short messaging service (SMS) 1
Signal to Interference plus noise Ratio (SINR) 2, 325

Signal-to-Noise Ratio (SNR) 26, 151, 165
Single channel frequency division multiple access (SC-FDMA) 8
Single channel sensing 325
Sleep mode 137
Slotted Aloha 59
Smart Cognitive-Femto Network (SCFN) 65
Solid angle 192
SON (Self-Organized Networks) 128
Sounding Reference Signal (SRS) 65, 71
Specific Absorption Rate (SAR) 204–205, 290, 300, 302, 303, 308, 316, 329
  The Federal Communications Commission's (FCC) 204–205
  International Commission on Non-Ionising Radiation Protection (ICNIRP) 204–205, 302
Spectral efficiency 322
Spectrum holes 50
Stacked patches 261
Static permittivity 308
SU *see* Secondary user (SU)
Subcarrier (SC) 27
Subframes (SF) 26
Surface current distribution 222, 240, 248, 269
Synchronization 131

t

Third Generation partnership project (3GPP) 1, 322
Time division duplex (TDD) 11, 200, 324
Time division multiple access (TDMA) 3
Tri-sector 90
Truncated annular slots 263
TV white spaces (TVWS) 64, 67, 71

u

UEs *see* User Equipment (UE)
Universal mobile Telecommunications system / standard (UMTS) 4
Universal Terrestrial Radio Access (UTRA) 8, 106
Universal Terrestrial Radio Access Network (UTRAN) 8
Unused RBs 134
Uplink (UL) 86, 200
User Equipment (UE) 21, 326

v

Varactor 232–233
Varying load 126
Vector network analyser 266, 279, 296
Vehicle-to-Vehicle (V2V) 326, 327
Vendors 108
Virtual Clusters (VCs) 69
Visible Human Project 292
Voice over LTE (VoLTE) 12
Voltage standing wave ratio (VSWR) 189, 201, 218, 220, 226
Voxel 291–293, 307

w

Waveform detection 160
Wavelet detection 161
White spaces 172
Wideband CDMA (WCDMA) 4

WiFi *see* Wireless Fidelity (Wi-Fi)
WINNERII+, 90
Wireless Application Protocol (WAP) 3
Wireless Body Area Networks (WBAN) 289, 290
Wireless Fidelity (Wi-Fi) 23, 323
Wireless Personal Area Networks (WPAN) 289, 290
Wireless World Research Forum (WWRF) 290
WLAN 214–216, 305
Worldwide Interoperability of Microwave Access (WiMAX) 1

X

X2 interface 125